Progress in Probability
Volume 29

Series Editors
Thomas Liggett
Charles Newman
Loren Pitt

Seminar on Stochastic Processes, 1991

E. Çınlar P. Fitzsimmons

K. L. Chung S. Port

M. J. Sharpe T. Liggett

Editors *Managing Editors*

1992

Birkhäuser
Boston · Basel · Berlin

E. Çınlar
Dept. of Civil Engineering
and Operations Research
Princeton University
Princeton, NJ 08544

P. J. Fitzsimmons
(Managing Editor)
Dept. of Mathematics
University of California-San Diego
La Jolla, CA 92093

K. L. Chung
Dept. of Mathematics
Stanford University
Stanford, CA 94305

S. Port
T. Liggett
(Managing Editors)
Dept. of Mathematics
University of California
Los Angeles, CA 90024

M. J. Sharpe
Dept. of Mathematics
University of California-San Diego
La Jolla, CA 92093

Library of Congress Cataloging-in-Publication Data

Seminar on stochastic processes, 1991 / edited by E. Çınlar, K. L.
Chung, M. J. Sharpe.
      p.   cm. -- (Progress in probability ; v. 29)
    ISBN 0-8176-3628-5
    1.   Stochastic processes--Congresses.   I. Çınlar, E. (Erhan),
1941-   II.   Chung, Kai Lai, 1917-   . III. Sharpe, M. J., 1941-
    IV. Series: Progress in probability : 29.
    QA274.A1S443 1992   91-47703
    519.2--dc20   CIP

ISBN 0-8176-3628-5
ISBN 3-7643-3628-5

Camera-ready copy prepared by the Authors in TEX.
Printed and bound by Quinn-Woodbine, Woodbine, N.J.
Printed in the U.S.A.

9 8 7 6 5 4 3 2 1

This Volume is Dedicated to the Memory of

STEVEN OREY

LIST OF PARTICIPANTS

K. Alexander

R. Banuelos

M. Barlow

R. Bass

K. Burdzy

D. Burkholder

R. Carmona

E. Çinlar

Z. Q. Chen

M. Cranston

R. Dalang

R. Darling

S. Evans

N. Falkner

R. E. Feldman

P. Fitzsimmons

R. Getoor

B. Hambly

T. Harris

H. Hughes

D. Khoshnevisan

F. Knight

G. Lawler

T. Liggett

P. March

M. Marcus

P. McGill

T. Mountford

C. Mueller

C. Neuhauser

X. Pei

R. Pemantle

J. M. Penrose

Y. Peres

E. Perkins

M. Perman

J. Picard

J. Pitman

L. Pitt

S. Port

J. Rosen

T. Salisbury

M. Sanz

R. Schonmann

M. J. Sharpe

C. T. Shih

H. Sikic

R. Song

D. Stroock

G. Swindle

M. Talagrand

L. Taylor

E. Toby

Z. Vondracek

X. Wang

J. Watkins

R. Williams

R. Wu

Z. Zhao

TABLE OF CONTENTS

FOREWORD

The 1991 Seminar on Stochastic Processes was held at the University of California, Los Angeles, from March 23 through March 25, 1991. This was the eleventh in a series of annual meetings which provide researchers with the opportunity to discuss current work on stochastic processes in an informal and enjoyable atmosphere. Previous seminars were held at Northwestern University, Princeton University, the University of Florida, the University of Virginia, the University of California, San Diego, and the University of British Columbia. Following the successful format of previous years there were five invited lectures. These were given by M. Barlow, G. Lawler, P. March, D. Stroock, M. Talagrand. The enthusiasm and interest of the participants created a lively and stimulating atmosphere for the seminar. Some of the topics discussed are represented by the articles in this volume.

P. J. Fitzsimmons
T. M. Liggett
S. C. Port

Los Angeles, 1991

In Memory of Steven Orey

M. CRANSTON

The mathematical community has lost a cherished colleague with the passing of Steven Orey. This unique and thoughtful man has left those who knew him with many pleasant memories. He has also left us with important contributions in the development of the theory of Markov processes. As a friend and former student, I wish to take this chance to recall to those who know and introduce to those who do not a portion of his lifework.*

Steven was born in Berlin and at an early age fled with his family first to Lisbon then to New York. His university studies, both undergraduate and graduate, were completed at Cornell University. It was there that he met his wife Delores who was a student in philosophy. Upon graduation in 1953, Steven took a position at the University of Minnesota. Here he remained throughout his career. In addition to his research contributions, he has authored two books [32] and [46] and directed several Ph.D. students. Among these are Robert Anderson, Carol Bezuidenhout, Tzuu-Shuh Chiang, Dean Isaacson, Peter March, Timo Seppalainen, Ananda Weerasinghe, Albert Wang and myself.

A common theme runs through much of the early work of Steven Orey, namely the ergodic behavior of Markov chains. His work grew to include many related topics in the theory of Markov processes including central limit theorems, renewal theory, tail σ-fields, and large deviations. He also wrote about Gaussian processes as well as control theory and optimization. Still this does not give a complete account of his work as his first interest was logic, a subject on which he

---

* I apologize to those whose work has been overlooked in this account. Omissions are due solely to my limited knowledge. Finally, Peter March has provided me with valuable assistance.

has written several papers. However, I would like to confine myself to a summary of what I have referred to as the common theme, the ergodic theory of Markov chains, to related topics and to some of his other work which has been influential.

Steven's Ph.D. thesis was written at Cornell in logic under J. B. Rosser. Before graduating he attended lectures by William Feller on one-dimensional diffusions. A year spent at Berkeley early in his career further propelled him in the direction of probability. Evidently, he was inspired by the ideas of Doeblin and this influence appears often in Steven's work.

His first paper on probability theory [4] established the central limit theorem for a sequence of m-dependent random variables. The condition put forward there reduces to the usual Lindeberg condition in the case $m = 0$, a feature that had been lacking in previous theorems of this type.

What might fairly be called the principal interest of his career appeared in his next work [5]. Here he is interested in examining the relation between Harris recurrence and Doeblin's condition for Markov chains. Among the main results are a ratio ergodic theorem and a very pretty central limit theorem. The latter concerns the additive functional $\sum_{k=0}^{n-1} f(X_k)$, of values of a given function f of the first n positions of the chain. Doeblin had handled this functional in the case of a recurrent chain by noting that excursions from a given state are independent and identically distributed. Steven observed that with Harris recurrent chains the excursions from a set form a Markov chain satisfying Doeblin's condition and with sufficient independence for a central limit result to hold.

Other highlights from his work of the early 1960's are [12], [7] and [11]. The first of these contains (in approximately two pages) perhaps his most well-known result. Here he considers a recurrent, aperiodic, irreducible Markov chain on a countable state space. Using a coupling or cancellation idea due to Doeblin, he proves that the total variation of the difference between the distributions of two copies of the chain at time n, starting from any two probability distributions, goes to zero as n tends to infinity. The clever idea used here is to partition the integers into two classes: those times when the mass from the first distribution exceeds the mass from the second at a fixed state and the complementary set of times. The chain must visit the fixed state infinitely often on one of these sets of times. This allows cancellation to exhaust both masses on the set of times when they both visit the fixed state. In the second paper, a necessary and sufficient condition is given for the strong ratio limit property to hold for a recurrent, aperiodic, irreducible Markov chain on a countable state space. The strong ratio limit property holds when there are positive constants π_h and π_k for each h and k in the state space so that

$$\lim_{n \to \infty} \frac{p_{i,h}^{n+m}}{p_{j,k}^{n}} = \frac{\pi_h}{\pi_k}.$$

An equivalent condition is that

$$\lim_{n \to \infty} \frac{p_{0,0}^{n+1}}{p_{0,0}^{n}} = 1.$$

The third result, obtained jointly with William Feller [11], is a renewal theorem at a level of generality which had been sought previously by several mathematicians.

The renewal theorem of Feller and Orey and an article of David Blackwell and David Freedman, which was inspired by Steven's ergodic theorem, naturally led him to consider tail and invariant σ-fields. His first result in this area [21] was to characterize the tail σ-field for sums of independent (not necessarily identically distributed) random variables. Following up on the Blackwell-Freedman results, Steven and Benton Jamison [24] considered Harris recurrent chains and proved the state space of the chain decomposes into disjoint cyclically moving subsets, $C_1, \ldots, C_n$, whose union is the entire state space and that the tail σ-field is generated by events of the form $\{X_0 \in C_i\}$. In a joint article with Bert Fristedt [40], the tail σ-field for one dimensional diffusions was specified. Here a colorful path crossing argument was used in conjunction with a refinement of the martingale convergence theorem. His work in this area was concluded with two articles done jointly with Uwe Rosler and myself [42], [47]. These papers found the tail and invariant σ-fields and Martin boundary for higher dimensional analogues of the Ornstein-Uhlenbeck process.

Steven also had an interest in sample path propeties and he obtained some elegant results in this field. There are three among these results which well illustrate his sense of aesthetics. The first of these [36], is joint work with S. J.

Taylor. Here they consider a question which is quite natural in view of Levy's modulus of continuity for $X(\cdot)$, the Brownian path. Namely, how often along the Brownian path will the law of the iterated logarithm fail? Setting, for $0 \leq \alpha \leq 1$,

$$E(\alpha) = \left\{ t : \limsup_{h \to 0} \frac{X(t+h) - X(t)}{\sqrt{2h \log h^{-1}}} \geq \alpha \right\},$$

they prove the tidy relationship that almost surely $E(\alpha)$ has Hausdorff dimension $1 - \alpha^2$. Second is a joint result with Naresh Jain [26]. Here they consider random walk on the integer lattice Z^d. Letting f_k be the probability that the first return to the origin occurs at time k, it is assumed that

$$\sum_{n=1}^{\infty} \sum_{k=n+1}^{\infty} f_k < \infty.$$

Now set $p = \sum_{k=1}^{\infty} f_k$, and take R_n to be the number of distinct lattice points visited by the walk up to time n. Breaking the path up appropriately enables them to establish a central limit result, namely $(R_n - np)/\sqrt{n}\sigma$ tends in law to a standard normal distribution. Lastly, with William Pruitt, in a paper which proved to be fundamental in its area [33], they consider path properties of the R^d-valued N-parameter Brownian sheet. Included here are generalizations of Strassen's law of the iterated logarithm and an integral test for upper functions of the path, both for global and local continuity. In addition, they give an exact account for interval and point recurrence, depending on the size of d and N: interval (point) recurrence holds for $d \leq 2N$ ($d < 2N$).

In the late 1970s, Steven's interest in large deviations was aroused by the articles of Donsker and Varadhan. Among his earliest works in this area was a joint paper [38] with his former student Robert Anderson on a version of the Ventcel-Freidlin theory for the Neumann problem. Included in this paper is a novel path-by-path construction of reflecting Brownian motion. In another work in the area [48], Steven considers an intriguing version of Schilder's theorem. Here he determines the asymptotics as $t \to 0$ with ϵ fixed, or $\epsilon \to 0$ with t fixed, of the probability that a Brownian path will stay for time t within ϵ of an independent Brownian path. Steven then pursued a program in the direction the field took as it moved from Cramér's Theorem to large deviation principles for Markov processes. Namely, to see what sort of path dependence would admit of a level three large deviation principle. One such result was obtained jointly with Hans Föllmer: in [53] a large deviation principle was shown to hold for stationary Gibbs measures on Z^d and the corresponding rate function was identified as specific relative entropy.

In another vein, he examined by himself [50], and jointly with Stephan Pelikan [54] [57], large deviation principles for dynamical systems. As he stated in the first of these articles, "Our primary goal is to find a result which comes as close as possible to solving the proportion X : ergodic theorem = Cramér's theorem : law of large numbers." Indeed, in [50], with the aid of his own generalization of the Shannon-McMillan theorem, he succeeded in establishing a large deviation principle for certain shifts. The upper bound for this case follows from considering the associated chain whose state at time n is the entire history of the path up to time n. The arguments of Donsker and Varadhan for the upper bound apply to this chain and are transferred back to the dynamical system via homomorphism. The lower bound cannot be obtained in this way. However, the arguments of Donsker and Varadhan were pushed through using the above mentioned refinement of Shannon-McMillan. In the articles with Pelikan, they find the deviation function for shifts invariant under Gibbs measures. In a beautiful application of their result, and using the symbolic dynamics associated via Markov partitions with an Anosov diffeomorphism on a compact manifold, they were able to prove a large deviation principle for trajectory averages of the diffeomorphism and to identify the rate function.

In his last work [58], Steven seems to have come 'full circle' and considered ergodic theoretic questions which occupied him early in his career, but for Markov chains whose transition probabilties are selected from a stationary stochastic process.

A Correlation Inequality for Tree-Indexed Markov Chains

ITAI BENJAMINI[1] AND YUVAL PERES[2]

§1. Introduction

Let T be a tree, i.e. an infinite, locally finite graph without loops or cycles. One vertex of T, designated 0, is called the root; we assume all other vertices have degree at least 2. Attach a random variable S_σ to each vertex σ of T as follows. Take $S_0 = 0$ and for $\sigma \neq 0$ choose S_σ randomly, with equal probabilities, from $\{S_{\tilde{\sigma}} - 1, S_{\tilde{\sigma}} + 1\}$ where $\tilde{\sigma}$ is the "predecessor" of σ in T (see §2 for precise definitions). We call the process $\{S_\sigma : \sigma \in T\}$ a T-walk on $\mathbb{Z}$; note that taking $T = \{0, 1, 2, \cdots\}$ with consecutive integers connected, we recover the (ordinary) simple random walk on $\mathbb{Z}$. Allowing richer trees T we may observe quite different asymptotic behaviour. One can study this behaviour either by considering the levels $\{\sigma : |\sigma| = n\} = T_n$ of T ($|\sigma|$ is the distance from 0 to σ) or by observing the rays (infinite non-self-intersecting paths) in T. The first approach, which we adopt here, was initiated by Joffe and Moncayo [JM] who gave conditions for asymptotic normality of the empirical measures determined by $\{S_\sigma : |\sigma| = n\}$. The second approach is used in [E], [LP] and [BP1]. To elucidate both approaches, we quote a theorem which relates three notions of "speed" for a T-walk, to dimensional properties of the boundary ∂T of T (∂T is the collection of rays in T, emanating from 0). Equip ∂T with the metric ρ given by $\rho(\xi, \eta) = e^{-n}$ if $\xi, \eta \in \partial T$ intersect in a path of length precisely n from 0.

Part (i) of the next theorem is a special case of the results of Lyons and Pemantle in [LP]; parts (ii), (iii) are from [BP2].

[1]Math. Institute, the Hebrew University, Jerusalem. Partially sponsored by a grant from the Edmund Landau Center for research in Mathematical Analysis, supported by the Minerva Foundation (Germany).
[2]Math. Dept., Stanford University, Stanford, California 94305. Supported by a Weizmann Postdoctoral fellowship.

Theorem 1([LP], [BP2])
Let T be a tree and $\{S_\sigma\}$ the T-walk on $\mathbb{Z}$.

(i) The sustainable speed

(1.1)
$$\sup_{\xi \in \partial T} \ \liminf_{\sigma \in \xi} \frac{1}{|\sigma|} S_\sigma$$

(which is a.s. constant by Kolmogorov's 0-1 law) is positive if ∂T has positive Hausdorff dimension.

(ii) The burst speed

$$\sup_{\xi \in \partial T} \ \limsup_{\sigma \in \xi} \frac{1}{|\sigma|} S_\sigma$$

is positive iff ∂T has positive packing dimension (see [TT] for the definition of packing dimension and [BP1] for its adaptation to trees).

(iii) The level burst speed $\limsup_{n \to \infty} \frac{1}{n} \sup_{|\sigma|=n} S_\sigma$ is positive iff T has exponential growth, i.e., the level cardinalities

$$A_n = \#\{\sigma \in T : |\sigma| = n\}$$

satisfy $\limsup_{n \to \infty} \frac{1}{n} \log A_n > 0$.

Concentrating on part (iii) of the theorem, observe that among all trees with n'th level T_n of cardinality A_n, the random variable $\sup_{\sigma \in T_n} S_\sigma$ is stochastically greatest for the "independent" tree consisting of A_n disjoint rays. This is a consequence of of the FKG inequality (See [G], §2.2) since the events $\{S_\sigma \leq y\}$ for $\sigma \in T_n$ are always decreasing events in a product space and therefore

(1.2)
$$P[\bigcap_{\sigma \in T_n} \{S_\sigma \leq y\}] \geq \prod_{\sigma \in T_n} P\{S_\sigma \leq y\}.$$

The point of the next section, §2, is that a correlation inequality like (1.2) holds for occupation of sets different from a halfline; in these cases the FKG inequality is no longer available. This is established in the general setting of tree-indexed Markov chains. In §3 this inequality is applied to a problem about T-walks on $\mathbb{Z}$: For which trees is the origin occupied from some level on (a.s.)?

§2. The correlation inequality

For vertices τ, σ of tree T, we write $\tau \leq \sigma$ if τ is on the unique path connecting σ to the root 0. The predecessor of a vertex $\sigma \neq 0$ is the unique vertex $\tilde{\sigma}$ adjacent to σ which satisfies $\tilde{\sigma} \leq \sigma$.

Definition
Let $\{p(x,y) : x, y \in G\}$ be transition probabilities on a countable set G (i.e., $\sum_y p(x,y) = 1$ for each x). A (T,p)-walk on G, with initial state $x_0 \in G$, is a collection $\{S_\sigma : \sigma \in T\}$ of G-valued random variables, where $S_0 \equiv x_0$ and for $\sigma \neq 0$:

(2.1) $P[S_\sigma = y | S_{\tilde{\sigma}} = x] = P[S_\sigma = y | S_{\tilde{\sigma}} = x, S_\tau \text{ for } \tau \neq \sigma, |\tau| \leq |\sigma|] = p(x,y).$

The first equality in (2.1) is the Markov property for (T,p)−walks.

Theorem 2

With the notation above, fix a subset Q of the state space G. Then among all trees with nth level T_n of cardinality A_n, the probability

$$\mathbf{P}_{x_0}[\forall \sigma \in T_n, S_\sigma \in Q]$$

is minimized by the tree consisting of A_n disjoint rays (the subscript x_0 indicates the initial state).

Proof:

Denote by $\{p^n(x,y)\}$ the n-step transition probabilities on G and let $p^n(x,Q) = \sum_{y \in Q} p^n(x,y)$. The assertion of the theorem may be restated as

$$(2.2) \qquad \mathbf{P}_{x_0}[\forall \sigma \in T_n, S_\sigma \in Q] \geq [p^n(x,Q)]^{A_n},$$

for T-walks where the nth level T_n of T has cardinality A_n. We verify (2.2) by induction on n; for $n = 1$ it clearly holds with equality. Next, let us pass from (2.2) to its analogue for $n + 1$.

For each vertex $\tau \in T_1$, let $T^{(\tau)}$ be the subtree $\{\sigma \geq \tau\}$ of T, rooted at τ. Denote by $A_n(\tau)$ the cardinality of the nth level $T_n^{(\tau)}$ of $T^{(\tau)}$, and observe that

$$(2.3) \qquad A_{n+1} = \sum_{\tau \in T_1} A_n(\tau).$$

Finally, let $\{S_\sigma^{(\tau)} : \sigma \in T^{(\tau)}\}$ be the $(T^{(\tau)}, p)$-walk on G. The Markov property (2.1) implies that the $(T^{(\tau)}, p)$-walks for $\tau \in T_1$ are mutually independent. Therefore, utilizing the induction hypothesis,

$$\mathbf{P}_{x_0}[\forall \sigma \in T_{n+1}, S_\sigma \in Q] = \prod_{\tau \in T_1} \left[\sum_{z \in G} p(x_0, z) \mathbf{P}_z\{\forall \sigma \in T_n^{(\tau)}, S_\sigma^{(\tau)} \in Q\} \right]$$

$$(2.4) \qquad\qquad \geq \prod_{\tau \in T_1} \left[\sum_z p(x_0, z)(p^n(z,Q))^{A_n(\tau)} \right].$$

Since $t \to t^{A_n(\tau)}$ is a convex function for $t \geq 0$, we have

$$(2.5) \qquad \sum_z p(x_0, z)(p^n(z,Q))^{A_n(\tau)} \geq \left[\sum_z p(x_0, z)p^n(z,Q) \right]^{A_n(\tau)}.$$

Combining (2.3), (2.4) and (2.5) gives

$$\mathbf{P}_{x_0}[\forall \sigma \in T_{n+1}, S_\sigma \in Q] \geq \left[\sum_z p(x_0, z)p^n(z,Q) \right]^{A_{n+1}} = [p^{n+1}(x_0,Q)]^{A_{n+1}},$$

completing the induction step.

$$\square$$

Remark:

Many variations of the simple convexity argument above are possible. As noted by the referee, the argument is shorter when the Markov chain is a function of an i.i.d. process.

§3. The local profile of a T-walk on Z.

Here we apply the correlation inequality from the previous section to show that if the levels T_n of T do not grow faster than $n^{\frac{1}{2}}$, then a.s. the origin is free of T-walk particles at infinitely many even times. This and a (weak) converse indicate that the occupation behavior of a T-walk is largely determined by the growth of T, rather than finer parameters like Hausdorff measure.

Theorem 3

Consider a T-walk $\{S_\sigma : \sigma \in T\}$ on the integers and denote by V_{2n} the random variable

$$V_{2n} = \sum_{\sigma \in T_{2n}} 1_{[S_\sigma = 0]}.$$

(i) If the level cardinalities $\{A_n\}$ of T satisfy

$$\liminf_n A_n n^{-\frac{1}{2}} < \infty$$

then

$$\limsup_n \ \mathbf{P}[V_{2n} = 0] > 0$$

and furthermore

$$\mathbf{P}[V_{2n} = 0 \ i.o.] = 1,$$

where i.o. stands for "infinitely often".

(ii) If $\liminf\limits_{n \to \infty} A_n n^{-\frac{1}{2}} = \infty$ then

(3.1) $$V_{2n} \overset{n \to \infty}{\longrightarrow} \infty \text{ in probability.}$$

(iii) If

(3.2) $$\forall n \quad A_n > C n^{\frac{1}{2}} (\log n)^{3/2}$$

for a sufficiently large constant $C > 0$, then $V_{2n} \overset{n \to \infty}{\longrightarrow} \infty$ almost surely.

Proof

We apply theorem 2 to the T-walk on Z with $Q = Z \setminus \{0\}$. Since

$$p^{2n}(x, 0) \le \binom{2n}{n} 2^{-2n} \le n^{-\frac{1}{2}},$$

that theorem gives

(3.3) $$\mathbf{P}[V_{2n} = 0] = \mathbf{P}[\forall \sigma \in T_{2n}, S_\sigma \neq 0] \ge [1 - n^{-\frac{1}{2}}]^{A_n}.$$

Our hypothesis and the monotinicity of $\{A_n\}$ imply that there is an increasing sequence $\{n_j\}$ and a constant $c > 0$ such that for all $j \ge 1$

(3.4)
$$A_{2n_j} \le cn_j^{\frac{1}{2}}.$$

In conjunction with (3.3) this implies

(3.5)
$$P[V_{2n} = 0 \ i.o.] \ge \limsup_n \ P[V_{2n} = 0] > 0$$

An additional argument is required to show that the probability on the left-hand-side of (3.5) is actually 1. By passing to a subsequence we may assume that the sequence $\{n_j\}$ satisfying (3.4) also has $n_j \ge 2n_{j-1}$ for all j. For each $\tau \in T_{2n_{j-1}}$ denote by $a(\tau)$ the number of vertices $\sigma \in T_{2n_j}$ which satisfy $\sigma \ge \tau$. By applying Theorem 2 to the subtree $\{\sigma \in T : \sigma \ge \tau\}$ of T we infer

$$P[\forall \sigma \in T_{2n_j}, \sigma \ge \tau \Rightarrow S_\sigma \ne 0 | S_\tau] \ge [1 - n_j^{-\frac{1}{2}}]^{a(\tau)}.$$

Combining this information for all $\tau \in T_{2n_{j-1}}$ gives

(3.6)
$$P[V_{2n_j} = 0 | S_\tau \ \forall \tau \in T_{2n_{j-1}}] \ge [1 - n_j^{-\frac{1}{2}}]^{A_{2n_j}},$$

since $\sum_{\tau \in T_{2n_{j-1}}} a(\tau) = A_{2n_j}$.

The Markov property of the T-walk allows us to infer from (3.6) that

$$P[V_{2n_j} = 0 | \forall i < j, V_{2n_i} > 0] \ge [1 - n_j^{-\frac{1}{2}}]^{A_{2n_j}}.$$

Utilizing (3.4) this shows that

$$\sum_{j=2}^\infty P[V_{2n_j} = 0 | \forall i < j, \ V_{2n_i} > 0] = \infty,$$

so the assertion

$$P[V_{2n} = 0 \ i.o.] = 1$$

follows from the conditional variant of the Borel-Cantelli lemma (see [Bil]).

(ii) For $k, R \ge 1$ denote by $M(k, R)$ the random variable

$$M(k, R) = \sum_{\tau \in T_k} 1_{[|S_\tau| < R]}$$

and observe that

$$P[M(k, R) < \frac{1}{2}A_k] \le \frac{2}{A_k} E[A_k - M(k, R)] = 2P[|S_\tau| \ge R]$$

where $|\tau| = k$.

Now the standard large-deviations bound for ordinary random walk [H] implies

(3.7)
$$P[M(k, R) < \frac{1}{2}A_k] \le 2e^{-R^2/2k}.$$

To each $\lambda > 0$ there corresponds some $\beta_\lambda > 0$ such that if $\sigma \in T_{2n}$ is a descendant of $\tau \in T_k$ and $k \le n$ then

$$P[S_\sigma = 0 \| |S_\tau| \le \lambda n^{-\frac{1}{2}}] \ge \beta_\lambda n^{-\frac{1}{2}}.$$

Therefore, by choosing for each $\tau \in T_k$ one descendent $\sigma \in T_{2n}$ we see that for $k \le n$:

$$P[V_{2n} \le v | M(k, \lambda n^{\frac{1}{2}}) = M] \le \sum_{j=1}^{v} \binom{M}{j} [\beta_\lambda n^{-\frac{1}{2}}]^j [1 - \beta_\lambda n^{-\frac{1}{2}}]^{M-j} \le$$

$$(3.8) \qquad\qquad\qquad \le \sum_{j=1}^{v} [\beta_\lambda M n^{-\frac{1}{2}}]^j \exp[-\beta_\lambda (M - j) n^{-\frac{1}{2}}]$$

Note that the right hand side is decreasing in M for $M > v/\beta_\lambda n^{\frac{1}{2}}$.
Next, to prove (ii), we set $k = n$ in (3.8) and $R = \lambda n^{\frac{1}{2}}$ in (3.7) and find

$$P[V_{2n} \le v] \le P[M(n, \lambda n^{\frac{1}{2}}) < \frac{1}{2} A_n] + P[V_{2n} \le v | M(n, \lambda n^{\frac{1}{2}}) \ge \frac{1}{2} A_n] \le$$

$$(3.9) \qquad\qquad \le 2e^{-\lambda^2/2} + \sum_{j=1}^{v} [\frac{1}{2} \beta_\lambda A_n n^{-\frac{1}{2}}]^j \exp[-\beta_\lambda (\frac{An}{2} - j) n^{-\frac{1}{2}}].$$

The assumption $A_n n^{-\frac{1}{2}} \to \infty$ implies

$$\limsup_{n \to \infty} P[V_{2n} \le v] \le 2e^{-\lambda^2/2} \; ;$$

since λ is arbitrary, this establishes (3.1).

(iii) We assume (3.2) with the constant $C > 0$ specified later. Employ (3.7) and (3.8) with $\lambda = 1$, i.e., $R = n^{\frac{1}{2}}$ and infer

$$P[V_{2n} \le v] \le P[M(k, n^{\frac{1}{2}}) < \frac{1}{2} A_k] + P[V_{2n} \le v | M(k, n^{\frac{1}{2}}) \ge \frac{1}{2} A_k] \le$$

$$(3.9) \qquad\qquad \le 2e^{-n/2k} + v[\beta_1 A_k n^{-\frac{1}{2}}]^v exp[-\frac{1}{2} \beta_1 A_k n^{-\frac{1}{2}}].$$

For $k = \lfloor \frac{n}{4 \log n} \rfloor$ and large n, (3.2) implies

$$A_k > \frac{1}{3} C n^{\frac{1}{2}} \log n .$$

Inserting this into (3.9) yields

$$P[V_{2n} \le v] \le \frac{2}{n^2} + v[\frac{1}{3} \beta_1 C \log n]^v \exp[-\frac{1}{6} \beta_1 C \log n],$$

which is summable if $C > 6/\beta_1$. Invoking Borel-Cantelli concludes the proof.

$\square$

Remarks

1. The same proof shows that under the condition of Theorem 3(i), a.s. every finite interval is vacant from T-walk particles infinitely often, while under the condition in (iii), a.s. all even points in any fixed finite interval are occupied at all sufficiently large even times.

2. The proof applies to any bounded step-size distribution of lattice type with mean zero (if one uses the local central limit theorem).

3. As observed by the referee, the proof of Theorem 3 easily implies that the almost sure convergence in part (iii) of that theorem does not hold under the weaker assumption made in part (ii). Indeed, let $A_n = c_1 n^{1/2} \log(\log(n))$ for some small constant c_1. Then by (3.6), we have

$$P[V_{2^n} = 0 | S_\sigma \text{ for all } |\sigma| \le 2^{n-1}] \ge [1 - 2^{-(n-1)/2}]^{A_{2^n}} \ge e^{-c_2 \log n}$$

which is divergent for c_2 small enough. By the conditional Borel-Cantelli Lemma,

$$P[V_{2^n} = 0 \ i.o] = 1.$$

Question

Motivated by the results of [JM] and Theorem 3 above, it seems one should be able to prove a local limit theorem for the empirical measures of T-walks on $\mathbf{Z}$, when T is sufficiently "nice". Explicitly, for which trees T does almost sure convergence

$$\frac{\sqrt{n\pi}}{A_{2n}} \sum_{\sigma \in T_{2n}} 1_{[S_\sigma = 0]} \overset{n \to \infty}{\longrightarrow} 1$$

hold?

Acknowledgement

We are grateful to Amir Dembo for helpful comments on a previous version of this note, and to the referee for valuable criticism.

References

[Bil] P. Billingsley, *Probability and Measure*,
Wiley, New York (1979).

[BP1] I. Benjamini and Y. Peres, *Markov chains indexed by trees*, preprint (1991).

[BP2] I. Benjamini and Y. Peres, *Tree-indexed random walks and first passage percolation*, preprint (1991).

[E] S. Evans, *Polar and non polar sets for a tree indexed process*, preprint (1990).

[G] G. Grimmett, *Percolation*, Springer Verlag, New York (1989).

[H] W. Hoeffding, *Probability inequalities for sums of bounded random variables*, J. Amer. Statist. Assoc. 58, no. 301 (1963) 13-30.

[JM] A. Joffe and A.R. Moncayo, *Random variables, trees and branching random walks.* Advances in Math. 10 (1973) 401-416.

[LP] R. Lyons and R. Pemantle, *Random walk in a random environment and first passage percolation*, preprint (1990), to appear in the Annals of Probability.

[TT] S.J. Taylor and C. Tricot, *Packing measure, and its evaluation for a Brownian path*, Trans. Amer. Math. Soc. 288 (1985) 679-699.

Itai BENJAMINI
Institute of Mathematics
Hebrew University
Givat Ram 91904, Jerusalem

Yuval PERES
Department of Mathematics
Yale University
New Haven CT 06520

On Specifying Invariant σ-fields

by

M. CRANSTON[1]

The purpose of this paper is to use coupling to specify invariant σ- fields. Suppose then that $(X, P^x, \Omega, F_t, S_t)$ is a strong Markov process on some nice state space E. S_t is the shift on paths: $X_s(S_t\omega) = X_{t+s}(\omega)$. Denote by $\mathcal{I} = \mathcal{I}(X)$ the σ-field generated by events $\Lambda \in F_\infty$ for which $S_t\Lambda = \Lambda$ a.s. $\forall t > 0$. This is the invariant σ-field. $\mathcal{T}$ is the tail σ-field for X, $\mathcal{T} = \bigcap_{t>0} \sigma(X_s : s \geq t)$. A couple of things are well-known about these σ-fields. First, if $\tilde{X}_t = (t + \xi_0, X_t)$ is the space-time process, then $\mathcal{I}(\tilde{X}) = \mathcal{T}(X)$. Thus our arguments will apply to tail σ-fields as easily as to invariant σ-fields. Second, is the fact that $\mathcal{I}$ gives all the bounded harmonic functions for X. From here on we shall say simply harmonic, the "for X" being understood. Briefly put, the connection is that for $\Lambda \in \mathcal{I}$, $h(x) = P^x(\Lambda)$ is harmonic. On the other hand, if h is bounded and harmonic, $h(X_t)$ is a bounded, hence convergent, martingale with limit $H \in \mathcal{I}$. Moreover, $h(x) = E^x H$. Thus to specify $\mathcal{I}$ is to characterize all bounded harmonic functions. Typically, there is some $\mathcal{I}$-measurable random variable Z for which one suspects $\sigma(Z) = \mathcal{I}$. The inclusion $\sigma(Z) \subseteq \mathcal{I}$ is automatic, yet for equality it must be shown there isn't more information in $\mathcal{I}$ than is provided by Z. This is the role played by Theorem 1 below.

One of the main tools of this paper is coupling. By a coupling for (X, P^x), we shall mean any process $((X, Y), P^{(x,y)})$ on $E \times E$ such that $(X, P^{(x,y)})$ and

---

[1]Research supported by a grant from NSA/NSF.

$(Y, P^{(x,y)})$ are copies of the given process with $P^{(x,y)}(X_0 = x) = P^{(x,y)}(Y_0 = y) = 1$. In addition, if we define the coupling time

$$T(X, Y) = \inf\{t > 0 : X_t = Y_t\}$$

then we set $Y_t = X_t$ for $t > T(X, Y)$. It is fairly classical, that if for every $x, y \in E$ there is a successful coupling, i.e. $P^{(x,y)}(T(X, Y) < \infty) = 1$ then $\mathcal{I}$ must be trivial. Indeed, if h is bounded and harmonic

$$|h(x) - h(y)| \leq E^{(x,y)}[|h(X_t) - h(Y_t)|,\ T > t]$$
$$\leq 2\|h\|_{L^\infty(E)} P^{(x,y)}(T > t)$$
$$\to 0 \qquad \text{as } t \to \infty$$

so h is constant. We shall see later that a uniform bound $P^{(x,y)}(T(X, Y) < \infty) \geq \epsilon > 0$ will give the same conclusion.

As mentioned above, there are many occasions when the invariant σ-field is nontrivial and seems to be generated by some random variable Z. By 'localizing' the above argument to each value in the range of Z, we get the following theorem. This theorem is motivated by the argument in Fristedt, Orey (1978) and Jamison, Orey (1978). Our result differs from their ideas insofar as coupling is brought into the picture. This allows us to deal with higher dimensional situations where paths do not meet so easily, if at all. The work here is an extension of the author's thesis which was written under the direction of Steven Orey.

Theorem 1. *Suppose (X, P^x) is a strong Markov process with invariant σ-field $\mathcal{I}$. Suppose $Z : (\Omega, \mathcal{I}) \to (\Gamma, \mathcal{B})$ is measurable and that $\mathcal{B}$ is the largest σ- field for which Z is $\mathcal{I}$-measureable. Assume there is an $\epsilon_0 > 0$ and a set $\Omega' \in \mathcal{I} \subset \Omega$ with $P^x(\Omega') = 1$ for some $x \in E$ and if $\omega, \omega' \in \Omega'$ with $Z(\omega) = Z(\omega')$ then there exist times $T_n, S_n \uparrow \infty$ and couplings $P^{(\cdot, \cdot)}$ with*

$$P^{(X_{T_n}(\omega), X_{S_n}(\omega'))}(T(X, Y) < \infty) > \epsilon_0 .$$

Then $\sigma(Z) = \mathcal{I}$.

This conclusion also holds if there are times U_n, V_n and couplings $P_1^{(\cdot, \cdot)}$ and $P_2^{(\cdot, \cdot)}$ such that

$$P_1^{(X_{T_n}(\omega), X_{S_n}(\omega'))} \left(P_2^{(X_{U_n}, Y_{V_n})}(T(X, Y) < \infty \geq \epsilon_0 \right) \geq \epsilon_0 .$$

The proof of Theorem 1 relies on a refinement of the martingale convergence theorem known as Hunt's lemma (Hunt (1966)). This lemma was used in Fristedt-Orey (1978). The following description of Hunt's Lemma holds for discrete as well as continuous time strong Markov processes. Suppose $\Lambda \in \mathcal{I}$ and define $h(x) = P^x(\Lambda)$. Then by martingale convergence, $h(X_t) = P^{X_t}(\Lambda) = P^x(S_t \Lambda | F_t) = P^x(\Lambda | F_t) \to 1_\Lambda$ a.s. when $t \to \infty$. Define for $\frac{1}{2} > \epsilon > 0$ fixed

$$R(\epsilon) = \{x : h(x) > 1 - \epsilon\}$$

$$B(\epsilon) = \{x : h(x) < \epsilon\}$$

$$DR(\epsilon) = \{x : P^x(X_t \in R(\epsilon), \forall t > 0) > 1 - \epsilon\}$$

$$DB(\epsilon) \doteq \{x : P^x(X_t \in B(\epsilon), \forall t > 0) > 1 - \epsilon\} .$$

The original terminology is a point x is called red (of degree ϵ) if $h(x) > 1 - \epsilon$ and blue if $h(x) < \epsilon$, dark red if $x \in DR(\epsilon)$ and dark blue if $x \in DB(\epsilon)$. Now the martingale convergence of $h(X_t) \to 1_\Lambda$ may be expressed that either X_t is eventually red or X_t is eventually blue. Now set

$$\mathcal{R}_t(\epsilon) = \bigcap_{s > t} \{X_s \in R(\epsilon)\}$$

$$\mathcal{B}_t(\epsilon) = \bigcap_{s > t} \{X_s \in B(\epsilon)\} .$$

Then $x \in DR(\epsilon)$ says $P^x(\mathcal{R}_0(\epsilon)) > 1 - \epsilon$ and similarly for $x \in DB(\epsilon)$. The above mentioned refinement of martingale convergence is that either X_t is eventually dark red or X_t is eventually dark blue. To see this, notice that $1_{\mathcal{R}_t(\epsilon)} \uparrow 1_\Lambda$ and thus

$$P^{X_t}(\mathcal{R}_0(\epsilon)) = P^x(S_t \mathcal{R}_0(\epsilon) | F_t)$$

$$= P^x(\mathcal{R}_t(\epsilon) | F_t)$$

$$\to 1_\Lambda \qquad \text{a.s.}$$

Thus a.s. if $h(X_t(w)) \to 1$ then $X_t(w) \in DR(\epsilon)$ for sufficiently large t and if $h(X_t(w)) \to 0$ then $X_t(w) \in DB(\epsilon)$ for sufficiently large t.

We shall make use of one further extension of the above. Namely, set

$$DDR(\epsilon) = \{x : P^x(X_t \in DR(\epsilon), \forall t > 0) > 1 - \epsilon\}$$
$$DDB(\epsilon) = \{x : P^x(X_t \in DB(\epsilon), \forall t > 0) > 1 - \epsilon\} .$$

Thus a point $x \in DDR(\epsilon)$ has most of its grandchildren of color red. Now it turns out that a.s. either $X_t \in DDR(\epsilon)$ eventually or $X_t \in DDB(\epsilon)$ eventually. This is seen as follows, set

$$\mathcal{DR}_t(\epsilon) = \bigcap_{s>t} \{X_s \in DR(\epsilon)\}$$
$$\mathcal{DB}_t(\epsilon) = \bigcap_{s>t} \{X_t \in DB(\epsilon)\} .$$

Then a.s. $\mathcal{DR}_t(\epsilon) \uparrow 1_\Lambda$ as $t \to \infty$ so

$$P^{X_t}(\mathcal{DR}_0(\epsilon)) = P^x(S_t \mathcal{DR}_0(\epsilon)|F_t)$$
$$= P^x(\mathcal{DR}_t(\epsilon)|F_t)$$
$$\to 1_\Lambda \qquad \text{a.s.} .$$

The same holds for dark blue and these results are summarized by

Lemma 2 (Hunt). *The following hold a.s.*

$$\Omega = \{X_t \in R(\epsilon) \text{ eventually }\} \cup \{X_t \in B(\epsilon) \text{ eventually }\}$$
$$= \{X_t \in DR(\epsilon) \text{ eventually }\} \cup \{X_t \in DB(\epsilon) \text{ eventually }\}$$
$$= \{X_t \in DDR(\epsilon) \text{ eventually }\} \cup \{X_t \in DDB(\epsilon) \text{ eventually }\} .$$

If $\epsilon_k \downarrow 0$ as $k \to \infty$ then these a.s. hold with ϵ replaced by ϵ_k for any k and the null set does not depend on k. Moreover, the events on the right hand sides belong to $\mathcal{I}$.

We now turn to the proof of Theorem 1.

Proof (Theorem 1). It suffices to show $\sigma(Z) \supseteq \mathcal{I}$ so begin with $\Lambda \in \mathcal{I}$ and define $h(x) = P^x(\Lambda)$. Then take Ω'' to be the intersection of Ω' with the set where Hunt's lemma holds. Note that $P^x(\Omega'') = 1$ for one x and hence all x by the maximum principle since the left hand side is a harmonic function. Define

$$\mathcal{R} = \{z \in \text{ range } Z : \exists \omega \in \Omega'', \ Z(\omega) = z, \ \lim_{t \to \infty} h(X_t(\omega)) = 1\} .$$

Note that $\mathcal{R}$ is measurable in the range of Z by our assumption- construction. Then we claim $\Lambda = \{Z \in \mathcal{R}\}$, P^x a.s.

For this it suffices to eliminate the possibility that for some $\omega' \in \Omega''$, $Z(\omega') \in \mathcal{R}$ but $\lim_{t \to \infty} h(X_t(\omega')) = 0$. Since then there is also some $\omega \in \Omega''$ for which $Z(\omega) = Z(\omega')$ yet $\lim_{t \to \infty} h(X_t(\omega)) = 1$. We may take the ϵ in Hunt's lemma to be $\epsilon = (\epsilon_0/4) \wedge 1/8$ with ϵ_0 as in the statement of Theorem 1. Then for n large enough, both $X_{T_n}(\omega) \in DR(\epsilon)$ and $X_{S_n}(\omega') \in DB(\epsilon)$. This gives,

$$P^{(X_{T_n}(\omega), X_{S_n}(\omega'))}(T(X, Y) < \infty, X_t \in R(\epsilon), Y_t \in B(\epsilon), \forall t > 0) \geq \epsilon/2$$

which is a contradiction since on this event

$$h(X_{T(X,Y)}) > 1 - \epsilon, \ h(Y_{T(X,Y)}) < \epsilon, \ h(X_{T(X,Y)}) = h(Y_{T(X,Y)})$$

yet $\epsilon \leq 1/8$. Thus no such ω' exists. This shows $\Lambda = \{Z \in \mathcal{R}\} P^x$ a.s. for some x. But $P^x(\Omega') = 1$ for one x implies for all x so that $\Lambda = \{Z \in \mathcal{R}\}$ for all x.

For the second part, wait until n is sufficiently large that $X_{T_n}(\omega) \in DDR(\epsilon/2)$, $X_{S_n}(\omega') \in DDB(\epsilon/2)$. Then argue almost as before, observing that with probability at least $1 - \epsilon$,

$$X_{U_n} \in DR(\epsilon/2), \ Y_{V_n} \in DB(\epsilon/2)$$

and arrive at the same contradiction as in the first part of the proof, i.e. a red path meets a blue path. This completes the proof. $\square$

We turn to applications of Theorem 1.

Example 1. This example is almost trivial but well illustrates the idea of Theorem 1. Suppose (X_n, P^x) is the renewal process on the nonnegative integers,

$$P^x(X_{n+1} = k + 1 | X_n = k) = 1 - p_k$$

$$P^x(X_{n+1} = 0 | X_n = k) = p_k$$

and select the sequence $\{p_k\}$ so that X_n is transient. Now the field $\mathcal{I}$ for the space-time process (X_n, n) is the tail field $\mathcal{T}$ for (X_n). Turning attention first to the tail field for X_n set

$$Z = \lim_{n \to \infty} n - X_n .$$

Since X_n is transient, it visits 0 for some last time so Z exists and in fact $n - X_n = Z$ for n sufficiently large. Clearly, Z is a tail random variable but notice it is not invariant for (X_n). For finding $\mathcal{T}$, take Ω' from Theorem 1 to be the set where the limit defining Z exists. Now if $Z(\omega) = Z(\omega')$, then for n sufficiently large, $n - X_n(\omega) = n - X_n(\omega')$, i.e. $(n, X_n(\omega)) = (n, X_n(\omega'))$ and trivially coupling occurs with probability one when the two processes are commenced at the same point. Thus the tail field for (X_n) or invariant field for (n, X_n) is generated by Z.

Turning now to the invariant field for $\mathcal{I}$ simply take $Z \equiv 1$ and $\Omega' = \{\lim_{n \to \infty} X_n = \infty\}$. Then if $Z(\omega) = Z(\omega') = 1$, take $T_n = S_n$ defined by $T_n = \inf\{k > 0 : X_k = n\}$. Then $X_{T_n(\omega)}(\omega) = X_{S_n(\omega')}(\omega') = X_{T_n(\omega')}(\omega')$ and

$$P^{(X_{T_n(\omega)}(\omega), X_{S_n(\omega')}(\omega'))}(T < \infty) = 1 .$$

Thus $Z = 1$ generates $\mathcal{I}$ so the invariant field is trivial.

The next examples will rely on a coupling from Lindvall-Rogers (1986). They were interested in producing a successful, i.e. probability one coupling, for diffusions in $\mathcal{R}^d$ of the form $dX_t = \sigma(X_t)dB_t + b(X_t)dt$ where B is $BM(\mathcal{R}^d)$. We will only need the coupling they have developed in the case $\sigma(x) \equiv I$, the identity matrix. Also, we do not seek a probability one coupling. For our diffusions generally this will not exist as we are interested in diffusions with nontrivial invariant fields. Another, though different use of coupling in function theory appeared in Lyons-Sullivan (1984).

The Lindvall-Rogers idea is coupling by reflection. Given two starting points x and y, set

$$L_{x,y} = \{u : \langle u - (x+y)/2, x - y \rangle = 0\}$$

and reflect a Brownian motion B commenced at x in $L_{x,y}$ to get a Brownian motion B' commenced at y. This goes on until time

$$T(B, B') = \inf\{t > 0 : B_t = B'_t\}$$

after which the two move as one, i.e. set $B'_t = B_t$, $t \geq T(B, B')$. Note that $T(B, B') = \sigma_{L_{x,y}}(B) = \inf\{t > 0 : B_t \in L_{x,y}\} = \sigma_{L_{x,y}}(B')$.

Example 2. Consider now diffusions with generator $L^\alpha = \frac{1}{2}\Delta + r^\alpha \frac{\partial}{\partial r}$, on $\mathcal{R}^d$, where α may be $-1, 0$ or 1. Let $\theta(x)$ be the angular part of x in polar coordinates. For $\alpha > -1, \lim_{t\to\infty} \theta(X_t) = \Theta$ exists P^x a.s. which is not difficult to show. The following result holds for $-1 \leq \alpha \leq 1$ but will only be proved for $\alpha = -1, 0, 1$. More precisely, the conclusion of the Theorem for $\alpha = 0$ or 1 holds as well for $-1 < \alpha \leq 1$. The Martin boundary for these operators has been previously computed by Murato (1986). In the next example we will show how to get a new result by perturbing the operator L^0.

Theorem 2. For $\alpha = 0$ or 1, the diffusion with generator L^α has $\sigma(\Theta) = \mathcal{I}$. For $\alpha = -1$, $\mathcal{I}$ is trivial. Consequently, for $\alpha = 0$ or 1, every bounded L^α-harmonic function h may be represented $h(x) = E^x g(\Theta)$ for some $g \in L^\infty(S^{d-1})$.

For the diffusion with generator $L^\alpha, \alpha = -1$ or 0, we shall use a skew product representation. Let b and θ be independent $BM(\mathcal{R}^1)$ and $BM(S^{d-1})$, respectively. Define the process r_t by

$$r_t = r + b_t + \int_0^t \left(\frac{d-1}{2r_s} + r_s^\alpha\right) ds ,$$

Also, define the clock

$$\ell_t = \int_0^t r_s^{-2} ds \, .$$

Then $X_t = (r_t, \theta_{\ell_t})$ is a diffusion with generator L^α.

We now describe a modification of the Lindvall-Rogers coupling in terms of skew products. For that matter, it is also a special case of the coupling of W. Kendall (1986) which generalized the Lindvall-Rogers coupling to manifolds. This coupling relies on a skew product representation of θ which may be found in Itô-McKean (1966).

The Laplacian on S^{d-1} may be written

$$\Delta_{S^{d-1}} = (\sin \varphi)^{2-d} \frac{\partial}{\partial \varphi} (\sin \varphi)^{d-2} \frac{\partial}{\partial \varphi} + (\sin \varphi)^{-2} \Delta_{S^{d-2}}$$

with φ the so-called colatitude and Δ_1 is just $\frac{\partial^2}{\partial \theta^2}$. This means θ_t is $(\varphi_t, \gamma_{m_t})$ where φ is what is called a $LEG(2)$, a Legendre process, i.e. φ solves the s.d.e.

$$d\varphi_t = dW_t + \frac{d-2}{2} \cot \varphi_t dt$$

with W_t a $BM(\mathcal{R}^1)$, independent of γ,

$$m_t = \int_0^t (\sin \varphi_s)^{-2} ds$$

and γ_t is $BM(S^{d-2})$. Our interest is in a coupling $((X, Y), P^{(x,y)})$ of two copies X and Y of a diffusion with generator L^α started at x and y respectively in the special case where $\|x\| = \|y\|$. With this last assumption, both X and Y can and will be run by the same r process. Now it is also possible via rotation to start with x and y such that $\frac{\pi}{2} - \varphi(x) = \varphi(y) - \frac{\pi}{2}$. Using the fact that $\cot \varphi$ is an odd function with respect to $\pi/2$, given one $LEG(2)$ process φ which is independent of γ, a $BM(S^{d-2})$ we obtain $X_t = (r_t, \varphi_{\ell_t}, \gamma_{m \circ \ell_t})$ and $Y_t = (r_t, \varphi'_{\ell_t}, \gamma_{m \circ \ell_t})$ with $\varphi'_t = \pi - \varphi_t$ which gives automatically

$$m_t = \int_0^t (\sin \varphi_s)^{-2} ds = \int_0^t (\sin \varphi'_s)^{-2} ds = m'_t \, .$$

This is our coupling. Notice that the coupling time $T(X,Y) = T = \inf\{t > 0 : X_t = Y_t\}$ is equal to $\sigma_{\frac{\pi}{2}}(\varphi) = \inf\{t > 0 : \varphi_t = \frac{\pi}{2}\}$. When $\alpha > -1$, $P^{(x,y)}(T = \infty) > 0$ whenever $x \neq y$ and in some instances this is called an unsuccessful coupling.

We first consider the extreme cases $\alpha = 1$ and $\alpha = -1$. The case $\alpha = 1$ is contained in COR(1980) but we now give a different analysis using coupling. Actually, what is done here can be applied to the operators $L = \frac{1}{2}\Delta + Ax \cdot \nabla$, where A is a constant matrix, which were treated in COR(1980). In fact, new results may be obtained with minor modifications by perturbing the drift term A by adding a sufficiently small drift. How this may be done when $\alpha = 0$ will be outlined later. In this case $L^1 = \frac{1}{2}\Delta + x \cdot \nabla$ has a diffusion given by

$$X_t = e^t(x + \int_0^t e^{-s} dB_s)$$

where B is $BM(\mathcal{R}^d)$. It is clear that $\theta(X_t) = \frac{X_t}{\|X_t\|} = \frac{x + \int_0^t e^{-s} dB_s}{\|x + \int_0^t e^{-s} dB_s\|} \to \Theta$ since $\int_0^t e^{-s} dB_s$ has components which are L^2-convergent martingales. In this case the coupling is a time change away from that of Lindvall-Rogers for Brownian motion. Notice that the process $x + W_t = x + \int_0^{\sigma_t} e^{-s} dB_s$, $\sigma_t = \ell n(1 - 2t)^{-\frac{1}{2}}, 0 \leq t < \frac{1}{2}$, is d-dimensional Brownian motion. Now reflect $x + W_t$ in the hyperplane $L_{x,y}$ to obtain the Brownian motion $y + W'$. Then $Y_t = e^t(y + W'_{\sigma_t^{-1}})$ is a diffusion with generator L^1. Moreover, $T(X,Y) = \sigma_{L_{x,y}} = \inf\{t > 0 : x + W_{\sigma_t^{-1}} \in L_{x,y}\}$.

Lemma 3. *With the coupling just described, we have the lower bound*

$$P^{(x,y)}(T(X,Y) < \infty) \geq c\left(\frac{1}{\|x - y\|} - \frac{1}{\|x - y\|^3}\right) e^{-\frac{\|x-y\|^2}{2}}$$

for some positive constant c.

Proof. Selecting coordinates so that the first component is parallel to $x - y$,

$$
\begin{aligned}
P^{(x,y)}(T(X,Y) < \infty) &= P^{(x,y)}(\sigma_{L_{x,y}} < \infty) \\
&= P^{(x,y)}\left(W^1_{\sigma_t^{-1}} > \frac{\|x - y\|}{2}, \quad \text{for some } t > 0\right) \\
&= 2P^{(x,y)}\left(W^1_{\sigma_\infty^{-1}} > \frac{\|x - y\|}{2}\right) \\
&= 2P^{(x,y)}\left(W_{\frac{1}{2}} > \frac{\|x - y\|}{2}\right) \\
&= \sqrt{\frac{2}{\pi}} \int_{\frac{\sqrt{2}}{2}\|x-y\|}^{\infty} e^{-\frac{u^2}{2}} \, du \cdot \\
&\geq c\left(\frac{1}{\|x - y\|} - \frac{1}{\|x - y\|^3}\right) e^{\frac{-\|x-y\|^2}{2}}. \quad \square
\end{aligned}
$$

The next order of business is to establish an oscillation result telling how close the diffusion approaches its limiting ray and with what frequency. For this, define $Z_t \equiv x + \int_0^t e^{-s} dB_s = x + W_{\sigma_t^{-1}}$ and set $A_n = \{\|X_n - e^n Z_\infty\| < \kappa\}$ where κ will be selected later. Notice that $\theta(X_n) = \theta(Z_n)$ and $\theta(X_n) \to \theta(Z_\infty)$ so $\|X_n - e^n Z_\infty\|$ is measuring the distance from X_n to its limiting trajectory (ray).

Lemma 4. *Given κ sufficiently large, there is a constant $c(\kappa) > \frac{1}{2}$ such that*

$$
\lim_{N \to \infty} \frac{1}{N} \sum_{n=1}^{N} 1_{A_n} = c(\kappa) \qquad a.s. \quad .
$$

Proof. $X_n = e^n Z_n$ so consider $Z_\infty - Z_n = \int_n^\infty e^{-s} dB_s = W_{\sigma_\infty^{-1}} - W_{\sigma_n^{-1}}$. Notice that $\sigma_n^{-1} = \frac{1}{2}(1 - e^{-2n})$ and consequently, dropping the superscript (x, y) from the notation,

$$
P(A_n) = P(\|W_{\frac{1}{2}e^{-2n}}\| < \kappa e^{-n}) = P(\|W_1\| < \sqrt{2}\kappa) = c(\kappa).
$$

Furthermore, for $n > m$,

$$P(A_m \cap A_n) = P(\|Z_m - Z_\infty\| < \kappa e^{-m}, \|Z_n - Z_\infty\| < \kappa e^{-n})$$
$$\leq P(\|Z_m - Z_n\| < \kappa e^{-m}(1 + e^{m-n}), \|Z_n - Z_\infty\| < \kappa e^{-n})$$
$$= P(\|Z_m - Z_n\| < \kappa e^{-m}(1 + e^{m-n}))P(\|Z_n - Z_\infty\| < \kappa e^{-n})$$
$$= P\left(\|W_{\frac{1}{2}(e^{-2m}-e^{-2n})}\| < \kappa e^{-m}(1 + e^{m-n})\right) P\left(\|W_{\frac{1}{2}e^{-2n}}\| < \kappa e^{-n}\right)$$
$$= P\left(\|W_{\frac{1}{2}e^{-2m}}\| < \kappa e^{-m}\frac{1 + e^{m-n}}{\sqrt{1 - e^{2m-2n}}}\right) P(A_n)$$
$$= c_{m,n}P(A_m)P(A_n)$$
$$= c_{m,n}c(\kappa)^2$$

with $c_{m,n} = \dfrac{P\left(\|W_1\| < \sqrt{2}\kappa\frac{(1+e^{m-n})}{\sqrt{1-e^{2m-2n}}}\right)}{P(\|W_1\| < \sqrt{2}\kappa)}$.

A straightforward estimate gives

$$|c_{m,n} - 1| \leq ce^{m-n}, \text{ for } m < n.$$

This shows the rate of L^2-convergence of $\frac{1}{N}\sum_{n=1}^{N} 1_{A_n}$ is

$$E\left[\frac{1}{N}\sum_{n=1}^{N} 1_{A_n} - c(\kappa)\right]^2 = \frac{1}{N^2}\sum_{n=1}^{N} P(A_n) + \frac{2}{N^2}\sum_{1 \leq m < n \leq N} P(A_n \cap A_m)$$
$$- \frac{2c(\kappa)}{N}\sum_{n=1}^{N} P(A_n) + c(\kappa)^2$$
$$\leq \frac{1}{N} + \frac{2c(\kappa)^2}{N^2}\sum_{1 \leq m < n \leq N} |c_{m,n} - 1|$$
$$\leq \frac{1}{N} + \frac{2c(\kappa)^2}{N^2}\sum_{1 \leq m < n \leq N} e^{m-n}$$
$$\leq \frac{c}{N}.$$

Using Borel-Cantelli it follows that $\frac{1}{N^2}\sum_{n=1}^{N^2} 1_{A_n} \to c(\kappa)$, a.s. . The entire sequence may be filled in, if $N^2 < m < (N+1)^2$ then

$$\frac{1}{N^2}\sum_{n=1}^{N^2} 1_{A_n} \leq \frac{m}{N^2} \cdot \frac{1}{m}\sum_{n=1}^{m} 1_{A_n} \leq \left(\frac{N+1}{N}\right)^2 \frac{1}{(N+1)^2}\sum_{n=1}^{(N+1)^2} 1_{A_n}$$

so $\dfrac{1}{N}\sum_{n=1}^{N} 1_{A_n} \to c(\kappa)$ a.s.   as desired. $\square$

Proof (Theorem 2 for $\alpha = 1$).

Let Ω' be the a.s. set from Lemma 4 intersected with the set where $\Theta(X_t) \to \Xi$. Suppose that $\Theta(\omega) = \Theta(\omega')$ for some $\omega, \omega' \in \Omega'$. Take $T_n = S_n \equiv n$. Now by the oscillation result Lemma 4, there is a sequence n_k such that both

$$\|X_{n_k}(\omega) - e^{n_k} Z_\infty(\omega)\| \leq \kappa \text{ and } \|X_{n_k}(\omega') - e^{n_k} Z_\infty(\omega')\| \leq \kappa.$$

Assume, as we might, that $\|Z_\infty(\omega)\| > \|Z_\infty(\omega')\|$ and set $a = \ell n \frac{\|Z_\infty(\omega)\|}{\|Z_\infty(\omega')\|}$. Consider the diffusion $(X, P^{X_{n_k}(\omega')})$. These paths may be represented

$$X_t = e^t \left(X_{n_k}(\omega') + \int_0^t e^{-s} dB, \right).$$

Observe that at time $t = a$, X_a has normal distribution with mean $e^a X_{n_k}(\omega')$ and covariance $\frac{1}{2}(e^{2a} - 1)I$. In order to apply Theorem 1, we select $U_{n_k} = 0$, $V_{n_k} = a$. With this choice of V_{n_k},

$$P^{X_{n_k}(\omega')} \left(\|X_a - e^a X_{n_k}(\omega')\| \leq \sqrt{\frac{1}{2}(e^{2a} - 1)} \right) = c > 0$$

where c is independent of n_k and a. Moreover, for $x \in B(e^a X_{n_k}(\omega'), \sqrt{\frac{1}{2}(e^{2a} - 1)})$

$$\|x - X_{n_k}(\omega)\| \leq \|x - e^a X_{n_k}(\omega')\| + \|e^a X_{n_k}(\omega') - e^a e^{n_k} Z_\infty(\omega')\|$$
$$+ \|e^a e^{n_k} Z_\infty(\omega') - e^{n_k} Z_\infty(\omega')\| + \|e^{n_k} Z_\infty(\omega) - X_{n_k}(\omega)\|$$
$$\leq \sqrt{\frac{1}{2}(e^{2a} - 1)} + e^a \kappa + \kappa$$
$$= c(a, \kappa).$$

Thus, we can find an $\epsilon_0 = \epsilon_0(a, \kappa)$ such that

$$P^{(X_{n_k}(\omega), X_{n_k}(\omega'))} \left(P^{(X_0, X_a)}(T(X, Y) < \infty) \geq \epsilon_0) \right)$$
$$\geq P^{(X_{n_k}(\omega), X_{n_k}(\omega'))} \left(P^{(X_0, X_a)}(T(X, Y) < \infty) \geq \epsilon_0, X_a \in B(X_{n_k}(\omega), c(a, \kappa)) \right)$$
$$\geq \epsilon_0, \text{ by Lemma 3 and the above calculations.}$$

Thus, by part 2 of Theorem 1 we are done. $\square$

We now show the random variable Z_∞ generates the tail σ- field for X or what is the same Z_∞ generates the invariant field for (t, X_t). Let Ω' be the a.s. set from Lemma 4 intersected with the set where $Z_\infty(\omega)$ exists. When $\omega, \omega' \in \Omega'$ with $Z_\infty(\omega) = Z_\infty(\omega')$, Lemma 4 gives infinitely many n for which $\|X_n(\omega) - X_n(\omega')\| < 2\kappa$.

Writing $\tilde{X}_t = (t, X_t)$, $\|\tilde{X}_n(\omega) - \tilde{X}_n(\omega')\| < 2\kappa$ for these same values of n. Taking $T_n = S_n = n$ and using Lemma 3

$$P^{(\tilde{X}_n(\omega), \tilde{X}_n(\omega'))}(T(\tilde{X}, \tilde{Y}) < \infty) \geq c(\kappa)$$

again for those values of n for which $\|X_n(\omega) - X_n(\omega')\| < 2\kappa$. By Theorem 1, Z_∞ generates the invariant field for $\tilde{X}$ or the tail σ-field for X.

The case $\alpha = -1$ is covered by the elegant analysis in P. March (1986). In this instance,

$$dr_t = db_t + \frac{d+1}{2r_t} dt .$$

The one-dimensional diffusion r_t has scale

$$s(r) = \int_1^r \rho^{-d-1} d\rho$$

and speed measure

$$m(dr) = 2r^{d+1} dr .$$

Thus one finds with γ_t defined by $\int_0^{\gamma_t} r_s^{-2} ds = t$ that r_{γ_t} is a diffusion with speed $2r^{d-1} dr$ and scale $s(r)$.

Also, if $\tau_R = \inf\{t : r_{\gamma(t)} = R\}$ then $P(\tau_\infty < \infty) = P(\ell_\infty < \infty)$. But $\int_1^\infty r^{d-1} dr \int_r^\infty \rho^{-d-1} d\rho = \infty$ so ∞ is a nonexit boundary for r_γ. This means $P(\tau_\infty < \infty) = P(\ell_\infty < \infty) = 0$. Thus θ_{ℓ_t} has no limit as θ is recurrent on S^{d-1} and $\ell_t \to \infty$ a.s. with t.

We now show that $X_t = (r_t, \theta_{\ell_t})$ has what W. Kendall (1986) refers to as the "Coupling Property." That is, we can make a coupling $((X, Y), P^{(x,y)})$ of the diffusion such that $P^{(x,y)}(T < \infty) = 1$ and this is independent of x and y.

Actually, this will only be needed for our purposes in the instance $\|x\| = \|y\|$. In that case, use the coupling we have already described for the processes with generator L^α. Since the clock $\ell_t = \int_0^t r_s^{-2} ds$ runs forever, i.e. $\ell_t \to \infty$, it is clear that φ_{ℓ_t} will hit $\frac{\pi}{2}$ eventually at which time the two paths will couple. Also, $T = T(X,Y)$ is a stopping time, so if h is a bounded solution in $\mathcal{R}^d$ of $L^{-1}h = 0$, by optional sampling,

$$|h(x) - h(y)| = |E^{(x,y)}[h(X_T) - h(Y_T)]| = 0 \ .$$

Thus h is constant on $\{x : \|x\| = r\}$ for each r and thus h is constant on $\mathcal{R}^d$. This proves the $\alpha = -1$ portion of Theorem 1.

We now treat the case $\alpha = 0$.

Lemma 5. Given any δ, $\frac{1}{2} > \delta > 0$ define $1 - \delta_t = t^{-\frac{1}{2}+\delta}$, $\eta_{t,r} = 1 + \frac{d-1}{r} + t^{-\frac{1}{2}+\delta}$. Then there is a $T = T(\omega)$ such that for $t > T$ a.s.

(5.1a) $r + t\delta_t \le r_t$

(5.1b) $\ell_\infty - \ell_t \le (\delta_t(r + t\delta_t))^{-1}$

If $E_r = \{t : r_t > \frac{r}{2}, \ \forall t > 0\}$ then $P^r(E_r^c) \le ce^{-cr}$ and on E_r a.s. one has for $t > T$,

(5.1c) $r_t \le r + t\eta_{t,r}$

(5.1d) $\ell_\infty - \ell_t \ge (\eta_{t,r}(r + t\eta_{t,r}))^{-1}$

Proof. Let T be the random time after which $|b_t| \le t^{\frac{1}{2}+\delta}$ holds for all t. Then for $t > T$,

$$r_t \ge r + t - t^{\frac{1}{2}+\delta}$$

and so again for $t > T$

$$\ell_\infty - \ell_t = \int_t^\infty r_s^{-2} ds$$

$$\leq \int_t^\infty (r + s(1 - s^{-\frac{1}{2} + \delta}))^{-2} ds$$

$$\leq \int_t^\infty (r + s\delta_t)^{-2} ds$$

$$= (\delta_t(r + t\delta_t))^{-1}.$$

The estimate for $P^r(E_r^c)$ is obtained by solving the Dirichlet problem $\frac{1}{2} u'' + \left(\frac{d-1}{2r} + 1\right) u' = 0$ on $[\frac{r}{2}, \infty)$, $u(\frac{r}{2}) = 1$, $u(\infty) = 0$ and evaluating $u(r)$ which gives the desired probability. On E_r, $\frac{d-1}{2} \int_0^t r_s^{-1} ds < \frac{d-1}{r} t$ holds trivially and the bound (5.1c) follows as before using the lower bound $b_t > -t^{\frac{1}{2} + \delta}$ for $t > T$ for b_t. The bound (5.1d) follows from (5.1c). $\square$

Let ρ denote the arclength metric on S^{d-1} and define $A_n = \{\rho(\theta_{\ell_{2^n}}, \theta_{\ell_\infty}) \leq \kappa 2^{-\frac{n}{2}}\}$. We now establish the analog of Lemma 4 for the case $\alpha = 0$.

Lemma 6. *If κ is selected sufficiently large*

$$\varliminf_{N \to \infty} \frac{1}{N} \sum_{n=1}^N 1_{A_n} > \frac{1}{2}, \quad a.s. .$$

Proof. Keeping in mind that θ and ℓ are independent, we compare $\rho(\theta_{\ell_{2^n}}, \theta_{\ell_\infty})$ with $\|W_{\ell_{2^n}} - W_{\ell_\infty}\|$ where W is $BM(\mathcal{R}^{d-1})$ independent of ℓ. By means of a comparison theorem it can be arranged, since the plane is flat and the sphere round (positively curved), that

$$\|W_{\ell_{2^n}} - W_{\ell_\infty}\| \geq \rho(\theta_{\ell_{2^n}}, \theta_{\ell_\infty}).$$

This implies that if $B_n = \{\|W_{\ell_{2^n}} - W_{\ell_\infty}\| \leq \kappa 2^{-\frac{n}{2}}\}$ then $B_n \subseteq A_n$. Thus the $\varliminf$ bound will follow once it is shown that

$$\varliminf_{N \to \infty} \frac{1}{N} \sum_{n=1}^N 1_{B_n} \geq c(\kappa) \quad a.s.$$

for a constant $c(\kappa) > \frac{1}{2}$.

Noting that for some $a > 1$ and $\delta > 0$

$$P\left(|b_t| \le t^{\frac{1}{2}+\delta}, \quad \forall t \ge 2^n\right) \ge 1 - a^{-n}$$

on defining

$$C_n = \{r + t - t^{\frac{1}{2}+\delta} \le r_t \le r + t + t^{\frac{1}{2}+\delta}, \quad \forall t \ge 2^n\}$$

it follows that

$$P(C_n) \ge 1 - ca^{-n}$$

for some positive value of c.

Finally, set

$$D_n = B_n \cap C_n .$$

Then since for almost any ω, $\omega \in \bigcap_{m>T(\omega)} C_m$, for some $T(\omega)$, the whole thing comes down to showing

$$\varliminf_{N \to \infty} \frac{1}{N} \sum_{n=1}^{N} 1_{D_n} \ge c(\kappa) .$$

On C_n, using Lemma 5 for the upper bound and an entirely analogous argument for the lower bound one gets

$$(\gamma_n(r + 2^n \gamma_n))^{-1} \le \ell_\infty - \ell_{2^n} \le (\beta_n(r + 2^n \beta_n))^{-1}$$

where

$$\beta_n = \delta_{2^n} = 1 - 2^{n(-\frac{1}{2}+\delta)}, \; \gamma_n = \epsilon_{2^n} = 1 + 2^{n(-\frac{1}{2}+\delta)} .$$

These bounds on $\ell_\infty - \ell_{2^n}$ imply

$$\begin{aligned}
P(D_n) &\ge P\left(\|W_{(\beta_n(r+2^n \beta_n))^{-1}}\| \le \kappa 2^{-\frac{n}{2}}\right) P(C_n) \\
&= P\left(\|W_1\| \le \kappa \sqrt{\beta_n(r2^{-n} + \beta_n)}\right) P(C_n) \\
&\equiv c_n(\kappa) P(C_n)
\end{aligned}$$

and

$$P(D_n) \leq P\left(\|W_{(\gamma_n(r+2^n\gamma_n))^{-1}}\| \leq \kappa 2^{-\frac{n}{2}}\right) P(C_n)$$

$$= P\left(\|W_1\| \leq \kappa\sqrt{\gamma_n(r2^{-n} + \gamma_n)}\right) P(C_n)$$

$$\equiv d_n(\kappa)P(C_n).$$

If $c(\kappa) \equiv \lim_{n\to\infty} c_n(\kappa) \equiv \lim_{n\to\infty} d_n(\kappa)$ then

$$|d_n(\kappa) - c(\kappa)| \leq cb^{-n}, \text{ for some } b > 1$$

with a similar estimate for $c_n(\kappa)$.

Now recalling that $P(C_n) = 1 - c_n$ with $c_n \leq a^{-n}$ for some $a > 1$, we have now for some $a, b > 1$ that

$$c(\kappa)(1 + b^{-n}) \geq P(D_n) \geq c(\kappa)(1 - b^{-n})(1 - a^{-n}).$$

Using similar arguments for $m < n$, we have

$$P(D_n \cap D_m) \leq P\left(\|W_{\ell_{2^m}} - W_{\ell_{2^n}}\| \leq \kappa 2^{-\frac{m}{2}}\left(1 + 2^{\frac{m-n}{2}}\right), \|W_{\ell_{2^n}} - W_{\ell_\infty}\| \leq \kappa 2^{-\frac{n}{2}}, C_m\right)$$

$$\leq P\left(\|W_{(\gamma_m(r+2^m\gamma_m))^{-1}}\| \leq \kappa 2^{-\frac{m}{2}}(1 + 2^{\frac{m-n}{2}})\right) \times$$

$$P\left(\|W_{(\gamma_n(r+2^n\gamma_n))^{-1}}\| \leq \kappa 2^{-\frac{n}{2}}\right) P(C_m)$$

$$= c_{m,n} d_m(\kappa) d_n(\kappa) P(C_m)$$

$$\leq c_{m,n}(c(\kappa) + cb^{-m})(c(\kappa) + cb^{-n}) P(C_m)$$

with

$$c_{m,n} = \frac{P\left(\|W_{(\gamma_m(r+2^m\gamma_m))^{-1}}\| \leq \kappa 2^{-\frac{m}{2}}(1 - 2^{\frac{m-n}{2}})\right)}{P\left(\|W_{(\gamma_m(r+2^m\gamma_m))^{-1}}\| \leq \kappa 2^{-\frac{m}{2}}\right)}.$$

A simple estimate shows there is a $c > 1$ such that

$$|1 - c_{m,n}| \leq c^{m-n} \qquad \text{for } m < n.$$

Thus,

$$
E\left[\frac{1}{N}\sum_{n=1}^{N} 1_{D_n} - c(\kappa)\right]^2
$$

$$
= \frac{1}{N^2}\sum_{n=1}^{N} P(D_n) + \frac{2}{N^2}\sum_{1 \le m < n \le N} P(D_n \cap D_m)
$$

$$
- \frac{2c(\kappa)}{N}\sum_{n=1}^{N} P(D_n) + c(\kappa)^2
$$

$$
\le \frac{1}{N} + \frac{2}{N^2}\sum_{1 \le m < n \le N} |P(D_n \cap D_m) - c(\kappa)^2|
$$

$$
+ \frac{2c(\kappa)}{N}\sum_{n=1}^{N} [c(\kappa) - P(D_n)]
$$

$$
\le \frac{1}{N} + \frac{2}{N^2}\sum_{1 \le m < n \le N} [d_{m,n}c_n(\kappa)c_m(\kappa)P(C_m) - c(\kappa)^2]
$$

$$
+ \frac{2c(\kappa)}{N}\sum_{n=1}^{N} [c(\kappa) - c(\kappa)(1 - a^{-n})(1 - b^{-n})]
$$

$$
= O\left(\frac{1}{N}\right).
$$

This implies, as we showed in the case $\alpha = 0$, that $\frac{1}{N}\sum_{n=1}^{N} 1_{D_n} \to c(\kappa)$ a.s. and the lemma is proved. □

The next result says that if two points are sufficiently close then the probability of a successful coupling has a lower bound.

Lemma 7. *Given* $\kappa > 0$ *there is a constant* $c_3(\kappa) > 0$ *such that for* x, y *with* $\|x\| = \|y\|$ *and* $\rho(\theta(x), \theta(y)) < 3\kappa\|x\|^{-\frac{1}{2}}$ *one has*

$$
P^{(x,y)}(T(X,Y) < \infty) \ge c_3(\kappa).
$$

Proof. Take the coupling described above. Then

$$
P^{(x,y)}(T(X,Y) < \infty) = P^{(x,y)}\left(\frac{\pi}{2} - \varphi_{\ell_t} = 0 \text{ for some } t > 0\right).
$$

Note that $\frac{\pi}{2} - \varphi = \rho(\theta(x), \theta(y))$. By simple comparison of drift, if W_t is the $BM(\mathcal{R}^1)$ such that

$$
d\varphi_t = dW_t + \frac{d-2}{2}\cot \varphi_t dt, \quad \varphi_0 = \varphi(x) < \frac{\pi}{2}
$$

then $W_{\ell_t} + \varphi \leq \varphi_{\ell_t}$ holds until $W_{\ell_t} + \varphi$ hits $\frac{\pi}{2}$. Define $F_{\|x\|} = \{b_t \leq \frac{1}{2}\|x\| + t^{\frac{1}{2}+\delta}$ for all $t > 0\}$. Then on $F_{\|x\|}, \ell_\infty \geq \frac{c}{\|x\|}$ and

$$P^{(x,y)}(T(X,Y) < \infty) \geq P^{(x,y)}(W_{\ell_t} > \frac{\pi}{2} - \varphi \text{ for some } t > 0)$$

$$= 2P^{(x,y)}(W_{\ell_\infty} > \frac{\pi}{2} - \varphi), \text{ by independence of } \ell \text{ and } W$$

$$\text{and reflection principle}$$

$$\geq 2P^{(x,y)}\left(W_{c\|x\|^{-1}} > \frac{3}{2}\kappa\|x\|^{-\frac{1}{2}}, E_{\|x\|}, F_{\|x\|}\right), \text{ by (5.1d)}$$

$$\geq 2P^{(x,y)}(W_1 > c\kappa) - 2P^{(x,y)}(E^c_{\|x\|}) - 2P^{(x,y)}(F^c_{\|x\|})$$

$$\geq c_3(\kappa) > 0, \text{ provided } \|x\| \text{ is large. } \square$$

We now prove Theorem 2 for $\alpha = 0$.

Proof (Theorem 2 for $\alpha = 0$).

Take Ω' to be the a.s. set in Lemma 6. Suppose $\theta_{\ell_\infty}(\omega) = \theta_{\ell_\infty}(\omega')$ for $\omega, \omega' \in \Omega'$ and select a sequence $\{2^{n_k}\}$ using Lemma 6 such that

$$\rho(\theta_{\ell_{2^{n_k}}}(\omega), \theta_{\ell_{2^{n_k}}}(\omega')) < 2\kappa 2^{-\frac{n_k}{2}}.$$

¿From now on we drop the subscript k but work with this same subsequence. If t is sufficiently large, there is a $\delta > 0$ so that by the law of the iterated logarithm we may assume $r_t(\omega), r_t(\omega') \in (r + (1-\delta)t, r + (1+\delta)t)$. Now suppose all the n are large enough for this bound to hold at $t = 2^n$. Then if $r_{2^n}(\omega) < r_{2^n}(\omega')$, $r_{2^n}(\omega') - r_{2^n}(\omega) < 2\delta 2^n$ and setting $U_n = \inf\{t > 0 : r_t = r_{2^n}(\omega')\}$ we have

$$P^{X_{2^n}(\omega)}(\rho(\theta_{\ell_{U_n}}, \theta_{\ell_{2^n}}(\omega)) < \kappa 2^{-n/2})$$

$$\geq P^{X_{2^n}(\omega)}\left(\rho\left(\theta_{\ell_{\frac{2\delta}{1-\delta}2^n}}, \theta_{\ell_{2^n}}(\omega)\right) < \kappa 2^{-\frac{n}{2}}, U_n \leq \frac{2\delta}{1-\delta}2^n\right)$$

$$\geq c_1(\kappa)P^{X_{2^n}(\omega)}\left(U_n \leq \frac{2\delta}{1-\delta}2^n\right)$$

$$\geq c_2(\kappa).$$

Thus, on setting $V_n = 0$, by Lemma 7 and the last bound, there is an $\epsilon_0 > 0$ so that

$$P^{(X_{2^n}(\omega), X_{2^n}(\omega'))}\left(P^{(X_{U_n}, X_{V_n})}(T(X,Y) < \infty) \geq \epsilon_0\right) \geq \epsilon_0$$

and by Theorem 1, $\sigma(\theta_{\ell_\infty}) = \mathcal{I}$. □

Example 3. In the case of $\alpha = 0$, we show how to get away from the radially symmetric assumption on L^0. Namely, consider the operator

$$L = \frac{1}{2}\Delta + \frac{\partial}{\partial r} + \sum_i \frac{b^i(r, \theta)}{r} \frac{\partial}{\partial \theta^i}$$

where the $\frac{\partial}{\partial \theta^i}$ are a basis of unit vectors on the tangent space to the unit sphere S^{d-1}. Assume, that each b^i is Lipschitz continuous and that for some positive constants c, ϵ,

$$|b^i(r, \theta)| \leq \frac{c}{(r + 1)^{1+\epsilon}} \quad , r > 0 .$$

The diffusion X with generator L almost has a skew-product representation. Take the same radial process as for the diffusion with generator L^0,

$$dr_t = db_t + \left(1 + \frac{d - 1}{2r_t}\right) dt \quad , r_0 = r$$

and an independent $BM(S^{d-1})$, θ_t. Set $\ell_t = \int_0^t r_s^{-2} ds$ and solve for

$$\psi_t = \theta_{\ell_t} + \int_0^t \frac{b^i(r_s, \psi_s)}{r_s} ds \quad , \psi_0 = \theta .$$

Then $X_t = (r_t, \psi_t)$ is a diffusion with generator L starting at $x = X_0 = (r, \theta)$.

First, it is easy to see that $\Theta = \lim_{t \to \infty} \psi_t$ exists in this case. Also, Lemma 5 holds, as the radial process is unchanged when considering L instead of L^0. The analog of Lemma 6 holds with A_n replaced by $A'_n = \{\rho(\psi_{2^n}, \Theta) < 2\kappa 2^{-\frac{n}{2}}\}$. This is an easy consequence of Lemma 6, the bounds on $|b^i(r, \psi)|$ and the triangle inequality, where one uses that on the set E_r of Lemma 5,

$$\int_0^\infty \frac{|b^i(r_s, \psi_s)|}{r_s} ds \leq cr^{-(1+\epsilon)} .$$

Finally, the analog of Lemma 7 also remains valid. To complete the argument for the analog of Lemma 7, we need to follow the development in Lindvall-Rogers (1986). The coupling is complicated a bit by the presence of the drift term $b(r, \theta)$.

This time define

$$z = x - y$$

$$H = I - \frac{z}{|z|} \cdot \frac{z^T}{|z|} .$$

Then H is unitary, it is reflection in the hyperplane $L_{x,y}$, set $dB'_t = H_t dB_t$, and $H_t = H(X_t, Y_t)$,

$$dX_t = dB_t + \frac{X_t}{|X_t|} dt + b(|X_t|, \theta(X_t)) dt, \ X_0 = x,$$

$$dY_t = dB'_t + \frac{Y_t}{|Y_t|} dt + b(|Y_t|, \theta(Y_t)) dt, \ Y_0 = y .$$

This defines a coupling (X, Y) of two diffusions with generator L begun at x and y. In addition, if $\|x\| = \|y\|$ then for all $t > 0$, $\|X_t\| = \|Y_t\| = r_t$. Using Itô's formula on $\rho_t = \|X_t - Y_t\|$, with $Z_t = X_t - Y_t$,

$$d\rho_t = 2dW_t + \left(\frac{Z_t}{\rho_t}, \frac{X_t - Y_t}{r_t} \right) dt + \left(\frac{Z_t}{\rho_t}, b(X_t) - b(Y_t) \right) dt$$

$$= 2dW_t + \frac{\rho_t}{r_t} dt + \left(\frac{Z_t}{\rho_t}, b(X_t) - b(Y_t) \right) dt$$

where W is a one-dimensional Brownian motion, and by a slight abuse of notation, we consider b as a vector in $\mathcal{R}^d$ which has zero component in the radial direction.

In the case of L^0, denoting the coupled diffusions for this operator by X^0 and Y^0, $X_0^0 = x^0, Y_0^0 = y^0$ and putting $\eta_t = \|X_t^0 - Y_t^0\|$ one notices $r_t = \|X_t^0\| = \|Y_t^0\| = \|X_t\| = \|Y_t\|$ if $\|x^0\| = \|y^0\| = r$ and

$$d\eta_t = 2dW_t + \frac{\eta_t}{r_t} dt .$$

The drift term for ρ is larger than for η but by the bound on b, on the set E_r one has

$$\int_0^\infty \left| \left(\frac{Z_t}{\rho_t}, b(X_t) - b(Y_t) \right) \right| dt \leq c .$$

Recall $P^x(E_r) \geq 1 - ce^{-cr}$. Thus, by simple comparison, i.e. noting that the contribution on E_r of $\int_0^\infty \left(\frac{Z_s}{\rho_s}, b(X_s) - b(Y_s) \right) ds$ is bounded, we have

$$P^{(x,y)}(T(X, Y) < \infty) \geq P^{(x_0, y_0)}(T(X^0, Y^0) < \infty) - 2ce^{-cr},$$

provided $\|x\| = \|y\| = \|x^0\| = \|y^0\| = r$, $\|x - y\| \le c\sqrt{r}$, $\|x^0 - y^0\| \le 2c\sqrt{r}$. This gives the analog of Lemma 7 and the conclusion $\mathcal{I} = \sigma(\Theta)$ follows as in the radially symmetric case.

It should be possible to extend these results to

$$L = \frac{1}{2}\Delta + c(r,\theta)\frac{\partial}{\partial r} + \sum_i \frac{b^i(r,\theta)}{r}\frac{\partial}{\partial \theta^i}$$

when $0 < c_1 < c(r,\theta) < c_2$ for all r, θ, since the radial process for this diffusion will behave approximately as in the case $c(r,\theta) \equiv 1$.

Also, for the operator L^0 we can show that $\mathcal{T} = \mathcal{I}$. Since $\mathcal{T} \supsetneq \mathcal{I}$ for L^1 the tail and invariant σ-fields pull apart for some value of α between 0 and 1. I conjecture this happens at $\alpha = \frac{1}{3}$. This comes from an examination of the rate of growth of r_t. For $\alpha > \frac{1}{3}$, the martingale component grows too slowly in comparison with the bounded variation term for a coupling of the radial processes to occur.

REFERENCES

(1) M. Cranston (1983). Invariant σ-fields for a class of diffusions, *Z.f. Wahr. verw. Geb.*, **65**, 161–180.

(2) M. Cranston, S. Orey, U. Rösler (1980). Exterior Dirichlet problems and the asymptotic behavior of diffusions, *Lect. Notes in Control and Inform. Sci.*, **15**, Springer, New York.

(3) B. Fristedt, S. Orey (1978). The tail σ-field of one dimensional diffusions, *Proc. Sympos. Stochastic Diff. Eqn's*. Springer, New York.

(4) G. A. Hunt (1966). *Martingales et Processus de Markov*, Dunod, Paris.

(5) K. Itô, H.P. McKean (1966). *Diffusion Processes and Their Sample Paths*, Springer, New York.

(6) B. Jamison, S. Orey (1967). Markov chains recurrent in the sense of Harris, *Z.f. Wahr*, **8**, 41–48.

(7) W. Kendall (1986). Nonnegative Ricci curvature and the Brownian coupling property, *Stochastics*, Vol. 19, 111–129.

(8) T. Lindvall, L.C.G. Rogers (1986). Coupling of multidimensional diffusions by reflection, *Annals of Prob.* 11, No. 3, 860–872.

(9) T. Lyons and D. Sullivan (1984). Function theory, random paths and covering spaces, *Jour. Diff. Geom.* **19**, no. 2, 299-323.

(10) P. March (1986). Brownian motion and harmonic functions on rotationally symmetric manifolds, *Annals of Prob.*, **11**, No. 3, 793–801.

(11) Murata (1986). Structure of positive solutions to $(-\Delta + V)u = 0$ in $\mathcal{R}^n$, Duke Math. Jour., Vol.53, No.4, 869–943.

Mathematics Department
University of Rochester
Rochester, NY 14627

On the Martingale Problem for

Measure-Valued Markov Branching Processes

by

P. J. FITZSIMMONS*

1. Introduction

It was stated as Theorem (4.1)(b) in [F] that the martingale problem associated with the (ξ, ϕ)-superprocess X discussed in that paper has a unique solution. The theorem is true, but the proof supplied in [F] is inadequate since the assertion on p. 355 that $t \mapsto V_t f(x)$ is continuously differentiable is false in general. Our purpose in this note is to give a complete proof of this result. Owing to the weakness of the hypotheses employed in [F], the proof is rather involved, but since the theorem has found recent application (e.g. [DP]), we thought a thorough treatment was warranted. Also, the main auxiliary result (Theorem (2.22)) concerning space-time martingales may be of independent interest. In the remainder of this section we recall some notation from [F] and state the theorem under discussion. All unexplained notation is as in [F].

Let $\xi = (\Omega, \mathcal{F}, \mathcal{F}_t, \theta_t, \xi_t, P^x)$ be a Borel right Markov process, with semigroup (P_t) and resolvent (U^α). The state space $(E, \mathcal{E})$ of ξ is a Borel subspace of some compact metric space, and each P_t is a Markov operator on $b\mathcal{E}$ (the class of bounded $\mathcal{E}$-measurable functions on E). In particular, $P_t 1 = 1$ so that ξ has infinite lifetime. We fix a "branching mechanism" ϕ of the form

$$(1.1) \qquad \phi(x, \lambda) = -b(x)\lambda - c(x)\lambda^2 + \int_0^\infty (1 - e^{-\lambda u} - \lambda u)\, n(x, du).$$

---

* Research supported in part by NSF Grant DMS 87−21237.

where $c \geq 0$ and b are bounded and $\mathcal{E}$-measurable, and n is a (positive) kernel from $(E, \mathcal{E})$ to $(]0, \infty[, \mathcal{B}_{]0,\infty[})$ satisfying $\int_0^\infty (u \vee u^2) n(\cdot, du) \in b\mathcal{E}$. For each $f \in bp\mathcal{E}$ the integral equation

$$(1.2) \qquad V_t f(x) = P_t f(x) + \int_0^t P_s(x, \phi(\cdot, V_{t-s}f)) \, ds.$$

has a unique solution. Let $M(E)$ denote the class of finite measures on $(E, \mathcal{E})$, and let $\mathcal{M}(E)$ be the σ-field on $M(E)$ generated by the mappings

$$\mu \mapsto \langle \mu, f \rangle := \int_E f \, d\mu, \quad f \in bp\mathcal{E}.$$

Writing $e_f(\nu) = \exp(-\langle \nu, f \rangle)$, the formula

$$Q_t(\mu, e_f) = \exp(-\langle \mu, V_t f \rangle)$$

uniquely determines a Markov semigroup of kernels on $(M(E), \mathcal{M}(E))$. Let $M_o(E)$ (resp. $M_r(E)$) denote the topological space obtained by endowing the set $M(E)$ with the weak* topology induced by the bounded continuous function on E in its original topology (resp. the Ray topology associated with ξ). In either space the Borel σ-field coincides with $\mathcal{M}(E)$. Then by §§2,3 of [F] there is a strong Markov process $X = (X_t; \mathbb{P}_\mu, \mu \in M(E))$ with transition semigroup (Q_t) and state space $M(E)$. X is a Borel right process when viewed as a process in $M_o(E)$, and a Hunt process when viewed as a process in $M_r(E)$. In particular $t \mapsto X_t$ is càdlàg in $M_r(E)$, and we shall write X_{t-}^r for the left limit process. Thus we can (and do) take X to be the coordinate process on the space W of càdlàg paths from $[0, \infty[$ into $M_r(E)$ that are also right continuous in $M_o(E)$. The natural σ-fields on W are $\mathcal{G}_t^\circ := \sigma\{X_s : 0 \leq s \leq t\}$ and $\mathcal{G}^\circ := \sigma\{X_s : s \geq 0\}$.

Let $(A, \mathbf{D}(A))$ denote the weak infinitesimal generator of ξ as defined in §4 of [F], and let $\mathbf{D}(L)$ denote the class of functions on $M(E)$ of the form

$$F(\mu) = \psi(\langle \mu, f_1 \rangle, \ldots, \langle \mu, f_n \rangle),$$

where $\psi \in C_0^\infty(\mathbb{R}^n)$, $f_i \in \mathbf{D}(A)$, and $n \geq 1$. Define L on $\mathbf{D}(L)$ by

$$(1.3)$$

$$L(F)(\mu) = \int_E \mu(dx)c(x)F''(\mu; x) + \int_E \mu(dx)[AF'(\mu; \cdot)(x) - b(x)F'(\mu; x)]$$

$$+ \int_E \mu(dx) \int_0^\infty n(x, du)[F(\mu + u\epsilon_x) - F(\mu) - uF'(\mu; x)].$$

Here $F'(\mu; x)$ and $F''(\mu; x)$ are the first and second variational derivatives of F (e.g., $F'(\mu; x) = \lim_{\delta \downarrow 0}(F(\mu + \delta \cdot \epsilon_x) - F(\mu))/\delta)$. In [**F**, (4.1)(a)] it was shown that for each $F \in \mathbf{D}(L)$ the process

$$(1.4) \qquad M_t^F := F(X_t) - F(X_0) - \int_0^t LF(X_s)\,ds, \quad t \geq 0,$$

is a càdlàg $\mathbb{P}_\mu$-martingale. Our object is to prove the following converse assertion, which sharpens [**F**, (4.1)(b)]:

(1.5) Theorem. *For each $\mu \in M(E)$ the (L, μ) martingale problem has a unique solution. More precisely, let $\mathbb{P}$ be a probability measure on $(W, \mathcal{G}^\circ)$ such that*

(i) $\mathbb{P}(X_0 = \mu) = 1$;

(ii) for each $F \in \mathbf{D}(L)$ of the form $F(\nu) = \exp(-\langle \nu, f \rangle)$ the process M^F is a càdlàg local martingale over the system $(W, \mathcal{G}^\circ, \mathcal{G}_t^\circ, \mathbb{P})$.

Then $\mathbb{P} = \mathbb{P}_\mu$.

In section 2 we shall prove Theorem (1.5) under the hypotheses of [**F**]. We then show in section 3 how the integrability hypothesis in [**F**] on the jump kernel $n(x, du)$ can be weakened from a second moment to a first moment condition.

The martingale methods used in this note are not new, and the reader can consult [**RC**], [**EK-RC**], and [**M-RC**] for more information; our debt to these papers will be obvious to anyone familiar with them.

Acknowledgement: Thanks are due to Ed Perkins for pointing out the error in [**F**], and to E. B. Dynkin for discussions concerning the integrability conditions on $n(x, du)$.

2. Main Result

Throughout this section $\mathbb{P}$ is a probability measure satisfying conditions (i) and (ii) of Theorem (1.5). Evidently any right continuous martingale over the system $(W, \mathcal{G}^\circ, \mathcal{G}_t^\circ, \mathbb{P})$ is also a martingale over $(W, \mathcal{G}, \mathcal{G}_t, \mathbb{P})$, where $(\mathcal{G}_t)$ is the usual $\mathbb{P}$-augmentation of $(\mathcal{G}_{t+}^\circ)$. In the sequel the term "martingale" will mean a càdlàg (a.s. $\mathbb{P}$) process that is a martingale over the system $(W, \mathcal{G}, \mathcal{G}_t, \mathbb{P})$. The terms "local martingale" and "semimartingale" have similar interpretations.

If $\lambda > 0$, $f \in \mathbf{D}(A)$, and $F_\lambda(\nu) := \exp(-\lambda\langle\nu, f\rangle)$, then M^{F_λ} is a local martingale. A formal differentiation with respect to λ leads to the conclusion stated below. The rigorous proof is a standard application of Gronwall's lemma (cf. [**EtK**, Lemma 4.1]) and is left to the reader.

(2.1) Proposition. *If $f \in \mathbf{D}(A)$ then*

$$M_t^f := \langle X_t, f\rangle - \langle X_0, f\rangle - \int_0^t \langle X_s, Af - bf\rangle\, ds$$

is a martingale, and

$$(2.2) \qquad\qquad \mathbb{P}(\langle X_t, |f|\rangle) \leq \mu(1)\|f\|_\infty \exp(t\|b\|_\infty).$$

Define a semigroup (P_t^b) of bounded operators on $b\mathcal{E}$ by

$$P_t^b f(x) = P^x\left(f(X_t)\exp(-\int_0^t b(X_s)\, ds)\right),$$

and note that $P_t^b 1 \leq \exp(t\|b\|_\infty)$.

(2.3) Corollary. *If T is a bounded $(\mathcal{G}_t)$-stopping time, then for all $f \in b\mathcal{E}$,*

$$(2.4) \qquad\qquad \mathbb{P}(\langle X_{T+t}, f\rangle | \mathcal{G}_T) = \langle X_T, P_t^b f\rangle, \quad \forall t \geq 0.$$

If, in addition, T is predictable and > 0, then

$$(2.5) \qquad\qquad \mathbb{P}(\langle X_T, f\rangle | \mathcal{G}_{T-}) = \langle X_{T-}^r, f\rangle.$$

Proof. Fix $B \in \mathcal{G}_T$, and for $t \geq 0$ define measures on E by $\mu_t := \mathbb{P}(\langle X_{T+t}, \cdot\rangle; B)$. Note that if N is a bound for T, then $\mu_t(1) \leq \mu(1)\exp((t+N)\|b\|_\infty)$. To prove (2.4) we must show that $\mu_t = \mu_0 P_t^b$ for all $t \geq 0$. Because of the evident right continuity, it suffices to prove

$$(2.6) \qquad\qquad \hat{\mu}^\alpha(f) = \mu_0 U^{b+\alpha} f, \quad \forall f \in b\mathcal{E}, \alpha > \|b\|_\infty,$$

where $\hat{\mu}^\alpha := \int_0^\infty e^{-\alpha t}\mu_t\,dt$ and $U^{b+\alpha}f := \int_0^\infty e^{-\alpha t}P_t^b f\,dt$. (Note that both sides of (2.6) are $\leq \mu(1)\|f\|_\infty e^{N\|b\|_\infty}/(\alpha - \|b\|_\infty) < \infty$.) Given $g \in \mathcal{R}$ (the Ray cone associated with ξ; see [F, p.342]) and $\alpha > \|b\|_\infty$, put $f = U^\alpha g$, so that $f \in \mathbf{D}(A)$ and $Af = \alpha f - g$. Because of Proposition (2.1) and the boundedness of T we have

$$\mu_t(f) = \mu_0(f) + \int_0^t \mu_s(\alpha f - g - bf)\,ds, \quad \forall t \geq 0,$$

whence (upon passing to Laplace transforms)

$$(2.7) \qquad\qquad \hat{\mu}^\alpha(g + bU^\alpha g) = \mu_0 U^\alpha g,$$

first for all $g \in \mathcal{R}$ and then for all $g \in b\mathcal{E}$ by a monotone class argument. Now if $f \in b\mathcal{E}$ and $g := f - bU^{b+\alpha}f$, then $U^\alpha g = U^{b+\alpha}f$, so (2.7) implies (2.6), and (2.4) is proved. Proceeding to (2.5), let $g \in \mathcal{R}$ and put $f = U^1 g$. Then $f \in \mathbf{D}(A) \cap \mathcal{R}$, so $t \mapsto \langle X_t, f \rangle$ is right continuous with left limit process $\langle X_{t-}^r, f \rangle$. On the other hand, the predictable projection of the martingale M^f is (M_{t-}^f). It follows that $\langle X_{t-}^r, f \rangle$ is the predictable projection of $\langle X_t, f \rangle$, and this is precisely the content of (2.5), at least for f of the form $U^1 g$, $g \in \mathcal{R}$. The extension to $f \in b\mathcal{E}$ is routine. $\square$

(2.8) **Remark.** The projection argument used above shows that if $\varphi \in b(\mathcal{B}_{[0,\infty[} \otimes \mathcal{E})$ and if there is a left continuous $(\mathcal{G}_t)$-adapted process C_t such that $\langle X_t, \varphi_t \rangle - C_t$ is a local martingale, then

$$\lim_{s \uparrow t}\langle X_s, \varphi_s \rangle = \langle X_{t-}^r, \varphi_t \rangle, \quad \forall t > 0, \text{ a.s. } \mathbb{P}.$$

The following result (describing the Lévy system of $(X_t, \mathbb{P})$) is an immediate consequence of the implication $(2) \Rightarrow (3)$ in Théorème 7 in [EK-RC]. We write $\mathcal{P}$ for the predictable σ-field on $]0,\infty[\times W$.

(2.9) **Proposition.** Let $\Delta X_t := X_t - X_{t-}^r$, $t > 0$. If $((t,w),u) \in [0,\infty[\times W \times \mathbb{R} \mapsto G(t,w,u)$ is positive and $\mathcal{P} \otimes \mathcal{B}_{\mathbb{R}}$-measurable, and $G(t,w,0) = 0$, then for all $f \in p\mathcal{E}$ and $t > 0$,

$$\mathbb{P}\sum_{0 < s \leq t} G(s,\cdot,\langle\Delta X_s, f\rangle) = \mathbb{P}\left(\int_0^t ds \int_E X_s(dx) \int_0^\infty n(x,du)G(s,\cdot,uf(x))\right).$$

In particular, $\Delta X_t \geq 0$, for all $t > 0$ a.s. $\mathbb{P}$.

(2.10) Corollary. *If $f \in D(A)$ then the martingale M^f of (2.1) has quadratic variation*

$$\langle M^f, M^f \rangle_t = \int_0^t \langle X_s, \hat{c}f^2 \rangle \, ds,$$

where $\hat{c} := 2c + \int_0^\infty u^2 \, n(\cdot, du) \in bp\mathcal{E}$.

Proof. Fix $f \in D(A)$ so that by (2.1), $\langle X_t, f \rangle$ is a (special) semimartingale. Itô's formula thus yields the canonical semimartingale decomposition of $Z_t := \exp(-\langle X_t, f \rangle)$. But the fact that $Z = F(X)$, where $F(\nu) = \exp(-\langle \nu, f \rangle)$, leads via the martingale problem (1.5)(ii) to a second decomposition of Z. Comparison of these decompositions yields

$$\int_0^t Z_s \langle X_s, cf^2 \rangle \, ds + \sum_{0 < s \le t} Z_{s-} k(\Delta X_s, f)$$

$$= \frac{1}{2} \int_0^t Z_s \, d\langle N, N \rangle_s + \int_0^t Z_s \langle X_s, \psi(\cdot, f) \rangle \, ds + K_t,$$

where $k(\nu, f) = 1 - \exp(-\langle \nu, f \rangle) - \langle \nu, f \rangle$, N is the continuous martingale part of M^f, $\psi(x, \lambda) = \phi(x, \lambda) + b(x)\lambda + c(x)\lambda^2$, and K is a local martingale. But by (2.9) and (2.1),

$$\sum_{0 < s \le t} Z_{s-} k(\Delta X_s, f) - \int_0^t Z_s \langle X_s, \psi(\cdot, f) \rangle \, ds$$

is a martingale. It follows that

$$(2.11) \qquad \int_0^t Z_s \langle X_s, cf^2 \rangle \, ds = \frac{1}{2} \int_0^t Z_s \, d\langle N, N \rangle_s + \tilde{K}_t,$$

where $\tilde{K}$ is a local martingale. But (2.11) forces $\tilde{K}$ to be continuous and of finite variation, hence $\equiv 0$. Since (2.9) implies that the dual predictable projection of the increasing process $\sum_{0 < s \le t} (\Delta M_s^f)^2 = \sum_{0 < s \le t} \langle \Delta X_s, f \rangle^2$ is $\int_0^t \langle X_s, (\hat{c} - 2c)f^2 \rangle \, ds$, the result follows. $\square$

In view of (2.10), (2.2), and the fact that any element of $C_b(E)$ is the bounded pointwise limit of a sequence from $D(A)$, the linear map $f \in D(A) \mapsto$

M^f extends to a "martingale measure" in the sense of Walsh [W]. Indeed it follows from Corollary (2.10) that there is a (unique) orthogonal martingale measure $M = M(dx, ds)$ on $E \times [0, \infty[$ such that

$$M_t^f = \int_E \int_0^t f(x) M(dx, ds), \quad \forall f \in D(A).$$

See [W, Chapter 2]. In the sequel we shall write

$$M_t(\varphi) = \int_E \int_0^t \varphi_s(x) M(dx, ds), \quad \varphi \in b(\mathcal{B}_{[0,\infty[} \otimes \mathcal{E}).$$

By (2.10),

$$(2.12) \qquad \langle M(\varphi), M(\psi) \rangle = \int_0^t \langle X_s, \hat{c} \varphi_s \psi_s \rangle \, ds.$$

Given $f \in b\mathcal{E}$ and $t > 0$, consider the process

$$N_s := \langle X_s, P_{t-s}^b f \rangle, \quad 0 \le s \le t.$$

Because of (2.4), N is the optional projection of the constant (in time) process $s \mapsto \langle X_t, f \rangle$. Thus N is a (càdlàg) martingale, and by Remark (2.8), $N_{s-} = \langle X_{s-}^r, P_{t-s}^b f \rangle$, $0 < s \le t$.

(2.13) Proposition. *If $f \in b\mathcal{E}$ and $u > 0$, then*

$$(2.14) \qquad \begin{aligned} N_t^u :&= \langle X_t, P_{u-t}^b f \rangle - \langle X_0, P_u^b f \rangle \\ &= \int_E \int_0^t P_{u-s}^b f(x) M(dx, ds), \quad \forall t \le u, \text{ a.s. } \mathbb{P}. \end{aligned}$$

Moreover,

$$(2.15) \quad \mathbb{P}(\langle X_t, f \rangle \langle X_t, g \rangle) = (\mu P_t^b f)(\mu P_t^b g) + \int_0^t \mathbb{P}(\langle X_s, \hat{c}(P_{t-s}^b f)(P_{t-s}^b g) \rangle) \, ds,$$

and

$$(2.16) \qquad \mathbb{P}(\langle X_t, f \rangle M_t(\varphi)) = \int_0^t \mathbb{P}(\langle X_s, \hat{c}(P_{t-s}^b f)\varphi_s \rangle) \, ds$$

for all $f \in b\mathcal{E}$ and $\varphi \in b(\mathcal{B}_{[0,\infty[} \otimes \mathcal{E})$.

Proof. We extend the definition of N^u by setting $N_t^u = N_u^u$ for $t > u$. It suffices to consider $f \in C_b(E)$, and in this case both $u \mapsto N_t^u(w)$ and $u \mapsto$

$\varphi_s^u := 1_{\{s \le u\}} P_{u-s}^b f(x)$ are right continuous on $[0, \infty[$. Fix $\alpha > \|b\|_\infty$. Then $t \mapsto \int_E \int_0^t e^{-\alpha s} U^{b+\alpha} f(x) \, M(dx, ds)$ is a martingale and by Walsh's stochastic Fubini theorem [**W**, Thm. 2.6],

$$\int_E \int_0^t e^{-\alpha s} U^{b+\alpha} f(x) \, M(dx, ds) = \int_E \int_0^t \int_s^\infty e^{-\alpha s} P_{u-s}^b f(x) \, du \, M(dx, ds)$$
$$= \int_0^\infty e^{-\alpha u} M_t(\varphi^u) \, du.$$

On the other hand, an integration by parts shows that

$$\int_E \int_0^t e^{-\alpha s} U^{b+\alpha} f(x) \, M(dx, ds) = e^{-\alpha t} \langle X_t, U^{b+\alpha} f \rangle - \langle X_0, U^{b+\alpha} f \rangle + \int_0^t e^{-\alpha u} \langle X_u, f \rangle \, du$$
$$= \int_0^\infty e^{-\alpha u} N_t^u \, du,$$

the latter being finite since $\mathbb{P}(N_t^u) = \mathbb{P}(\langle X_u, f \rangle) \le C \exp(u\|b\|_\infty)$. It follows that there is a set $W_0 \subset W$ with $\mathbb{P}(W_0) = 1$ such that for all $w \in W_0$, $t \ge 0$, and $\alpha > \|b\|_\infty$,

$$\int_0^\infty e^{-\alpha u} N_t^u \, du = \int_0^\infty e^{-\alpha u} M_t(\varphi^u) \, du < \infty.$$

By Laplace inversion,

(2.17) $N_t^u = M_t(\varphi^u)$, a.e. $u \ge 0$, $\forall w \in W_0, t \ge 0$.

Now square both sides of (2.17) and integrate: for all $t \ge 0$, and $0 < a_1 < a_2$,

$$\int_{a_1}^{a_2} \mathbb{P}((N_t^u)^2) \, du = \int_{a_1}^{a_2} \mathbb{P}(M_t(\varphi^u)^2) \, du$$
$$= \int_{a_1}^{a_2} \mathbb{P}(\langle M(\varphi^u), M(\varphi^u) \rangle_t) \, du$$
$$= \int_{a_1}^{a_2} \mathbb{P}\left(\int_0^t \langle X_s, \hat{c}(\varphi_s^u)^2 \rangle \, ds \right) du.$$

Owing to the previously noted right continuity in u, we can safely conclude that

$$\mathbb{P}((N_t^u)^2) = \mathbb{P}\left(\int_0^t \langle X_s, \hat{c}(\varphi_s^u)^2 \rangle \, ds \right), \quad \forall t, u \ge 0.$$

Upon setting $u = t$ and polarizing we obtain (2.15), first for $f, g \in C_b(E)$, and then for $f, g \in b\mathcal{E}$ by a monotone class argument. A similar argument leads from (2.17) to (2.16). Finally, using (2.15) and (2.16) it is easy to compute:

$$\mathbb{P}((N_t^u - M_t(\varphi^u))^2) = 0, \quad \forall 0 \le t \le u,$$

so that (2.14) follows by right continuity. □

(2.18) Corollary. *Let* $\varphi, g \in b(\mathcal{B}_{[0,\infty[} \otimes \mathcal{E})$ *be related by*

$$(2.19) \qquad P_{t-s}^b \varphi_t = \varphi_s + \int_s^t P_{u-s}^b g_u \, du, \quad \forall 0 \leq s \leq t.$$

Then

$$N_t := \langle X_t, \varphi_t \rangle - \langle X_0, \varphi_0 \rangle - \int_0^t \langle X_s, g_s \rangle \, ds$$

is a (càdlàg) martingale. Indeed

$$(2.20) \qquad N_t = M_t(\varphi), \quad \forall t \geq 0, \text{ a.s. } \mathbb{P},$$

so that

$$(2.21) \qquad \langle N, N \rangle_t = \int_0^t \langle X_s, \hat{c}\varphi_s^2 \rangle \, ds, \quad \forall t \geq 0.$$

Proof. It follows from (2.4) and (2.19) that for each $u > 0$, $t \mapsto N_{t \wedge u}$ is the optional projection of $t \mapsto N_u$. Thus N is a càdlàg martingale. Using (2.15) and (2.16) it is easy to check that $\mathbb{P}((N_t - M_t(\varphi))^2) = 0$ for all $t \geq 0$. Points (2.20) and (2.21) now follow immediately. □

We are ready to state and prove the main result:

(2.22) Theorem. *Let* φ *and* g *be as in Corollary (2.18) and assume* $\varphi \geq 0$. *Define* $g_s^*(x) := g_s(x) + b(x)\varphi_s(x)$. *Then the process*

$$\exp\left(-\langle X_t, \varphi_t \rangle + \int_0^t \langle X_s, g_s^* + \phi(\cdot, \varphi_s) \rangle \, ds\right)$$

is a local martingale.

Proof. Let $Z_t := \exp(-\langle X_t, \varphi_t \rangle)$. In view of Corollary (2.18), Z is a semimartingale and by Itô's formula,

$$Z_t = Z_0 - \int_0^t Z_s \langle X_s, g_s \rangle \, ds - \int_0^t Z_{s-} \, dM_s(\varphi)$$
$$+ \frac{1}{2} \int_0^t Z_s \langle X_s, 2c\varphi_s^2 \rangle \, ds - \sum_{0 < s \leq t} Z_{s-} k(\Delta X_s, \varphi_s),$$

where $k(\nu, f) = 1 - \exp(-\langle \nu, f \rangle) - \langle \nu, f \rangle$ as before. Using (2.9) it now follows that

$$Y_t := Z_t - Z_0 + \int_0^t Z_s \langle X_s, g_s^* + \phi(\cdot, \varphi_s) \rangle \, ds$$

is a local martingale. But if $K_t := \exp(\int_0^t \langle X_s, g_s^* + \phi(\cdot, \varphi_s) \rangle \, ds)$, then $Z_t K_t = Z_0 K_0 + \int_0^t K_s \, dY_s$ is a local martingale, as desired. $\square$

(2.23) Corollary. *If $f \in p\mathcal{E}$ then the process $\exp(-\langle X_s, V_{t-s}f \rangle)$, $0 \le s \le t$, is a martingale. In particular, $\mathbb{P} = \mathbb{P}_\mu$, so that the (L, μ) martingale problem has a unique solution.*

Proof. The first assertion follows from (2.22) upon taking $\varphi_s = V_{t-s}f$ $(= f$ if $s > t)$ and $g_s^* = -\phi(\cdot, V_{t-s}f)$ $(= 0$ if $s > t)$, since $V_t f$ solves the integral equation (1.2). It now follows easily that X_t under $\mathbb{P}$ is a Markov process with transition semigroup Q_t (cf. [**EK-RC**, Thm.11]). Since $\mathbb{P}(X_0 = \mu) = 1$, we must have $\mathbb{P} = \mathbb{P}_\mu$. $\square$

3. An Extension.

In this section we show that all of the results in [**F**] remain valid under the condition

$$(3.1) \qquad \int_0^\infty (u \wedge u^2) \, n(\cdot, du) \in b\mathcal{E}.$$

This covers, for example, the case of "stable" branching: $\phi(x, \lambda) = -b(x)\lambda - c(x)\lambda^2 - k(x)\lambda^{\beta+1}$, where $k \in bp\mathcal{E}$ and $0 < \beta < 1$. The only real difficulty lies in proving that the (L, μ) martingale problem has a unique solution. The reader will note that the boundedness of $\int_0^\infty u^2 \, n(\cdot, du)$ was crucial to the proof given in §2. Although it is possible that a modification of the argument used in §2 will work under (3.1) alone, an easier path is provided by the Girsanov transformation discussed in [**EK-RC**, Prop.14]. On the other hand, *only* the boundedness of $\int_0^\infty u^2 \, n(\cdot, du)$ (as opposed to that of $\int_0^\infty (u \vee u^2) \, n(\cdot, du)$) was used explicitly in §2.

Hereafter the jump kernel n is assumed only to satisfy condition (3.1). The results in [F] concerning the existence and basic properties of the (ξ, ϕ)-superprocess $(X_t, \mathbb{P}_\mu)$ remain valid with only minor changes. The proof that the solution $V_t f$ of (1.2) is negative definite as a function of f becomes a bit more delicate when the condition $\int_0^1 u\, n(\cdot, du) \in b\mathcal{E}$ is relaxed to the condition $\int_0^1 u\, n(\cdot, du) \in b\mathcal{E}$; for a good discussion of this point the reader can consult Dynkin [D]. We shall now comment briefly on other points that require alteration.

(a) The second moment identity in [F, (2.7)] may be vacuous since $\hat{c}(x) = \infty$ is now possible.

(b) Formula (2.12) in the proof of [F, (2.11)] now fails (in general) but all that is needed at that point is $\sup_{0 \le r \le t,\, r \text{ rat.}} \langle Z_r, 1_{\bar{E}} \rangle < \infty$ a.s. $\mathbb{P}_\mu$, and this is true since $t \mapsto e^{-\alpha t} \langle Z_t, 1_{\bar{E}} \rangle$ is a supermartingale for $\alpha > 0$ sufficiently large.

(c) Since $t \mapsto e^{-\alpha t} \langle X_t, 1_{\bar{E}} \rangle$ is a supermartingale, if T is a stopping time bounded by N then $\mathbb{P}_\mu(\langle X_T, 1_{\bar{E}} \rangle) \le e^{\alpha N} \mu(1)$ by the optional sampling theorem. This suffices for the argument in [F, (2.15)] and subsequently.

(d) In the proof of [F, (2.20)], rather than looking at $(\langle X_T, \bar{f} \rangle - \langle X^r_{T-}, \bar{P}_0 \bar{f} \rangle)^2$ one should look at $(\exp(-\langle X_T, \bar{f} \rangle) - \exp(-\langle X^r_{T-}, \bar{P}_0 \bar{f} \rangle))^2$.

(e) [F, (3.10)] remains valid if the definitions of γ and $\hat{\beta}$ are altered by setting $\gamma = \|c + \frac{1}{2} \int_0^1 u^2\, n(\cdot, du)\|_\infty$ and $\hat{\beta} = \|b + \int_1^\infty u\, n(\cdot, du)\|_\infty$. The crucial point is that $\phi(x, \lambda) \ge \hat{\phi}(x, \lambda) := -\hat{\beta}\lambda - \gamma\lambda^2$.

(f) The boundedness of $\hat{c} = 2c + \int_0^\infty u^2\, n(\cdot, du)$ can be avoided in the proof of [F, (4.7)(b)] by an appeal to (2.9) of this paper, which is valid under (3.1) alone.

It remains to verify that [F, (4.1)(b)] holds under (3.1). To prove this we employ the Girsanov transformation of [EK-RC, Prop.14]. Define

$$\tilde{b} := b + 2c + \int_0^\infty (1 - e^{-u})\, u\, n(\cdot, du),$$

$$\tilde{c} := c. \quad \tilde{n}(x, du) := e^{-u} n(x, du),$$

and define $\tilde{\phi}(x,\lambda)$ as in (1.1) but with b, c, and n replaced by $\tilde{b}$, $\tilde{c}$, and $\tilde{n}$. In other words $\tilde{\phi}(x,\lambda) = \phi(x,\lambda+1) - \phi(x,1)$. Note that $\int_0^\infty u^2\,\tilde{n}(\cdot,du) \in b\mathcal{E}$. In view of the preceding remarks, the $(\xi,\tilde{\phi})$-superprocess $(\tilde{X}_t, \widetilde{\mathbb{P}}_\mu)$ is covered by [F] and §2 of this paper.

Now let $\mathbb{P}$ be a solution of the (L,μ) martingale problem. It is easy to check that if $f \in p\mathbf{D}(A)$ then

$$H_t(f) := \exp\left(-\langle X_t, f\rangle + \int_0^t \langle X_s, Af + \phi(\cdot,f)\rangle\,ds\right)$$

is a $\mathbb{P}$-local martingale which is reduced by the sequence of $(\mathcal{G}_{t+}^\circ)$ stopping times defined by

$$T_n := \inf(t \geq 0 : \langle X_t, 1\rangle \geq n) \wedge n, \quad n \in \mathbb{N}.$$

Let $\mathbb{P}^{(n)}$ (resp. $\mathbb{P}_\mu^{(n)}$, resp. $\widetilde{\mathbb{P}}_\mu^{(n)}$) denote the restriction of $\mathbb{P}$ (resp. $\mathbb{P}_\mu$, resp. $\widetilde{\mathbb{P}}_\mu$) to $\mathcal{G}_{T_n+}^\circ$, and define

$$\widetilde{\mathbb{P}}^{(n)} := H_{T_n}(1)\,\mathbb{P}^{(n)}.$$

Then $\widetilde{\mathbb{P}}^{(n)}$ is a probability measure on $(W, \mathcal{G}_{T_n+}^\circ)$ and since

$$H_t(f+1) = \tilde{H}_t(f)H_t(1),$$

where $\tilde{H}_t(f) := \exp(-\langle X_t, f\rangle + \int_0^t\langle X_s, Af + \tilde{\phi}(\cdot,f)\rangle\,ds)$, it is clear that

$$(\tilde{H}_{t\wedge T_n}(f))_{t\geq 0} \text{ is a } \widetilde{\mathbb{P}}^{(n)}\text{-local martingale}$$

for all $f \in p\mathbf{D}(A)$ and $n \in \mathbb{N}$. Writing $F(\nu) = \exp(-\langle \nu, f\rangle)$, and using the obvious notation, an integration by parts shows that there is a predictable process $K_t(f)$ such that

$$\tilde{M}_{t\wedge T_n}^F = \int_0^{t\wedge T_n} K_s(f)\,d\tilde{H}_s(f)$$

is also a $\widetilde{\mathbb{P}}^{(n)}$-local martingale. By the obvious localization of the argument used in §2 we see that $\widetilde{\mathbb{P}}_\mu^{(n)}$ is the unique solution of the $(\tilde{L},\mu)$ martingale problem on $[0,T_n]$. It follows that $\widetilde{\mathbb{P}}^{(n)} = \widetilde{\mathbb{P}}_\mu^{(n)}$. In the same way $H_{T_n}(1)\mathbb{P}_\mu^{(n)} = \widetilde{\mathbb{P}}_\mu^{(n)}$, so $H_{T_n}(1)\mathbb{P}^{(n)} = \widetilde{\mathbb{P}}_\mu^{(n)} = H_{T_n}(1)\mathbb{P}_\mu^{(n)}$, and finally $\mathbb{P}^{(n)} = \mathbb{P}_\mu^{(n)}$ since $H_{T_n}(1)(w) \in \,]0,\infty[$ for all $w \in W$. Consequently $\mathbb{P} = \mathbb{P}_\mu$ on $\sigma\{\cup_n\mathcal{G}_{T_n+}^\circ\}$, which coincides with $\mathcal{G}^\circ$ since $T_n(w) \to \infty$ as $n \to \infty$ for all $w \in W$. Thus Theorem (1.5) of this paper holds under condition (3.1) alone.

References

[DP] D. A. DAWSON and E. A. PERKINS. Historical Processes. Memoirs Amer. Math. Soc. **93**, no. 454. Providence, 1991.

[D] E. B. DYNKIN. Superdiffusions and partial differential equations. Preprint 1991. (To appear in *Ann. Probab.*)

[EK-RC] N. EL KAROUI and S. ROELLY-COPPOLETTA. Propriétés de martingales, explosion et représentation de Lévy-Khintchine d'une classe de processus de branchement à valeurs mesures. Stoch. Proc. App. **38** (1991) 239–266.

[EtK] S. N. ETHIER and T. G. KURTZ. *Markov Processes: Characterization and Convergence.* Wiley, New York, 1986.

[F] P. J. FITZSIMMONS Construction and regularity of measure-valued Markov branching processes. Israel J. Math. **64** (1988) 337–361.

[M-RC] S. MELÉARD and S. ROELLY-COPPOLETTA. Discontinuous measure-valued branching processes and generalized stochastic equations. Preprint, (1989).

[RC] S. ROELLY-COPPOLETTA. A criterion of convergence of measure-valued processes: application to measure branching processes, Stochastics, **17** (1986) 43–65.

[W] J. B. WALSH An introduction to stochastic partial differential equations. Lecture Notes in Math. **1180**, pp. 265–439. Springer, Berlin, 1986.

P. J. FITZSIMMONS
Department of Mathematics
University of California, San Diego
La Jolla, California 92093-0112

Potential Densities
Of Symmetric Lévy Processes

JOSEPH GLOVER

MURALI RAO

1. Introduction

H. Cartan introduced Hilbert space methods into the study of Newtonian potential theory in the 1940's [2,3]. Many of his results were generalized immediately to symmetric translation invariant potential theories in R^d by Deny [5], and most of the results are valid for general symmetric Markov processes.

We discuss one of Cartan's most striking results for Brownian motion (X_t, P^x) in R^3, namely, his realization of balayage as projection in Hilbert space. Recall that the potential density for X_t is $u(y - x) = |y - x|^{-1}$. Let $\mathcal{M}^+$ be the collection of positive measures μ on R^3 such that

$$\int \int u(y - x) \, \mu(dx) \, \mu(dy) < \infty$$

If we set $\mathcal{M} = \mathcal{M}^+ - \mathcal{M}^+$, then $\mathcal{M}$ is a pre–Hilbert space with inner product

$$(\mu, \nu) = \int \int u(y - x) \, \mu(dy) \, \nu(dx)$$

Let $\mathcal{H}$ be the completion of $\mathcal{M}$ in this inner product. Let $K \subset R^3$ be compact, and let $T = \inf\{t > 0 : X_t \in K\}$. If $\mu \in \mathcal{M}^+$ has closed support contained in K^c, the balayage of μ onto K is the measure μP_K defined by $\mu P_K f = P^\mu[f(X_T) : T < \infty]$. It can be characterized in terms of $\mathcal{H}$ as follows. Let $\mathcal{M}_K^+$ be the collection of measures in $\mathcal{M}^+$ which are supported by K. Then Cartan showed that $\mathcal{M}_K^+$ is a closed convex subset of $\mathcal{H}$ and that μP_K is the orthogonal projection of μ onto $\mathcal{M}_K^+$.

Define $\mathcal{M}_K = \mathcal{M}_K^+ - \mathcal{M}_K^+$: $\mathcal{M}_K$ is not a closed subspace of $\mathcal{H}$, in general, and we let $\mathcal{H}_K$ denote the closure of $\mathcal{M}_K$ in $\mathcal{H}$. We can show that μP_K is the

* Research of the first author supported in part by NSA and NSF by grant MDA904–89–H–2037

linear projection of μ onto $\mathcal{H}_K$, which displays clearly the fact that $(\mu_1 + \mu_2)P_K = \mu_1 P_K + \mu_2 P_K$ for μ_1 and μ_2 positive measures, a linearity which is not evident from Cartan's representation. Since $\mathcal{M}_K$ is dense in $\mathcal{H}_K$, it is enough to show that $(\mu, \nu) = (\mu P_K, \nu)$ for every $\nu \in \mathcal{M}_K$. But if $\nu \in \mathcal{M}_K$, then $\nu = \nu P_G$ for every open set G containing K. By Hunt's switching identity (VI-1.16 in [1]), $(\mu, \nu) = (\mu, \nu P_G) = (\mu P_G, \nu)$. As G decreases to K, $(\mu P_G, \nu)$ converges to $(\mu P_K, \nu)$.

In this article, we characterize which kernels $u(y - x)$ are potential densities of a Markov process on a finite group. The characterization requires that $u(y - x)$ be symmetric and positive definite so that (μ, ν) is an inner product. Our one extra hypothesis is that the projection of the point mass at the group identity onto the measures supported by the complement of the identity be a *positive* measure. These conditions suffice to show that u is a potential density (see (2.2)). This is an extremely simple condition to check, since it amounts to computing a linear projection in Hilbert space (i.e. compute a few Fourier coefficients). In section 3, we use the result in sec. 2 to characterize continuous symmetric translation invariant potential kernels of Markov processes on the circle group.

2. Finite Groups

Let $E = \{0, 1, 2, \ldots, n\}$ be a finite group with an addition $+$ and an inverse $-$. In spite of the fact that we use additive notation, we do not assume the group is abelian. Let $w : E \to (0, \infty)$ be a *symmetric* function on E (so $w(x) = w(-x)$ for every $x \in E$), and let $\mathcal{M}$ be the collection of finite signed measures μ on E. If we let $\mu_j = \mu(j)$, then each $\mu \in \mathcal{M}$ may be identified with the vector $(\mu_j)_{j=0}^n$ in R^{n+1}, but regarding them as measures is more conducive to extending to locally compact spaces later. We say that w is *positive semi-definite* if

$$(\mu, \mu) = \sum_{i \in E} \sum_{j \in E} \mu_i \mu_j w(j - i) \geq 0$$

for every $\mu \in \mathcal{M}$. A positive semi-definite function w is *positive definite* if $(\mu, \mu) = 0$ only when $\mu_j = 0$ for every $j \in E$.

Fix such a symmetric positive definite function w, and define an inner product on $\mathcal{M}$ by setting

$$(\mu, \nu) = \sum_{i \in E} \sum_{j \in E} \mu_i \nu_j w(j - i)$$

Since $\mathcal{M}$ is complete in the inner product, it is a Hilbert space with norm $\|\mu\| = (\mu, \mu)^{1/2}$. For each $k \in E$, we define a closed linear subspace of $\mathcal{M}$ by setting

$\mathcal{M}_k = \{\mu \in \mathcal{M} : \mu_k = 0\}$. Let δ^k be the Dirac measure assigning unit mass to the point k, and let Q^k be the (linear) Hilbert space projection of δ^k onto $\mathcal{M}_k$. Our only other assumption is the following.

(2.1) Hypothesis. Q^0 *is a positive measure on* E.

By the translation invariance and symmetry of the inner product, it is clear that $Q_j^0 = Q_{-j}^0$ and $Q_j^0 = Q_{j+k}^k$ for every $k, j \in E$. It follows from this that $Q_j^k = Q_{j-k}^0 = Q_{k-j}^0 = Q_k^j$ for every $k, j \in E$. The rest of this section is devoted to proving the next theorem.

(2.2) Theorem. *If (2.1) holds, then* $w(j - i)$ *is the potential density of a Markov process on* E.

That is, there is a continuous time Markov process (X_t, P^x) on E such that

$$P^i \int_0^\infty f(X_t)\, dt = \sum_{j \in E} f(j) w(j - i)$$

for every function $f : E \to R$.

The measure Q^0 is the unique measure in $\mathcal{M}_0$ which attains the value

$$\min\{\|\delta^0 - \mu\| : \mu \in \mathcal{M}_0\}$$

Thus, if we define the function $n : \mathcal{M}_0 \to R^+$ by

$$n(\mu) = \|\delta^0 - \mu\|^2 = w(0) - 2\sum_{j \neq 0} \mu_j w(j) + \sum_{j \neq 0}\sum_{i \neq 0} \mu_j \mu_i w(j - i)$$

then n achieves its minimum at Q^0. Taking the gradient with respect to μ_i, we obtain for each $i \neq 0$, the equation

$$(2.3) \qquad\qquad -w(i) + \sum_{j \neq 0} Q_j^0 w(j - i) = 0$$

(2.4) Lemma. $\sum_{j \neq 0} Q_j^0 < 1$.

PROOF: Let $c = \sum_{j=0}^n w(j)$. By (2.3), we have for each $i \neq 0$,

$$w(i) = \sum_{j \neq 0} Q_j^0 w(j - i)$$

Summing both sides on $i \neq 0$, we obtain

$$(2.5) \qquad\qquad c - w(0) = \sum_{j \neq 0} Q_j^0 (c - w(j))$$

Let us observe that $w(0) > w(j)$ for each $j \neq 0$. To see this, let $\gamma \in \mathcal{M}$ be defined by $\gamma(0) = 1, \gamma(j) = -1, \gamma(i) = 0$ for $i \neq 0, j$. Then $0 < (\gamma, \gamma) = 2w(0) - 2w(j)$. Thus we obtain from (2.5) that

$$c - w(0) > (\sum_{j \neq 0} Q_j^0)(c - w(0))$$

Since $c - w(0) > 0$ (recall $w(j) > 0$ for every j), we obtain the desired result. ∎

As we observed earlier, $Q_k^j = Q_j^k$, so Q_k^j is the transition matrix of a symmetric Markov chain X_t. More precisely, for each $\alpha > 0$, we can construct a process X_t^α on E such that each point in E is an exponential holding point with parameter α. If $T = \inf\{t : X_t^\alpha \neq X_0^\alpha\}$, then $P^k[X_T^\alpha = j] = Q_j^k$. Since $Q_j^k = Q_{j-k}^0 = Q_k^j$, X_t^α is a symmetric Lévy process. For each $\alpha > 0$, there is a symmetric positive definite function v_α such that

$$P^k \int_0^\infty f(X_t^\alpha)\, dt = \sum_{j \in E} f(j) v_\alpha(k - j)$$

If we set

$$m(\mu) = v_\alpha(0) - 2\sum_{j \neq 0} \mu_j v_\alpha(j) + \sum_{j \neq 0}\sum_{i \neq 0} \mu_j \mu_i v_\alpha(j - i)$$

then it is a classical fact about energy of symmetric Markov processes that m also achieves its minimum at Q^0, and we obtain as before

$$(2.6) \qquad -v_\alpha(i) + \sum_{j \neq 0} Q_j^0 v_\alpha(j - i) = 0$$

If we write (2.3) in matrix form, we obtain $\widetilde{A}\overline{w} = \overline{0}$, where $\widetilde{A}$ is an $n \times (n+1)$-dimensional matrix whose j^{th} row is

$$(Q_j^0, Q_{j-1}^0, \ldots, Q_1^0, -1, Q_n^0, \ldots, Q_{j+1}^0)$$

and where $\overline{w}$ is the column vector $(w(0), \ldots, w(n))$. We create a matrix A by adding a 0th row to $\widetilde{A}$ consisting of $(-1, Q_n^0, \ldots, Q_1^0)$ to obtain the equation $A\overline{w} = \overline{d}$, where $\overline{d}$ is the column vector $(d, 0, \ldots, 0)$, and where $d = -w(0) + Q_n^0 w(1) + \cdots + Q_1^0 w(n)$. Similarly, we can write (2.6) in matrix form: $A\overline{v}_\alpha = \overline{e}_\alpha$, where $\overline{v}_\alpha$ is the column vector $(v_\alpha(0), \ldots, v_\alpha(n))$, $\overline{e}_\alpha = (e_\alpha, 0, \ldots, 0)$, and where $e_\alpha = -v_\alpha(0) + Q_n^0 v_\alpha(1) + \cdots + Q_1^0 v_\alpha(n)$. Since $v_\alpha = \alpha v_1$, $e_\beta = d$ for some $\beta > 0$. By (2.4), A is a diagonally dominant matrix [4], so A^{-1} exists, and we conclude that $\overline{v}_\beta = \overline{w}$. This completes the proof of (2.2).

A simple consequence of (2.2) is the following result about perturbations of potential densities.

(2.7) Corollary. *Let $u(y - x) > 0$ be the symmetric potential density of a Markov process on E such that $\inf\{Q_j^0 : j \neq 0\} > 0$. If $z(x)$ is a symmetric function on E, then there exists an $\epsilon_0 > 0$ such that $w_\epsilon(y - x) = u(y - x) + \epsilon z(y - x)$ is the potential density of a Markov process on E for every $\epsilon < \epsilon_0$.*

PROOF: There is an $\epsilon_1 > 0$ such that w_ϵ is positive definite for each $\epsilon < \epsilon_1$. If we let $Q(\epsilon)^0$ be the projection of δ^0 onto $\mathcal{M}_0$ with respect to the inner product generated by the function $u + \epsilon z$, then $Q(0)^0 = Q^0$ and $\epsilon \to Q(\epsilon)_j^0$ is continuous for each j. Therefore, there is an ϵ_0 with $0 < \epsilon_0 < \epsilon_1$ such that for every $\epsilon < \epsilon_0$, $Q(\epsilon)_j^0 > 0$ for every $j \neq 0$. ∎

3. The Circle Group

Let T be the circle group $\{e^{2\pi i\theta} : 0 \leq \theta < 1\}$ under multiplication, and let $w : T \to (0, \infty)$ be a *continuous* symmetric positive definite function. That is, $w(x) = w(-x)$ and

$$0 < (\mu, \mu) = \int \int w(y - x) \, \mu(dy) \, \mu(dx)$$

for every finite signed nonzero measure μ on T. For brevity, we write $e(\theta) = e^{2\pi i\theta}$. For each n, let $T_n = \{e(k/n) : 0 \leq k < n\}$ be the nth roots of unity, and let $\mathcal{M}^n$ be the signed measures on T_n. Let $\delta^{k,n}$ be the Dirac measure assigning unit mass to the point $e(k/n)$, and let $\mathcal{M}_k^n = \{\mu \in \mathcal{M}^n : \mu(e(k/n)) = 0\}$. As before, we let $Q^{n,k}$ be the linear Hilbert space projection of $\delta^{k,n}$ onto $\mathcal{M}_k^n$.

(3.1) Hypothesis. $Q^{n,0}$ *is a positive measure on T_n for each n.*

Define $Wf(x) = \int w(y - x)f(y)\lambda(dy)$, where λ is Lebesgue measure on T.

(3.2) Theorem. *There is a positive subMarkov resolvent (W^α) on T such that $W = W^0$.*

PROOF: From the results in section 2, we know there is a Markov process X_t^n on T_n such that

$$(3.3) \qquad P^x \int f(X_t^n)dt = n^{-1} \sum_{k=0}^{n-1} w(e(k/n) - x)f(e(k/n))$$

for each $x \in T_n$, for each n. We call the left side of (3.3) $W_n f(x)$, and note that W_n is the zero potential of a resolvent W_n^α. Since w is bounded and continuous, $W_n f(x)$ converges to $Wf(x)$ as n tends to infinity for each continuous function $f : T \to R$. How about W_n^α? Since $W_n 1(x) < w(0) = \sup\{w(x) : x \in T\}$ (see the discussion following formula (2.5)), (V–5.10) in [1] implies

$$(3.4) \qquad W_n^\alpha f(x) = \sum_{k=0}^{\infty} (-\alpha)^k (W_n)^{k+1} f(x)$$

for $0 \leq \alpha \leq w(0)^{-1}$. If $f : T \to R$ is continuous, the dominated convergence theorem applies to (3.4) to yield

$$(3.5) \qquad \lim_{n \to \infty} W_n^\alpha f(x) = \sum_{k=0}^{\infty} (-\alpha)^k (W)^{k+1} f(x)$$

since w is bounded and continuous. Call the right side of (3.5) $W^\alpha f(x)$. For $0 \leq \alpha, \beta \leq w(0)^{-1}$, we have

$$W_n^\alpha f(x) - W_n^\beta f(x) = (\beta - \alpha) W_n^\alpha W_n^\beta f(x)$$

As n tends to infinity, we obtain for f continuous,

$$W^\alpha f(x) - W^\beta f(x) = (\beta - \alpha) W^\alpha W^\beta f(x)$$

by the dominated convergence theorem since w is bounded and continuous. For $w(0)^{-1} \leq \alpha \leq 2w(0)^{-1}$,

$$W_n^\alpha f(x) = \sum_{k=0}^{\infty} (w(0)^{-1} - \alpha)^k (W_n^{w(0)^{-1}})^{k+1} f(x)$$

and a repetition of these arguments yields $W^\alpha f(x) = \lim_{n \to \infty} W_n^\alpha f(x)$ satisfies the resolvent equation for $\alpha \leq 2w(0)^{-1}$. The induction argument to extend to all $\alpha > 0$ is straightforward.

REFERENCES

[1] Blumenthal, R.M. and Getoor, R.K. *Markov Processes and Potential Theory* Academic Press, New York (1968).

[2] Cartan, H. Sur les fondements de la théorie du potentiel. *Bull. Soc. Math. France* **69** 71-96 (1941).

[3] Cartan, H. Théorie du potentiel newtonien: énergie, capacité, suites de potentiels. *Bull. Soc. Math. France* **73** 74-106 (1945).

[4] Conte, S.D.. and De Boor, C. *Elementary Numerical Analysis: An Algorithmic Approach* McGraw-Hill, New York (1980).

[5] Deny, J. Les potentiels d'énergie finie. *Acta Math.* **82** 107-183 (1950).

Joseph Glover, Murali Rao
Department of Mathematics
University of Florida
Gainesville, FL 32611

An Absorption Problem for Several Brownian motions

by

HARRY KESTEN[1]

Introduction. Let $B_0, B_1, \ldots, B_N$ be independent Brownian motions, starting at $0, b_1, \ldots, b_N$, respectively, with $0 < b_i \leq 1$, $1 \leq i \leq N$. We estimate the tail of the distribution of

$$\tau = \inf\{t > 0 : B_i(t) \leq B_0(t) \text{ for some } i\}.$$

In an elegant paper on coupling various stochastic processes, Bramson and Griffeath (1991) considered the analogue of the stopping time τ for continuous time random walks. It is very likely that the kind of tail estimates which we derive here are the same for Brownian motions as for continuous time random walks. However, for our purposes Brownian motions are easier to work with, so that we will stick with the setup described above.

Bramson and Griffeath raised the question for which N is $E\tau < \infty$. They showed that $E\tau = \infty$ for $N = 2$ or 3, and showed computer simulations which indicate that $E\tau < \infty$ for $N \geq 4$. Can one at least prove that $E\tau < \infty$ for large N? The last question is settled by the following theorem which we shall prove in the next section. P_b will be used for the probability measure of $B_0, B_1, \ldots, B_N$, given that $B_i(0) = b_i$ and E_b for expectation with respect to P_b.

Theorem 1. *Let $B_0, B_1, \ldots, B_N$ be independent Brownian motions starting at $0, b_1, \ldots, b_N$, respectively. Let*

(1.1) $$\tau = \inf\{t > 0 : B_i(t) = B_0(t) \text{ for some } 1 \leq i \leq N\}.$$

If $0 < b_i \leq a$, $1 \leq i \leq N$, then for each $\gamma > 0$ there exist constants $C = C(\gamma) > 0$ and $C_1(\gamma) < \infty$ (independent of N and a) such that

(1.2) $$P_b\{\tau > t\} \leq a^{2\gamma}t^{-\gamma} + 2^N a^{2\gamma N}t^{-\gamma N} + (2e^{3C_1})^N a^{2CN}t^{-CN}.$$

In particular $E_b\tau < \infty$ whenever $N > [C(\gamma)]^{-1}$ for some $\gamma > 1$.

The proof will actually give some estimate for $C(\gamma)$, which will lead to the following

---

[1] Research supported by the NSF through a grant to Cornell University.

Corollary 1. *There exists some constant $\alpha > 0$ such that for each N there is a* $t_0(N)$ *such that for* $0 < b_i \leq a$, $1 \leq i \leq N$

$$(1.3) \qquad P_b\{\tau > t\} \leq (a^{-2}t)^{-\alpha \log N}, \quad a^{-2}t \geq t_0(N).$$

Can we give a lower bound for $P_b\{\tau > t\}$? Let

$$(1.4) \qquad c = \min\{b_1, \ldots, b_N\}.$$

Then for $0 < \bar{c} < c$ we have the easy lower bound

$$(1.5) \qquad P_b\{\tau > t\} \geq P\{B_0(s) \leq \bar{c}, \; 0 \leq s \leq t\} \prod_{i=1}^{N} P\{B_i(s) > \bar{c}, \; 0 \leq s \leq t\}$$
$$\geq C(t \wedge 1)^{-(N+1)/2},$$

for some $C = C(b_1, \ldots, b_N) > 0$ (use Ito–McKean (1965, Sect. 1.7) for the second inequality). This bound is already too crude to show $E_b \tau = \infty$ for $N = 2$. The exponent $(N+1)/2$ in (1.5) is very far from the exponent $\alpha \log N$ in (1.3) for large N. In Section 3 we prove that (1.3) is indicative of the true behavior.

Theorem 2. *For each $\varepsilon > 0$ there exists an $N_0(\varepsilon)$ and for each N there exists a $t_1(N)$ such that for $b_0 = 0$ and $b_i \geq c \geq 1$, $1 \leq i \leq N$,*

$$(1.6) \qquad P_b\{\tau \geq t\} \geq (c^{-2}t)^{-(1+\varepsilon)\frac{1}{4}\log N}$$

for $N \geq N_0$ and $c^{-2}t \geq t_1(N)$.

Of course τ equals the first exit time by the $(N+1)$-dimensional Brownian motion $(B_0(t), \ldots, B_N(t))$ from the "wedge"

$$\{x = (x_0, \ldots, x_{N+1}) \in \mathbf{R}^{N+1} : x_i - x_0 > 0, \; 1 \leq i \leq N\}.$$

De Blassie (1987) has shown that

$$(1.7) \qquad P\{\tau > t \,|\, B_i(0) = b_i\} \sim C(b)t^{-\theta}, \quad t \to \infty,$$

when $b_0 = 0 < b_i$, $b = (b_0, \ldots, b_N)$, and where $\theta = \theta(N)$ is determined by the first eigenvalue of the Laplace–Beltrami operator on a subset of the unit sphere S^N. However, as Bramson and Griffeath point out, it seems very difficult to find θ explicitly by this approach, and it even seems difficult to show $\theta > 1$ by this method.

A closely related approach would be via the theory of large deviations. Breiman (1967) noted in a similar situation the usefulness of the processes

$$(1.8) \qquad U_i(t) := e^{-t}\left(B_i(e^{2t}) - b_i\right), \quad t \in \mathbf{R}, \qquad \text{(with } b_0 = 0\text{)}.$$

These are independent stationary Ornstein–Uhlenbeck processes. Each has mean zero and covariance function $\rho(s,t) = E\{U_i(s)U_i(t)\} = \exp(-|s-t|)$ and a standard normal distribution as its stationary distribution.   Clearly, for $0 < b_i \leq 1, 1 \leq i \leq N$,

$$\{\tau > t\} \subset \{U_i(s) > U_0(s) - 1 \text{ for } 0 \leq s \leq \frac{1}{2} \log t \text{ and } 1 \leq i \leq N\}.$$

Thus, if one defines

$$R = \{s : 0 \leq s \leq \frac{1}{2} \log t, \ U_i(s) > U_0(s) - 1 \text{ for all } 1 \leq i \leq N\},$$

and if we denote the Lebesgue measure of a set S by $|S|$, then

$$P\{\tau > t\} \leq P\{|R| \geq \frac{1}{2} \log t\}.$$

The theory of large deviations (cf. Donsker and Varadhan (1976)) gives that

(1.9) $$\log P\{|R| \geq \frac{1}{2} \log t\} \sim -I \log t, \quad t \to \infty$$

for some constant I, and hence, for any $I' < I$

$$P\{\tau > t\} \leq t^{-I'} \text{ eventually.}$$

However, it seems difficult to find I explicitly. Our approach to Theorem 1 can be viewed as a rather clumsy way to find a lower bound for I.

Our approach is roughly as follows. Define

(1.10) $$\sigma = \inf \{t > 0 : U_i(t) \leq U_0(t) - 1 \text{ for some } i\}.$$

For $a = 1$ (i.e., $0 < b_i \leq 1$) we then have

$$\{\tau > t\} \subset \{\sigma > \frac{1}{2} \log t\},$$

and it therefore suffices to prove for (1.2) that

(1.11) $$P\{\sigma > T\} \leq e^{-2\gamma T} + 2^N e^{-2\gamma NT} + 2^N e^{-2CNT}.$$

To prove (1.11) we introduce

(1.12) $$S = S(d) = \{t \geq 0 : U_0(t) > -d\}$$

and show that for each fixed $\gamma > 0$ and $\varepsilon > 0$ one has for sufficiently large d

(1.13) $$P\{|S \cap [0, T]| < (1 - \varepsilon)T\} \leq e^{-2\gamma T}.$$

(1.13) is a *one*–dimensional large deviation estimate, and is therefore much easier than finding I in (1.9). Even for this one–dimensional problem we do not calculate the exact exponential rate of decay. Once we have (1.13) we show that for any fixed (measurable) subset S_0 of $[0, T]$ with $|S_0| \geq (1 - \varepsilon)T$ one has

(1.14) $P\{U_i(t) \leq -d - 1 \text{ for some } t \in S_0\} \geq 1 - e^{-2\gamma T} - e^{-2CT}.$

(1.11) follows easily from (1.13) and (1.14) by conditioning on the set $S \cap [0, T]$ (see Section 2).

As for Theorem 2, we use

$$P_b\{\tau > t\} \geq P\left\{B_0(s) < \frac{1}{2} \text{ for } 0 \leq s \leq 1 \text{ and } B_0(s) \leq -d\sqrt{s} \text{ for } 1 \leq s \leq t\right\}$$

$$\prod_{i=1}^{N} P\left\{B_i(s) > \frac{1}{2} \text{ for } 0 \leq s \leq 1 \text{ and } B_i(s) > -d\sqrt{s} \text{ for } 1 \leq s \leq t\right\}$$

$$\geq C(b, d, N)P\left\{U_0(s) \leq -d \text{ for } 0 \leq s \leq \frac{1}{2}\log t \,\Big|\, U_0(0) = -d - 1\right\}$$

(1.15)

$$\left[P\left\{U_1(s) > -d \text{ for } 0 \leq s \leq \frac{1}{2}\log t\right\} \Big| U_1(0) = 0\right]^N$$

for some $C(b, d, N) > 0$ (see Lemmas 3 and 4). The factors on the right are essentially known, and we choose d as a suitable function of N to obtain (1.6).

To close this introduction we remark that in view of (1.3) and (1.6) it seems reasonable to conjecture that the $\theta = \theta(N)$ of (1.7) satisfies

(1.16) $\lim_{N \to \infty} \dfrac{\theta(N)}{\log N}$ exists.

2. Proof of the upper bound. The standard scaling relation of Brownian motion tells us that $\{a^{-1}B_i(t)\}_{t \geq 0}$ has the same distribution as $\{\tilde{B}_i(a^{-2}t)\}_{t \geq 0}$ for Brownian motions $\tilde{B}_i$ starting at $a^{-1}b_i$. We can therefore reduce the case with $0 < b_i \leq a$, $1 \leq i \leq N$ to the case with $0 < b_i \leq 1$. For the remainder of this section we shall therefore assume

$$0 < b_i \leq 1, \ 1 \leq i \leq N, \ \text{ and } \ b_0 = 0.$$

We define U_i and S as in (1.8) and (1.12). Trivially (1.8) shows that $U_0(0)$ has a standard normal distribution and that the covariance of $U_0(\cdot)$ is given by

$$E\{U_0(s)U_0(t)\} = e^{-|s-t|}.$$

Therefore, the conditional density of $U_0(t)$ at y, given $U_0(s) = x$ is given by

(2.1) $\left\{2\pi(1 - e^{-2|s-t|})\right\}^{-1/2} \exp\left\{-\left(2(1 - e^{-2|s-t|})\right)^{-1}(y - xe^{-|s-t|})^2\right\}.$

This means that the generator of U_0 is

$$Lf(u) = f''(u) - uf'(u)$$

for any bounded, twice continuously differentiable function f on $\mathbb{R}$.

Our first step is to prove (1.13). Even though this can be proved "by hand", we obtain a much shorter proof and better estimates by using a general estimate of Gross (1976), based on a logarithmic Sobolev inequality. We are grateful to L. Gross for his help with this improved proof of Lemma 1.

Lemma 1. *For each $\gamma > 0$ and $\varepsilon > 0$ one has*

$$(2.2) \qquad P\{|S(d) \cap [0,T]| \leq (1-\varepsilon)T\} \leq e^{-2\gamma T}$$

whenever

$$(2.3) \qquad \frac{\Phi(-d)}{\varepsilon} - \log \frac{\Phi(-d)}{\varepsilon} \geq 1 + \frac{4\gamma}{\varepsilon},$$

where

$$(2.4) \qquad \Phi(y) = \frac{1}{\sqrt{2\pi}} \int_{-\infty}^{y} e^{-x^2/2} dx.$$

Corollary 2. *(2.2) holds for*

$$(2.5) \qquad d \geq \frac{e}{\varepsilon\sqrt{2\pi}} \vee \left(\frac{8\gamma}{\varepsilon}\right)^{1/2}$$

Proof: By the standard Chebychev bound

$$(2.6) \qquad P\{|S| \leq (1-\varepsilon)T\} = P\{|[0,T] \setminus S| \geq \varepsilon T\}$$
$$\leq \inf_{\lambda \geq 0} e^{-\lambda \varepsilon T} E\{\exp(\lambda |[0,T] \setminus S|)\}.$$

Note that the probabilities and expectation in (2.2) and (2.6) are calculated for the stationary process U_0, i.e., when U_0 has a standard normal distribution. Now note that

$$|[0,T] \setminus S| = \int_0^T I_{(-\infty,-d)}(U_0(s))ds,$$

and define

$$u(x,T) = u(x,T,\lambda) = E\left\{\exp\left(\lambda \int_0^T I_{(-\infty,-d)}(U_0(s))ds\right) \Big| U_0(0) = x\right\}.$$

Define further the semigroup $Q_\lambda(t)$ by

$$(Q_\lambda(T)g)(x) = E\left\{\exp\left(\lambda \int_0^T I_{(-\infty,-d)}(U_0(s))ds\right) g(U_0(T)) \,|\, U_0(0) = x\right\}$$

for bounded measurable g. Then

$$u(x,T) = (Q_\lambda(T)\mathbf{1})(x)$$

and

$$(2.7) \qquad E\{\exp(\lambda|[0,T] \setminus S|)\} = \,<\mathbf{1}, Q_\lambda(T)\mathbf{1}>,$$

where $\mathbb{1}$ is the constant function equal to 1 everywhere, and the inner product $< , >$ is defined as

$$< f, g > = \int_{\mathbb{R}} f(x)\overline{g(x)} \frac{1}{\sqrt{2\pi}} e^{-x^2/2} dx$$

for bounded measurable f, g. Moreover, $< f, Q_\lambda(T)g > = < f, e^{-T(H+V)}g >$, where $-H$ is the generator in $L^2(\frac{1}{\sqrt{2\pi}}e^{-s^2/2}dx)$ of the Ornstein–Uhlenbeck process (i.e., $Hf(x) = -f''(x) + xf'(x)$ for a twice continuously differentiable function f with compact support) and V is the multiplication operator by $-\lambda I_{(-\infty,-d)}$ (compare Simon (1979, Theorem 6.2)). H satisfies the hypotheses of Theorem 7 in Gross (1976) with $c = 1$ (see also Theorem 4 of that article). Therefore, for any f in the domain of H

$$< f, (H + V)f > \geq -\frac{1}{2} \log \left\{ \frac{1}{\sqrt{2\pi}} \int e^{-2V(x)-x^2/2} dx \right\} < f, f >^{1/2}$$

$$= -\frac{1}{2} \log \left(1 + (e^{2\lambda} - 1)\Phi(-d)\right) < f, f >^{1/2},$$

with Φ as in (2.4). Then also (by the spectral theorem)

$$< \mathbb{1}, Q_\lambda(T)\mathbb{1} > = < \mathbb{1}, e^{-T(H+V)}\mathbb{1} >$$

$$\leq \exp \left\{ \frac{T}{2} \log(1 + (e^{2\lambda} - 1)\Phi(-d)) \right\}$$

$$\leq \exp \left\{ \frac{T}{2}(e^{2\lambda} - 1)\Phi(-d) \right\}.$$

Now the left hand side of (2.2) is at most

$$e^{-\lambda\varepsilon T}(\mathbb{1}, Q_\lambda(T)\mathbb{1})$$

for any $\lambda \geq 0$ (by (2.6) and (2.7)). The lemma follows by taking

$$\lambda = \frac{1}{2} \log \frac{\varepsilon}{\Phi(-d)}.$$

∎

The **Corollary** follows from the estimate

$$\Phi(-d) \leq \frac{1}{d\sqrt{2\pi}} e^{-d^2/2}, \quad d \geq 0$$

(see Feller (1968, Lemma VII.1.2)).   ∎

Another immediate consequence of the Lemma is that (by symmetry) (2.2) or (2.6) also imply

(2.8) $P\left\{ |\{t : 0 \leq t \leq T, \ U_i(t) < d\}| \leq (1 - \varepsilon)T \right\} \leq e^{-2\gamma T}.$

In the sequel we always take d such that (2.8) holds. We shall write

$$S_i = \{t : t \geq 0, \ U_i(t) < d\}, \quad 1 \leq i \leq N,$$

and, for some fixed L,

$$\tilde{S} = S - L = \{t \geq -L : U_0(t + L) > -d\}.$$

We write $\mathcal{F}_t$ for the σ–field generated by $\{U_i(s) : s \leq t, i = 1, \ldots, N\}$. We now turn to the **proof of (1.14)**. The proof rests on the simple observation that once S is given as a fixed set S_0, and $t \in S_i \cap (S_0 - L)$, then we know that $U_0(t + L) > -d$ and $U_i(t) < d$, so that (see (2.1))

$$(2.9) \qquad P\Big\{U_i(s) < U_0(s) - 1 \text{ for some } s \in [t, t + L] \,\big|\, \mathcal{F}_t\Big\}$$

$$\geq P\Big\{U_i(t + L) \leq -d - 1 \,\big|\, \mathcal{F}_t\Big\}$$

$$\geq P\Big\{U_1(L) \leq -d - 1 \,\big|\, U_1(0) = d\Big\}$$

$$= C_1(d, L) := \Phi\left(-\frac{1 + d + de^{-L}}{(1 - e^{-2L})^{1/2}}\right) > 0.$$

Lemma 2. Fix $\varepsilon < 1/2$. Let S_0 be a fixed measurable subset of $[0, T]$ which satisfies

$$(2.10) \qquad |[0, T] \cap S_0| \geq (1 - \varepsilon)T.$$

Then for each $i = 1, \ldots, d$

$$P\Big\{U_i(t) > -d - 1 \text{ for all } t \in S_0 \cap [0, T]\Big\}$$

$$(2.11) \qquad \leq \exp\left(-(1 - 2\varepsilon)\frac{C_1}{L}T + 3C_1\right) + \exp(-2\gamma T).$$

Proof: Fix i, define $\rho_0 = 0$ and

$$(2.12) \qquad \rho_{j+1} = \inf\big\{t \geq \rho_j + L : t \in S_i \cap S_0\big\}, \quad j \geq 0.$$

As usual we define $\rho_{j+1} = \infty$ if the set in the right hand side of (2.12) is empty. We further define

$$\nu = \max\big\{j : \rho_j \leq T - L\big\}.$$

Then, by (2.9), for $j \geq 1$ and on the set $\{\rho_j \leq T - L\}$,

$$P\Big\{U_i(s) \leq -d - 1 \text{ for some } s \in [\rho_j, \rho_j + L] \,\big|\, \mathcal{F}_{\rho_j}\Big\} \geq C_1.$$

Let E_j be the event

$$\Big\{U_i(s) \leq -d - 1 \text{ for some } s \in [\rho_j, \rho_j + L]\Big\} \cup \{\rho_j > T - L\}.$$

Note that $E_\ell \in \mathcal{F}_{\rho_\ell + L} \subset \mathcal{F}_{\rho_{\ell+1}}$. Thus $E_1, \ldots, E_{j-1} \in \mathcal{F}_{\rho_j}$. Consequently

$$P\{E_j \mid E_1^c, \ldots, E_{j-1}^c\} \geq C_1 \text{ for all } j \geq 1.$$

Thus for any M

$$(2.13) \qquad P\Big\{U_i(t) > -d - 1 \text{ for all } t \in S_0 \cap [0, T]\Big\}$$
$$\leq P\Big\{E_j^c \text{ for all } 1 \leq j \leq \nu\Big\}$$
$$\leq P\Big\{E_j^c \text{ for all } 1 \leq j \leq M\Big\} + P\{\nu < M\}$$
$$\leq (1 - C_1)^M + P\{\nu < M\}.$$

Finally we note that by definition of the ρ_j, $t \notin S_i \cap S_0$ on $(\rho_j + L, \rho_{j+1})$ for $j \geq 0$, so that

$$\Big|S_i \cap S_0 \cap [0, T]\Big| \leq \Big|\bigcup_{j=0}^{\nu+1} [\rho_j, \rho_j + L]\Big| \leq (\nu + 2)L.$$

This implies (by virtue of (2.10))

$$(\nu + 2) \geq L^{-1}\Big[T - \big|[0, T] \setminus S_i\big| - \big|[0, T] \setminus S_0\big|\Big]$$
$$\geq L^{-1}\Big[T - \varepsilon T - \big|[0, T] \setminus S_i\big|\Big].$$

Thus, if we take $M = -3 + (1 - 2\varepsilon)T/L$, then by (2.8)

$$P\{\nu < M\} \leq P\Big\{\big|[0, T] \setminus S_i\big| \geq \varepsilon T\Big\} \leq e^{-2\gamma T},$$

and (2.11) therefore follows from (2.13).   ∎

To complete the **proof of Theorem 1** we combine Lemmas 1 and 2. By Lemma 1

$$P\Big\{\big|[0, T] \cap S\big| < (1 - \varepsilon)T\Big\} \leq e^{-2\gamma T}.$$

Therefore, if we now regard $\{U_0(t)\}_{0 \leq t \leq T}$ as a random element of $C[0, T]$, the space of continuous functions on $[0, T]$, and $[0, T] \cap S$ as a function of $\{U_0(t)\}_{0 \leq t \leq T}$, then

$$P\{\sigma > T\} \leq P\Big\{\big|[0, T] \cap S\big| < (1 - \varepsilon)T\Big\}$$
$$+ \int_D P\Big\{\{U_0(t)\}_{0 \leq t \leq T} \in dw\Big\} \prod_{i=1}^N P\Big\{U_i(t) > -d - 1$$
$$\text{for all } t \in [0, T] \cap S(\omega)\Big\}$$
$$\leq e^{-2\gamma T} + \Big\{\exp\Big(-(1 - 2\varepsilon)\frac{C_1}{L}T + 3C_1\Big) + \exp(-2\gamma T)\Big\}^N,$$

where

$$D = \Big\{ \omega \in C[0,T] : \big| [0,T] \cap S(\omega) \big| \geq (1 - \varepsilon)T \Big\}.$$

(1.11), and hence (1.2), follows with

$$(2.14) \qquad C = C(\gamma) = (1 - 2\varepsilon) \frac{C_1(d, L)}{2L}.$$

∎

Proof of Corollary 1. We choose $\varepsilon = 1/4$ and $L = 1$ (we have no idea whether this choice is anywhere near optimal). This gives

$$C(\gamma) = \frac{1}{4} C_1(d, 1) = \frac{1}{4} \Phi \left(-\frac{1 + d(1 + e^{-1})}{(1 - e^{-2})^{1/2}} \right)$$

where we only have to choose d to satisfy (2.5). For large γ, $d = 4\sqrt{2\gamma}$ will do. Then, for large γ

$$C(\gamma) = \frac{1}{4} \Phi \left(-\frac{1 + 4(2\gamma)^{1/2}(1 + e^{-1})}{(1 - e^{-2})^{1/2}} \right) \geq C_2 \exp(-C_3 \gamma).$$

We substitute this for C in (1.2) and choose

$$\gamma = \frac{1}{C_3} \Big\{ \log N - \log \log N + \log(C_2 C_3) \Big\}.$$

This choice makes γ and $C(\gamma)N$ approximately equal, and yields (1.3). Any $\alpha < C_3^{-1}$ will do for N sufficiently large.

3. Proof of the lower bound. Again we can rescale the Brownian motions so that we only have to prove Theorem 2 for

$$b_0 = 0, \quad b_i \geq 1, \quad 1 \leq i \leq N.$$

We henceforth assume $c = 1$ and begin with the factor for B_i in the second member of (1.15).

Lemma 3. For $\varepsilon > 0$, $d \geq d_1(\varepsilon)$, and $t \geq t_2 = t_2(d, \varepsilon)$

$$P\Big\{ B_i(s) > \frac{1}{2} \text{ for } 0 \leq s \leq 1 \text{ and } B_i(s) > -d\sqrt{s} \text{ for } 1 \leq s \leq t \,\Big|\, B_i(0) = b_i \Big\}$$

$$(3.1) \qquad \geq t^{-(1+\varepsilon)\beta(d)},$$

with

$$(3.2) \qquad \beta(d) = \frac{d}{2\sqrt{2\pi}} e^{-d^2/2}.$$

Proof: Obvious monotonicity arguments show that the left hand side of (3.1) is at least

$$P\Big\{ B_i(s) > \frac{1}{2} \text{ for } 0 \leq s \leq 1 \,\Big|\, B_i(0) = b_i \Big\}$$

$$(3.3) \qquad \cdot P\Big\{ B_i(s) > -d\sqrt{s} \text{ for } 1 \leq s \leq t \,\Big|\, B_i(1) = 0 \Big\}$$

$$= C_4 P\Big\{ U_i(s) > -d \text{ for } 0 \leq s \leq \frac{1}{2} \log t \,\Big|\, U_i(0) = 0 \Big\}$$

for some universal $C_4 > 0$. Now let

$$\kappa = \kappa(d) = \inf \left\{ t \geq 0 : U_i(s) \leq -d \right\}.$$

Then Darling and Siegert (1953) showed that

$$E\left\{ e^{-\lambda \kappa} \mid U_i(0) = 0 \right\} = \frac{D_{-\lambda}(0)}{e^{d^2/4} D_{-\lambda}(-d)},$$

for $Re \lambda > 0$, where $D_s(z)$ is the Weber or parabolic cylinder function (see Whittaker and Watson (1952, Sect. 16.5). For $0 < Re \lambda < 1$ we can represent $e^{z^2/4} D_{-\lambda}(z)$ as

$$-\frac{\Gamma(1-\lambda)}{2\pi i} 2i \sin \pi(\lambda - 1) \int_0^\infty e^{-zt - t^2/2} t^{\lambda - 1} dt$$

(see Whittaker and Watson (1952, Sect. 16.6). This gives, by an integration by parts

$$E\left\{ e^{-\lambda \kappa} \mid U_i(0) = 0 \right\} = \frac{2^{\lambda/2 - 1} \Gamma(\frac{\lambda}{2})}{\int_0^\infty e^{dt - t^2/2} t^{\lambda - 1} dt} = \frac{2^{\lambda/2} \Gamma(\frac{\lambda}{2} + 1)}{\int_0^\infty e^{dt - t^2/2} t^\lambda (t - d) dt}.$$

This can be analytically continued in λ for $Re \lambda > (-\lambda_0) \vee (-1)$, where $-\lambda_0$ is the real part of the rightmost zero in the left half plane of

$$h(\lambda) := \int_0^\infty e^{dt - t^2/2} t^\lambda (t - d) dt.$$

In fact, the first singularity of $E\{ e^{-\lambda \kappa} \mid U_i(0) = 0 \}$ has to be on the real axis, (see Widder (1946, Theorem II.5b). $-\lambda_0$ will therefore be the rightmost zero on the negative real axis of $h(\cdot)$. We shall prove that

$$(3.4) \qquad \lambda_0 \sim \frac{d}{\sqrt{2\pi}} e^{-d^2/2} \text{ and } h'(-\lambda_0) \neq 0 \text{ as } d \to \infty.$$

Once we have (3.4) it will follow that for $s \to 0$ through the positive reals, one has

$$E\left\{ e^{(\lambda_0 - s)\kappa(d)} \mid U_i(0) = 0 \right\} = \frac{2^{(s - \lambda_0)/2} \Gamma(\frac{s - \lambda_0}{2} + 1)}{h(s - \lambda_0)} \sim \frac{2^{-\lambda_0/2} \Gamma(1 - \frac{\lambda_0}{2})}{s h'(-\lambda_0)},$$

and by the Tauberian theorem (cf. Widder (1946, Theorem V.4.3))

$$\int_{[0,v]} e^{\lambda_0 x} P\left\{ \kappa \in dx \mid U_i(0) = 0 \right\} \sim \frac{2^{-\lambda_0/2}}{h'(-\lambda_0)} \Gamma\left(1 - \frac{\lambda_0}{2} \right) v$$

as $v \to \infty$. Consequently, for fixed $s > 0$ and d,

$$P\left\{ \kappa(d) \geq v \mid U_i(0) = 0 \right\} \geq P\left\{ v \leq \kappa(d) \leq v(1 + \varepsilon) \mid U_i(0) = 0 \right\}$$

$$\geq e^{-\lambda_0(1 + \varepsilon)v} \int_{(v, (1 + \varepsilon)v]} e^{\lambda_0 x} P\left\{ \kappa \in dx \mid U_i(0) = 0 \right\}$$

$$\sim e^{-\lambda_0(1 + \varepsilon)v} \frac{2^{-\lambda_0/2} \Gamma(1 - \frac{\lambda_0}{2})}{h'(-\lambda_0)} \varepsilon v, \quad v \to \infty.$$

Finally

$$P\left\{U_i(s) > -d \text{ for } 0 \leq s \leq \frac{1}{2}\log t \,\middle|\, U_i(0) = 0\right\}$$

$$= P\left\{\kappa(d) > \frac{1}{2}\log t \,\middle|\, U_i(0) = 0\right\}$$

$$\geq e^{-\lambda_0(1+\varepsilon)\frac{1}{2}\log t}$$

$$= t^{-\lambda_0(1+\varepsilon)/2} \quad \text{for} \quad t \geq t_1(d, \varepsilon).$$

In view of (3.3) and (3.4) this will prove the lemma.

It remains to prove (3.4). To this purpose we expand $h(\cdot)$ around the origin. This gives

$$(3.5) \quad \left|h(\lambda) - h(0) - \lambda h'(0)\right| \leq \frac{|\lambda|^2}{2} \int_0^\infty e^{dt - t^2/2}(1 + t^{Re\lambda})|d - t|(\log t)^2 dt.$$

But

$$h(0) = \int_0^\infty e^{dt - t^2/2}(t - d)dt = 1,$$

$$h'(0) = \int_0^\infty e^{dt - t^2/2}(t - d)\left[\log d + \log\left(1 - \frac{d-t}{d}\right)\right]dt$$

$$= \log d - e^{d^2/2}\int_{-\infty}^d e^{-u^2/2}u \log\left(1 - \frac{u}{d}\right)du$$

$$= \log d - e^{d^2/2}\int_{|u|\leq\sqrt{d}} e^{-u^2/2}u\left\{-\frac{u}{d} + 0\left(\frac{u^2}{d^2}\right)\right\}du$$

$$\quad + 0\left(e^{d^2/2}\int_{\sqrt{d}<|u|<d-1} e^{-u^2/2}|u|(\log d)du\right)$$

$$\quad + 0\left(e^{d^2/2}de^{-(d-1)^2/2}(\log d)\right)$$

$$\sim \frac{\sqrt{2\pi}}{d}e^{d^2/2} \quad (d \to \infty).$$

By estimates of the same kind the right hand side of (3.5) is at most

$$C_5|\lambda|^2 e^{d^2/2}d^{1+(Re\lambda)^+}(\log d)^2 \quad \text{for} \quad Re\,\lambda \geq -\frac{1}{2}$$

for some $C_5 < \infty$, independent of λ, d. Consequently, for $|\lambda| < 1$, $-\frac{1}{2} \leq Re\,\lambda < 0$

$$\left|h(\lambda) - 1 - \frac{\lambda\sqrt{2\pi}}{d}(1 + o_d(1))e^{d^2/2}\right| \leq C_5|\lambda|^2 e^{d^2/2}d(\log d)^2,$$

from which it is immediate that

$$\lambda_0 \sim \frac{d}{\sqrt{2\pi}}e^{-d^2/2}.$$

Similar calculations show that

$$h'(-\lambda_0) = \int_0^\infty e^{dt-t^2/2} t^{-\lambda_0}(t-d)\log t\, dt$$

$$\sim \frac{\sqrt{2\pi}}{d} e^{d^2/2},$$

so that (3.4) holds. ∎

We turn to the factor involving B_0 in (1.15).

Lemma 4. For $\varepsilon > 0$, $d \geq d_2(\varepsilon)$, and $t \geq t_3(d,\varepsilon)$

$$(3.6) \qquad P\Big\{B_0(s) < \frac{1}{2} \text{ for } 0 \leq s \leq 1 \text{ and } B_0(s) \leq -d\sqrt{s}$$

$$\text{for } 1 \leq s \leq t \,\Big|\, B_0(0) = 0\Big\}$$

$$\geq t^{-(1+\varepsilon)\gamma(d)}$$

with

$$(3.7) \qquad \gamma = \gamma(d) = \frac{1}{8}(d+1)^2.$$

Proof: Analogously to (3.3), the left hand side of (3.6) is at least

$$(3.8) \qquad \int_{-\infty}^{-d-1} P\Big\{B_0(s) \in dx \text{ and } B_0(s) < \frac{1}{2} \text{ for } 0 \leq s \leq 1 \,\Big|\, B_0(0) = 0\Big\}$$

$$\cdot P\Big\{B_0(s) \leq -d\sqrt{s} \text{ for } 1 \leq s \leq t \,\Big|\, B_0(1) = -d-1\Big\}$$

$$= C_7 P\Big\{U_0(s) \leq -d \text{ for } 0 \leq s \leq \frac{1}{2}\log t \,\Big|\, U_0(0) = -d-1\Big\}$$

for some $C_7 > 0$. This time we define

$$\xi = \xi(d) = \inf\Big\{t \geq 0 : U_0(s) > -d\Big\}.$$

Then, for any integer k and $\Delta > 0$

$$(3.9)$$

$$P\Big\{U_0(s) \leq -d \text{ for } 0 \leq s \leq k\Delta \,\Big|\, U_0(0) = -d-1\Big\}$$

$$= P\Big\{\xi \geq k\Delta \,\Big|\, U_0(0) = -d-1\Big\}$$

$$\geq \int_{x \leq -d-1} P\Big\{U_0(\Delta) \in dx, \xi \geq \Delta \,\Big|\, U_0(0) = -d-1\Big\}$$

$$\cdot P\Big\{\xi \geq (k-1)\Delta \,\Big|\, U_0(0) = x\Big\}$$

$$\geq P\Big\{\xi \geq (k-1)\Delta \,\Big|\, U_0(0) = -d-1\Big\}$$

$$\cdot P\Big\{U_0(\Delta) \leq -d-1, \xi \geq \Delta \,\Big|\, U_0(0) = -d-1\Big\}$$

$$\geq \cdots \geq \Big[P\Big\{U_0(\Delta) \leq -d-1, \xi \geq \Delta \,\Big|\, U_0(0) = -d-1\Big\}\Big]^k.$$

The main point, therefore, is to find a lower bound for

(3.10)

$$P\left\{U_0(\Delta) \le -d-1, \xi \ge \Delta \,\middle|\, U_0(0) = -d-1\right\}$$
$$= P\left\{U_0(\Delta) \le -d-1 \,\middle|\, U_0(0) = -d-1\right\}$$
$$- \int_0^\Delta P\left\{\xi \in ds \,\middle|\, U_0(0) = -d-1\right\}$$
$$\cdot P\left\{U_0(\Delta-s) \le -d-1 \,\middle|\, U_0(0) = -d\right\}$$
$$= \Phi\left(\frac{-d-1+(d+1)e^{-\Delta}}{(1-e^{-2\Delta})^{1/2}}\right)$$
$$- \int_0^\Delta P\left\{\xi \in ds \,\middle|\, U_0(0) = -d-1\right\}\Phi\left(\frac{-d-1+de^{s-\Delta}}{(1-e^{2s-2\Delta})^{1/2}}\right)$$

(see (2.1)). Now take

$$e^{-\Delta} = \frac{d}{d+1} \quad \text{or} \quad \Delta = \log\left(\frac{d+1}{d}\right) = \frac{1}{d} + 0(d^{-2}).$$

It is easy to see that

$$\Phi\left(\frac{-d-1+de^{s-\Delta}}{(1-e^{2s-2\Delta})^{1/2}}\right)$$

is decreasing in s on $[0, \Delta]$. Therefore the left hand side of (3.10) is at least

$$\Phi\left(\frac{-d-1+(d+1)e^{-\Delta}}{(1-e^{-2\Delta})^{1/2}}\right) - \Phi\left(\frac{-d-1+de^{-\Delta}}{(1-e^{-2\Delta})^{1/2}}\right)$$
$$= \Phi\left(\frac{-d-1}{(2d+1)^{1/2}}\right) - \Phi\left(-(2d+1)^{1/2}\right)$$
$$\sim \left(\frac{2d+1}{2\pi}\right)^{1/2}(d+1)^{-1}\exp\left\{-\frac{1}{2}\frac{(d+1)^2}{(2d+1)}\right\}$$

as $d \to \infty$ by Feller (1968, Lemma VII.1.2). The lemma now follows easily from (3.8) and (3.9), by taking $k = \lfloor \frac{1}{2\Delta}\log t\rfloor + 1 \sim \frac{d}{2}\log t$ in (3.9). ∎

Proof of Theorem 2. From the first inequality in (1.15) (which is obvious) and Lemmas 3 and 4 we have for $t \ge t_2(d,\varepsilon) \vee t_3(d,\varepsilon)$

$$P\{\tau > t\} \ge t^{-(1+\varepsilon)(\gamma(d)+N\beta(d))}.$$

(1.6) now results by taking

$$d = (2\log N)^{1/2}.$$

REFERENCES

Bramson, M. and Griffeath, D., *Capture problems for coupled random walks*, Festschrift for Frank Spitzer, Birkhäuser–Boston, 1991.

Breiman, L., *First exit times from a square root boundary*, Proc. Fifth Berkeley Symp. Math. Stat. and Probab., vol. II, Part II, University of California Press, 1967, pp. 9–16.

Darling. D. A. and Siegert, A. J. F., *The first passage problem for a continuous Markov process*, Ann. Math. Statist. **24** (1953), 624–639.

De Blassie, R. D., *Exit times from cones in $\mathbb{R}^n$ of Brownian motion*, Prob. Theory Rel. Fields **74** (1987), 1–29.

Donsker, M. D. and Varadhan, S. R. S., *Asymptotic evaluation of certain Markov process expectations for large time – III*, Comm. Pure and Appl. Math. **29** (1976), 389–461.

Feller, W., *Introduction to Probability Theory and its Applications*, vol. I, 3rd ed., John Wiley and Sons, 1968.

Gross, L., *Logarithmic Sobolev inequalities*, Amer. J. Math. **97** (1976), 1061–1083.

Ito, K. and McKean, H. P. Jr., *Diffusion processes and their sample paths*, Springer Verlag, 1965.

Simon, B., *Functional integration and quantum physics*, Academic Press, 1979.

Whittaker, E. T. and Watson, G. M., *A course of modern analysis, 4th ed.*, Cambridge University Press, 1952.

Widder, D. V., *The Laplace transform*, Princeton University Press, 1946.

Cornell University
Department of Mathematics
Ithaca, NY 14853

FORMS OF INCLUSION BETWEEN PROCESSES

BY FRANK B. KNIGHT

One of the most salient features of the work of Professor Steven Orey, and a feature of which the present writer stands in considerable awe, is the almost uncanny dedication to logic and realism which it manifests. Time and again, one saw him depart from familiar settings to take up some new and innovative development, no matter how formidable it must have appeared at a first glance. This is already true in his change of field from logic to probability theory. We see it again, for example, in his joint paper with the present author [4]. At the time of its instigation (by Orey) the only precursor to [4] was the famous paper [1] of Blumenthal, Getoor, and McKean. This last, however, was basically analytic, in sharp distinction to the synthetic approach of Orey. Other examples could be found easily (even from the limited perspective of the writer) since Orey's work ran the gamut of many branches of probability from Markov chains to absolute continuity of diffusion processes to large deviations.

In the present work, while not proposing to simulate the style of Professor Orey, we are concerned with certain logical relationships which can exist between two stochastic processes X_t and Y_t, $-\infty < t < \infty$, on the same complete probability space $(\Omega, \mathcal{F}, P)$. We will propose various types of inclusion from X to Y, which become types of equivalence when assumed as well from Y to X, and we will investigate the logical implications between them. An important role will be played by the prediction processes Z^X and Z^Y of X and Y, as defined for instance in the author's books [6] and [7].

To proceed formally, let $(\Omega, \mathcal{F}, P)$ be a complete probability space and let X_t, Y_t be two real–valued stochastic processes, $-\infty < t < \infty$, such that the paths of each process are with probability one right–continuous with left limits at all t (we abbreviate this to r.c.l.l., P a.s.). Since our results are meant to exhibit general principles, and it is not clear what the setting of maximum generality would be, we do not aim for the most general type of path space for which results such as ours subsist. For any two σ–subfields $\mathcal{G}$ and $\mathcal{H}$ of $\mathcal{F}$, we write $\mathcal{G} \overset{\_}{\supseteq} \mathcal{H}$ to denote that $\mathcal{G}$ includes $\mathcal{H}$ up to P–nullsets in $\mathcal{F}$, i.e. for $E_1 \in \mathcal{G}$ there is an $E_2 \in \mathcal{H}$ with $P(E_1 \triangle E_2) = 0$ (so that the L^2–subspace measurable with respect to $\mathcal{G}$ can be viewed as containing that of $\mathcal{H}$). For any stochastic process W_t on

$(\Omega, \mathcal{F}, P)$, the past of W at time t is $\mathcal{M}_t^W \doteq \sigma(W_s, s \leq t)$, and the future is $\mathcal{N}_t^W \doteq \sigma(W_s, s \geq t)$. We emphasize that (the present) $\sigma(W_t)$ is contained in both past and future. Before turning to the prediction processes, we can indicate the type of inclusion result that we have in mind as follows.

Definition 1. We say that X is "past inclusive" with respect to Y (respectively, is "future inclusive") if, for all t, $\mathcal{M}_t^X \supseteq \mathcal{M}_t^Y$ (respectively, $\mathcal{N}_t^X \supseteq \mathcal{N}_t^Y$).

Theorem 2. *X is past inclusive with respect to Y (resp. future inclusive) if and only if, for each t, there is a borel function $g_t(x_1, \ldots, x_n, \ldots) \in \mathcal{B}^\infty / \mathcal{B}$ and a sequence (t_n), $t_n \leq t$ (resp. $t_n \geq t$) such that $Y_t \equiv g_t(X_{t_1}, \ldots, X_{t_n}, \ldots)$, where, for any two random variables, $X \equiv Y$ denotes $P\{X = Y\} = 1$.*

Proof. Except for the allowance of infinitely many t_n and the exceptional P-nullsets, this is a result of Doob [3, Supplement, Theorem 1.5]. However, our proof here is quite different from [3] (but an analogous proof is given later in Lemma 11). We will show by montone class argument that the class of all random variables $Y \equiv g(X_{t_1}, \ldots, X_{t_n}, \ldots)$, with g and t_n as asserted, equals all Y with $\sigma(Y) \subseteq \mathcal{M}_t^X$. Since clearly $\sigma(Y_t) \subset \mathcal{M}_t^X$, and the assertion for $\mathcal{N}_t^Y$ is entirely analogous, this will prove the existence of the representation. Now assuming that $\sigma(Y) \subseteq \mathcal{M}_t^X$, it is easy to construct a random variable $Y^* \equiv Y$ with $\sigma(Y^*) \subset \mathcal{M}_t^X$, hence it is enough to show the existence of g and (t_n) for which $Y^* = g(X_{t_1}, \ldots, X_{t_n}, \ldots)$. The class of Y^* so representable certainly includes all X_s, $s \leq t$, and it is obviously closed under composition with Borel functions $g(x_1, \ldots, x_n, \ldots)$, hence it is an algebra. Moreover, if $Y^1 \leq Y^2 \leq \cdots$ is a monotone sequence of elements in the class, which is uniformly bounded by K, then we have $\lim_{n \to \infty} Y^n = g^{\sup}(X_{t_i^*})$, where (t_i^*) is the combined sequence of all $t_{i,n}$ from the assumed representations $Y^n = g_n(X_{t_{i,n}})$, and $g^{\sup} = \limsup_{n \to \infty}(g_n \wedge K)$ (regarding each g_n as function of $(X_{t_i^*})$). It follows by the monotone class theorem [2, Chapter 1, Theorem 2.1] that the class contains all bounded random variables measurable over $\mathcal{M}_t^X$. The boundedness restriction is then easily removed by writing $Y^* = \tan \tan^{(-1)}(Y^*)$.

Conversely, from such a representation, setting $Y^* = g_t(X_{t_1}, \ldots, X_{t_n}, \ldots)$ we have $\sigma(Y^*) \subset \mathcal{M}_t^X$, and therefore $\sigma(Y_t) \subseteq \mathcal{M}_t^X$. Then for $s \leq t$ (resp. for $s \geq t$) we have $\sigma(Y_s) \subseteq \mathcal{M}_s^X \subset \mathcal{M}_t^X$, (resp. $\subseteq \mathcal{N}_s^X \subseteq \mathcal{N}_t^X$) and it follows that $\mathcal{M}_t^Y \subseteq \mathcal{M}_t^X$, as required.

The meaning of Theorem 2 may be paraphrased as a logical equivalence between an inclusion of filtrations, on the one hand, and a deterministic functional relationship between random variables on the other. It is this type of equivalence in which we shall be mainly interested below. But first we need to introduce the concept of the prediction process of a process X (for brevity of notation, we shall only define it here for r.c.l.l. processes).

Notation 3. Let $(\Omega^+, \mathcal{F}^+)$ denote the space of all r.c.l.l. functions $w^+ : R^+ \to R$, with $\mathcal{F}_t^+ = \sigma(w^+(s); 0 \leq s)$, and let $(H, \mathcal{H})$ denote the space of all probability measures $z(S)$ on $(\Omega^+, \mathcal{F}^+)$, with $\mathcal{H} = \sigma\{z(S); S \in \mathcal{F}^+\}$.

The prediction process of X is a process Z_t^X on $(\Omega, \mathcal{F}, P)$, $-\infty < t < \infty$, with state space $(H, \mathcal{H})$, such that $Z_t^X(S)$ is a version of $P(X_{t+(\cdot)} \in S | \mathcal{M}_{t+}^X)$, $S \in \mathcal{F}^+$, where as usual $\mathcal{M}_{t+}^X = \bigcap_{\epsilon > 0} \mathcal{M}_{t+\epsilon}^X$.

We may give a formal existence and uniqueness assertion as follows.

Definition 4. The prediction process Z_t^X of the (r.c.l.l.) process X is the unique (up to P–equivalence), $(H, \mathcal{H})$–valued process on $(\Omega, \mathcal{F}, P)$, such that the following two properties hold.

(a) For $\mathcal{M}_{t+}^X$–optional $T < \infty$, we have

$$P(X_{T+(\cdot)} \in S | \mathcal{M}_{T+}^X) = Z_T^X(S), \quad S \in \mathcal{F}^+,$$

and

(b) Z_t^X is $\mathcal{M}_{t+}^X$–optional.

In this formulation, uniqueness is an easy consequence of the optional section theorem of P.–A. Meyer. For, if another process satisfying (a) and (b) were not P–a.s. identical to Z^X, one could by the section theorem find an optional $T < \infty$ at which they differed with positive probability. Since $\mathcal{F}^+$ is countably generated, this would contradict property (a) at T. As to existence, it is known (see for example Theorem 1.7 of [6]) that such a process exists when $(\Omega, \mathcal{F}, P)$ is the canonical path space $(\Omega', \mathcal{F}', P')$ of all r.c.l.l. functions, X_t is the coordinate random variable, and $\mathcal{M}_{t+}^X$ is the canonical filtration $\mathcal{F}_{t+}'$ augmented by all P–nullsets. The mapping $w \to X_{(\cdot)}(w)$ from Ω into Ω' is $\mathcal{M}_{t+}^X / \mathcal{F}_{t+}'$–measurable for each t, and both filtrations may be augmented by the corresponding nullsets. Then optionality is preserved (along with r.c.l.l.) and hence the properties (a) and (b) on $(\Omega', \mathcal{F}', P')$ for a process $Z_t^{P'}$ define a process Z_t^X on $(\Omega, \mathcal{F}, P)$ having the same properties by the prescription

$$Z_t^X(S, w) = Z_t^{P'}(S, X_{(\cdot)}(w)).$$

Remarks. (a) Actually, this only yields optionality with respect to the augmented filtration. The $\mathcal{M}_{t+}^X$–optionality is shown in Exercise 1.8 of [7].

(b) In the present paper we do not require the 'left limit' process Z_{t-}^X, and hence topology on $(H, \mathcal{H})$ will not be needed. Nonetheless, it should be kept in mind that Z_t^X is r.c.l.l. in a suitable topology (such that $(H, \mathcal{H})$ is a Lusin space), and hence it is determined from a countable set of t.

The prediction process Z^X is a very useful object—in some ways it is more useful than X itself. One reason for this is that there is a functional relationship between Z^X and X:

Theorem 5. *There is a function* $\rho \in \mathcal{H}/\mathcal{B}$ *such that*

$$P\{X_t = \rho(Z_t^X), \quad -\infty < t < \infty\} = 1.$$

This is proved in Theorem 1.9 of [6], but it is more important here to recognize the underlying reason, namely that since $X_t \in \mathcal{M}_t^X \cap \mathcal{N}_t^X$, X_t is determined by its wide–sense conditional distribution $Z_t^X(S)$, $S \in \mathcal{F}_0^+ (\doteq \sigma(w^+(0)))$.

In fact, there is a further inclusion relationship between Z^X and X, but in the opposite direction and extended over the past. This is a consequence of what we may call the fundamental theorem concerning Z^X, which may be stated as follows.

Theorem 6. Let $q(t, z, A) = P^z\{Z_t^z \in A\}$, $0 \le t$, $z \in H$, $A \in \mathcal{H}$, where $P^z \doteq z$ and Z^z is the prediction process of P^z on $(\Omega^+, \mathcal{F}^+)$. Then for $\mathcal{M}_{t+}^X$-optional $T < \infty$, $P(Z_{T+t}^X \in A | \mathcal{M}_{T+}^X) = q(t, Z_T^X, A)$.

Comment. We shall omit the details concerning measurability of q, existence of Z^z, etc., which are all found in [6] or (in greater generality) in [7]. Suffice it to say that Z^X is a homogeneous strong–Markov process, with transition function q not depending on P.

The consequence of Theorem 6 which was referred to above is stated as

Corollary 7. For each t, we have $\mathcal{M}_{t+}^X \equiv \mathcal{M}_{t+}^{Z^X}$, where $\equiv$ between σ–fields denotes $\subseteq$ and $\supseteq$, i.e. equality up to P–nullsets.

Proof. The inclusion $\mathcal{M}_t^X \subseteq \sigma(Z_s^X, s \le t)$ is immediate from Theorem 5. Conversely, by definition of Z_s^X as a conditional probability (property (a) with $T = s$) we have $\sigma(Z_s^X, s \le t) \subseteq \mathcal{M}_{t+}^X$. Thus it only remains to replace $\mathcal{M}_t^X$ by $\mathcal{M}_{t+}^X$ in the former inclusion, and since $\mathcal{M}_{t+}^X = \mathcal{M}_{t+}^{Z^X}$ in view of the latter, this is a familiar application of the Blumenthal $0 - 1$ Law for the process Z^X.

The point of introducing Z^X for the present paper is that, although $\mathcal{M}_t^{Z^x}$ is not basically different from $\mathcal{M}_t^{X_t}$ (at least, when we take right–limits), $\sigma(Z_t^X)$ is not at all the same as $\sigma(X_t)$. By Theorem 5 we have $\sigma(X_t) \subseteq \sigma(Z_t^X)$, but the converse inclusion is entirely false in general, and the only general relationship is $\sigma(Z_t^X) \subset \mathcal{M}_{t+}^X$. The role of $\sigma(Z_t^X)$ is nevertheless fundamental, in view of the following characterization.

Theorem 8. (a) Let $\mathcal{G} \subset \mathcal{M}_{t+}^X$ be a σ–field such that, for $S_1 \in \mathcal{M}_{t+}^X$ and $S_2 \in \mathcal{N}_t^X$, $P(S_1 \cap S_2 | \mathcal{G}) \equiv P(S_1 | \mathcal{G})P(S_2 | \mathcal{G})$. Then $\sigma(Z_t^X) \subseteq \mathcal{G}$.

(b) Conversely, $\sigma(Z_t^X)$ has the property of part (a).

In other words, we may describe $\sigma(Z_t^X)$ as the (unique up to equivalence) minimal splitting field of $\mathcal{M}_{t+}^X$ and $\mathcal{N}_t^X$ in $\mathcal{M}_{t+}^X$.

Proof. We rely on the fact that the conditional independence assumption of (a) is equivalent to the assertion that

$$(1) \qquad P(S_2 | \mathcal{M}_{t+}^X) \equiv P(S_2 | \mathcal{G}), \quad S_2 \in \mathcal{N}_t^X.$$

(Of course, the meaning of $\equiv$ here is slightly different). A proof of this may be found in [8, 25.3]. Now it is easy to see that for $S \in \mathcal{F}^+$ we have $\{X_{t+(\cdot)} \in S\} \in \mathcal{N}_t^X$, and for this as S_2, (1) becomes

$$(2) \qquad Z_t^X(S) \equiv P(X_{t+(\cdot)} \in S | \mathcal{G}),$$

which proves (a). Conversely, since any $S_2 \in \mathcal{N}_t^X$ has the form $S_2 = \{X_{t+(\cdot)} \in S\}$ for some $S \in \mathcal{F}^+$, we have $P(S_2 | \mathcal{M}_{t+}^X) \equiv Z_t^X(S)$ which is clearly measurable over $\sigma(Z_t^X)$. Then we have $P(S_2 | \mathcal{M}_{t+}^X) \equiv P(S_2 | \sigma(Z_t^X))$ by the definition of conditional probability.

Remark. For purposes of symmetry, introducing $\mathcal{N}_{t-}^X = \bigcap_{\epsilon > 0} \mathcal{N}_{t-\epsilon}^X$, we can show that $\sigma(Z_t^X)$ is also the minimal splitting field of $\mathcal{M}_{t+}^X$ and $\mathcal{N}_{t-}^X$ in $\mathcal{M}_{t+}^X$. This follows from the above by applying (for example) Lemma 1 of [5]. On the other hand, it should be emphasized that the definition of Z_t^X is not symmetric with respect to past and future. Thus Corollary 7 has no counterpart for $\mathcal{N}_{t-}^X$ beyond the obvious $\mathcal{N}_{t-}^X \subseteq \mathcal{N}_{t-}^{Z^X}$. It then becomes of interest to examine the meaning of inclusions from Z^X to Z^Y, as well as from X to Y.

As we saw in Theorem 2, an inclusion of σ–fields gives rise to a functional dependence of random variables. For Z^X and Z^Y, however, there are two types of functional dependence, corresponding roughly to whether we regard Z_t^X as a state of a Markov process or as a conditional probability. Let us introduce formally

Definition 9. We say that Z_t^Y is nonlinearly functionally dependent on Z_t^X if there is an $f_t \in \mathcal{H}/\mathcal{H}$ with $Z_t^Y \equiv f_t(Z_t^X)$. We say that Z_t^Y is linearly functionally dependent on Z_t^X if there is a mapping $\psi_t : \mathcal{F}^+ \to \mathcal{F}^+$ such that, for $S \in \mathcal{F}^+$, $Z_t^Y(S) \equiv Z_t^X(\psi_t S)$.

Remarks. The term "linearly" derives from the fact that $z(S)$ is a linear functional on $(H, \mathcal{H})$ for $S \in \mathcal{F}^+$. Since $\mathcal{F}^+$ is countably generated, it is not hard to see that linear functional dependence implies nonlinear functional dependence.

Theorem 10. *If both $\mathcal{M}_{t+}^X \equiv \mathcal{M}_{t+}^Y$ and $\mathcal{N}_t^X \supseteq \mathcal{N}_t^Y$, for a single t then Z_t^Y is linearly functionally dependent on Z_t^X. The mapping ψ_t may be chosen to be a σ–homomorphism.*

Proof. This assertion depends on the following lemma, which extends Theorem 1.5 of [3].

Lemma 11. *Let X and Y be random variables on the same probability space with values in a Lusin spaces $(E_i, \mathcal{E}_i)$, $i = 1$ or 2 (i.e. E_i is a Borel subset of a compact metric space $(\bar{E}_i, \bar{\mathcal{E}}_i)$ with its Borel sets). If $\sigma(Y) \subseteq \sigma(X)$, then there is an $f \in \mathcal{E}_1 | \mathcal{E}_2$ with $Y \equiv f(X)$.*

Proof. Let $A_{n,j} \in \mathcal{E}_2$, $1 \leq n$, $1 \leq j \leq j_n$, be a sequence of partitions of E_2 such that (the maximum diameter) $\max_{j \leq j_n} |A_{n,j}| < n^{-1}$, and such that $\{A_{n+1,j}\}$ is a refinement of $\{A_{n,j}\}$ for each n. Such is easily consturcted, using the compactness of $\bar{E}_2$. Let $B_{n,j} \in \mathcal{E}_1$ be such that $\{X \in B_{n,j}\} \equiv \{Y \in A_{n,j}\}$. For each n, the sets on the left being disjoint in j, we can replace $B_{n,j}$ by $B_{n,j} \cap (\bigcup_{i<j} B_{n,i}^c)$, $j < j_n$, and then B_{n,j_n} by $E_1 - \bigcup_{i<j_n} B_{n,i}$, so that the $B_{n,j}$ also form a partition of E_1. Now let $a_{n,j} \in A_{n,j}$ be chosen arbitrarily, and set $f_n(x) = a_{n,j}$ for $x \in B_{n,j}$, $1 \leq j \leq n_j$. Then $\{f_n(X) \in A_{nj}\} \equiv \{Y \in A_{n,j}\}$, and therefore $P\{f_n(X) \text{ and } Y \text{ have distance} < n^{-1}\} = 1$. It follows that $P\{Y = \lim_{n\to\infty} f_n(X)\} = 1$. To complete the proof we need only set.

$$f(x) \doteq \begin{cases} \lim_n f_n(x) & \text{if the limit exists in } E_2 \\ x_0 & \text{elsewhere } (x_0 \in E_2 \text{ fixed}). \end{cases}$$

Now to prove Theorem 10, we apply Lemma 11 with $(E_i, \mathcal{E}_i) = (\Omega^+, \mathcal{F}^+)$ and $X = X_{t+(\cdot)}$, $Y = Y_{t+(\cdot)}$, to get an $f \in \mathcal{F}^+/\mathcal{F}^+$ with $Y_{t+(\cdot)} \equiv f(X_{t+(\cdot)})$.

This is justified since $(\Omega^+, \mathcal{F}^+)$ is a Lusin space [2, IV, 19]. Then $\psi_t = f^{(-1)}$, defines a σ–homomorphism $\mathcal{F}^+ \to \mathcal{F}^+$, and $\{Y \in S\} \equiv \{X \in \psi_t(S)\}$, $S \in \mathcal{F}^+$. Consequently $Z_t^Y \equiv Z_t^X(\psi_t(S))$, as needed (note that only a single t is involved).

For a partial converse to Theorem 10, we have the following.

Theorem 12. *If Z_t^Y is linearly functionally dependent on Z_t^X for all t, then $\mathcal{M}_{t+}^X \supseteq \mathcal{M}_{t+}^Y$ and $\mathcal{N}_t^X \supseteq \mathcal{N}_t^Y$ for all t.*

Proof. Since linear dependence implies nonlinear dependence, it is clear from Corollary 7 that the hypothesis implies $\mathcal{M}_{t+}^X \supseteq \mathcal{M}_{t+}^Y$. To prove $\mathcal{N}_t^X \supseteq \mathcal{N}_t^Y$, we begin by noting that, by Theorem 5, $\mathcal{N}_t^X \equiv \sigma(\rho(Z_s^X), t \le s)$. Now we claim, furthermore, that

$$(3) \qquad \mathcal{N}_t^X \equiv \sigma(Z_{t+s}^X(S) : P\{Z_{t+s}^X(S) = 0 \text{ or } 1\} = 1, \quad 0 \le s, \quad S \in \mathcal{F}^+).$$

To see this, note that for $B \in \mathcal{B}$, setting $S(B) = \{w^+(0) \in B\}$ we have $I_{\{X_{t+s} \in B\}} \equiv I_{\{Z_{t+s}^X(S(B))=1\}}$ by definition of $Z_{t+s}^X(S(B))$ as a conditional probability given $\mathcal{F}_{(t+s)+}^X$. Replacing B by B^c, we see that $P^X\{Z_{t+s}^X(S(B)) = 0 \text{ or } 1\} = 1$, and moreover since $Z_{t+s}^X(S(B)) \equiv I_{\{X_{t+s} \in B\}}$, we have $\mathcal{N}_t^X \subseteq \sigma\{Z_{t+s}^X(S(B)); 0 \le s, B \in \mathcal{B}\}$. This proves the claim (3) in one direction. Conversely, let $S \in \mathcal{F}^+$ be such that $P\{Z_{t+s}^X(S) = 0 \text{ or } 1\} = 1$, for some $s \ge 0$. Then by conditional probability

$$\begin{aligned} \{Z_{t+s}^X(S) = 1\} &\equiv \{X_{t+s+(\cdot)} \in S\} \\ &= \{X_{t+(\cdot)} \in \theta_s^{-1}S\} \\ &\in \mathcal{N}_t^X, \end{aligned}$$

as claimed.

Now if Z_t^Y is linearly functionally dependent on Z_t^X for all t, then $P\{Z_{t+s}^Y(S) = 0 \text{ or } 1\} = 1$ if and only if $P\{Z_{t+s}^X(\psi_{t+s}(S) = 0 \text{ or } 1\} = 1$, and so clearly $\mathcal{N}_t^Y \subseteq \mathcal{N}_t^X$ as asserted.

Combining Theorem 10 and 12, we obtain immediately a characterization of linear functional dependence.

Theorem 13. *If $\mathcal{M}_{t+}^X \equiv \mathcal{M}_{t+}^Y$ for all t, then Z_t^Y is linearly functionally dependent on Z_t^X for all t if and only if, for all t, $\mathcal{N}_t^X \supseteq \mathcal{N}_t^Y$. The mappings ψ_t of Definition 9 may be taken as σ–homomorphisms.*

The interpretation of nonlinear dependence of Z^X and Z^Y in terms of inclusion of σ–fields is quite straightforward. We have

Theorem 14. *Z_t^Y is nonlinearly dependent on Z_t^X for all t if and only if, for all t, $\mathcal{M}_{t+}^X \supseteq \mathcal{M}_{t+}^Y$ and $\mathcal{N}_t^{Z^X} \supseteq \mathcal{N}_t^{Z^Y}$.*

Proof. If Z_t^Y is nonlinearly dependent on Z_t^X for all t, then clearly $\mathcal{M}_t^{Z^X} \supseteq \mathcal{M}_t^{Z^Y}$, and hence by Corollary 7 we have $\mathcal{M}_{t+}^X \supseteq \mathcal{M}_{t+}^Y$. Obviously $\mathcal{N}_t^{Z^X} \supseteq \mathcal{N}_t^{Z^Y}$, proving the necessity (we note too, from Theorem 5, that $\mathcal{N}_t^X \supseteq \mathcal{N}_t^Y$, which in view of Theorem 13 is a stronger statement when $\mathcal{M}_{t+}^X \equiv \mathcal{M}_{t+}^Y$ for all t).

Conversely, as in Theorem 2 there are $\mathcal{H}^\infty/\mathcal{H}$–measurable functions $g_t(x_1,\ldots,x_n,\ldots)$ and $h_t(x_1,\ldots,x_n,\ldots)$, and sequences $(s_n \leq t)$, $(t_n \geq t)$, such that

(4) $$Z_t^Y \equiv g_t(Z_{s_1}^X,\ldots,Z_{s_n}^X,\ldots) \equiv h_t(Z_{t_1}^X,\ldots,Z_{t_n}^X,\ldots).$$

Actually, the proof of Theorem 2 (based on a monotone class argument) does not easily extend to $(H,\mathcal{H})$–valued processes. However, the existence of (s_n) for which $\sigma(Z_t^Y)$ is contained in $\sigma(Z_{s_n}^X, 1 \leq n)$ is a simple fact, and then Lemma 11 applies since $\mathcal{H}$ and $\mathcal{H}^\infty$ are Lusin spaces [6, Essay I, Proposition 1.3]. Suppose now that Z_t^X is given. By the Markov property of Z_t^X, the two terms on the right in (4) are then (conditionally) independent. Since they are equal P-a.s., they must each generate a σ–subfield of $\sigma(Z_t^X)$, i.e. in particular $\sigma(g_t(Z_{s_1}^X,\ldots, Z_{s_n}^X,\ldots))\subseteq\sigma(Z_t^X)$. In more detail, Z_t^Y is conditionally independent of itself. Then a simple partioning argument for $(H,\mathcal{H})$ (as in Lemma 11) shows that the conditional distribution of Z_t^Y given Z_t^X reduces to a unit point mass, P-a.s., and the assertion follows. Therefore, again by Lemma 11, there is an $f_t \in \mathcal{H}/\mathcal{H}$ with $Z_t^Y = f_t(Z_t^X)$, as required.

In the case that X and Y are $\mathcal{M}_{t+}$–equivalent, we can combine the above results in the following logical sequence, in which neither of the implications $\Longrightarrow$ may be replaced by $\Longleftrightarrow$ in general.

Summary Theorem 15. *If $\mathcal{M}_{t+}^X \equiv \mathcal{M}_{t+}^Y$ for all t (written $\forall t$, and understanding all functions below to exist and be Borel measurable on the corresponding spaces) then we have*

$$Y_t = f_t(X_t), \quad \forall t$$
$$\Longrightarrow \mathcal{N}_t^X \supseteq \mathcal{N}_t^Y, \quad \forall t$$
$$\Longleftarrow Z_t^Y \text{ is linearly functionally dependent on } Z_t^X, \quad \forall t$$
$$\Longrightarrow \mathcal{N}_t^{Z_X} \supseteq \mathcal{N}_t^{Z_Y}, \quad \forall t$$
$$\Longleftarrow Z_t^Y \text{ is nonlinearly functionally dependent on } Z_t^X, \quad \forall t$$

Counterexamples to the two possible $\Longleftarrow$ implications may be found in the Appendix of Chapter 6 of [7] (under stronger assumptions, so they suffice also here). As they are straightforward we shall not repeat them.

Remark. In the above Theorem we have chosen to work 'from the right', using $\mathcal{M}_{t+}^X$, $Z_t^X(= Z_{t+}^X)$, etc. This is only for simplicity, however. The assmption $\mathcal{M}_{t+}^X \equiv \mathcal{M}_{t+}^Y$, $\forall t$, is easily seen to be equivalent to $\mathcal{M}_{t-}^X \equiv \mathcal{M}_{t-}^Y$, $\forall t$. Moreover, as mentioned in the Remark to Definition 4, there also exist the left prediction processes Z_{t-}^X and Z_{t-}^Y defining the respective conditional futures given $\mathcal{M}_{t-}^X$ and $\mathcal{M}_{t-}^Y$. Actually, these could be used also in the statement of Theorem 15.

To sketch the argument, we note first that the same σ–homomorphisms ψ_t suffice to show that Z_{t-}^Y is linearly functionally dependent on Z_{t-}^X for all t. Conversely, if this holds, since it is known that $P\{Z_{t-}^X = Z_t^X\} = 1$ except for countably many t, it can be seen from the proof of Theorem 12 that $\mathcal{N}_t^X \supseteq \mathcal{N}_t^Y$, and therefore Z_t^Y is linearly functionally dependent on Z_t^X. Similarly, if Z_t^Y is only nonlinearly dependent on Z_t^X for all t, and if $P\{Z_{t-}^Y \neq Z_t^Y\} > 0$ (i.e. Z_{t-}^Y is a branch point with positive probability) then it follows from the equation $P(Z_t^X \in A | \mathcal{M}_{t-}^X) = q(0, Z_{t-}^X, A)$, $A \in \mathcal{H}$, that Z_{t-}^Y is nonlinearly dependent on Z_{t-}^X. Here the converse follows because $P\{Z_{t-}^X = Z_t^X\} = 1$ for almost every t, and hence every t is a right limit of times at which Z_t^Y is nonlinearly dependent on Z_t^X. But this implies the same dependence at t, because if Z_t^X is known it follows by the zero–or–one law that $Z_t^Y (= \lim_{s \to t+} Z_s^Y)$ is determined with probability 1. Hence Theorem 15 remains true if Z_t^X and Z_t^Y are replaced by Z_{t-}^X and Z_{t-}^Y.

As a final consideration, let us examine briefly how our results behave under reversal of time. It is to be emphasized that our whole approach is time-asymmetric, not only because the processes are r.c.l.l., but more basically because Z_t^X is defined asymmetrically in $\mathcal{M}_{t+}^X$ and $\mathcal{N}_t^X$. To make a sensible reversal of time, one must first redefine X_t and Y_t, replacing them by $\frac{1}{2}(X_{t-} + X_t)$ and $\frac{1}{2}(Y_{t-} + Y_t)$ for all t. We assume for the rest of the paper that this adjustment has been made, and we continue the notations X_t, Y_t, Z_t^X, Z_t^Y, etc., for the corresponding quantities (note that $(\Omega^+, \mathcal{F}^+)$ is again a Lusin space).

Definition 16. Introducing the processes $X_t^* \doteq X_{-t}$, $Y_t^* \doteq Y_{-t}$, $-\infty < t < \infty$, with prediction processes $Z_t^{X^*}$, $Z_t^{Y^*}$ respectively, we call the *postdiction* processes (or prediction processes with past and future interchanged) of X and Y the processes $Z_{-t}^{X^*}$ and $Z_{-t}^{Y^*}$, $-\infty < t < \infty$.

Because of the adjustment it is more symmetrical now to replace $\mathcal{N}_t^X$ and $\mathcal{N}_t^Y$ by $\mathcal{N}_{t+}^X$ and $\mathcal{N}_{t+}^Y$. Besides, we recover in this way the original $\mathcal{N}_t^X$ and $\mathcal{N}_t^Y$ before the change of definition, since $\mathcal{N}_{t+}^X$ contains only X_{t+} rather than the present $\frac{1}{2}(X_{t-} + X_{t+})$. This change of $\mathcal{N}_t^X$ is convenient for a time reversal. We thus consider $Z_t^X(S)$ as defined only for $S \in \sigma(w^+(s), 0 < s)$, which is a proper subfield of the new $\mathcal{F}^+$.

Now we are ready to reverse the direction of time, i.e. at time t the past and future are interchanged. Writing $\mathcal{M}_t^{*,X} \doteq \mathcal{N}_t^X$ and $\mathcal{N}_t^{*,X} \doteq \mathcal{M}_t^X$, with the analogous notation for Y, the assumption $\mathcal{M}_{t+}^X \equiv \mathcal{M}_{t+}^Y$ and $\mathcal{N}_{t+}^X \equiv \mathcal{N}_{t+}^Y$ for all t (in which we can replace $t+$ by $t-$ without loss since $\mathcal{M}_{t-}^X = \bigvee_{s<t} \mathcal{M}_{t+}^X$, etc.) is equivalent to $\mathcal{M}_{t-}^{*,X} \equiv \mathcal{M}_{t-}^{*,Y}$ and $\mathcal{N}_{t+}^{*,X} \equiv \mathcal{N}_{t+}^{*,X}$ for all t. Then by weakening Theorem 13 to an assertion about *equivalence* of past and future it becomes reversable in time, and we obtain immediately

Theorem 17. *The assumption $\mathcal{M}_{t-}^X \equiv \mathcal{M}_{t-}^Y$ and $\mathcal{N}_{t+}^X \equiv \mathcal{N}_{t+}^Y$ for all t is equivalent to the linear functional equivalence of Z_t^X and Z_t^Y for all t, and equivalently to that of $Z_{-t}^{X^*}$ and $Z_{-t}^{Y^*}$ for all t.*

Unfortunately, there seems to be no corresponding reversalibility for nonlinear equivalence of Z_t^X and Z_t^Y. Indeed, for any pair X, Y such that Z_t^Y is nonlinearly equivalent to Z_t^X for all t, but not linearly equivalent, since clearly $\mathcal{M}_{t-}^X \equiv \mathcal{M}_{t-}^Y$ for all t, Theorem 17 implies that $\mathcal{N}_{t+}^X$ and $\mathcal{N}_{t+}^Y$ cannot be equivalent for all t. But these are the same as $\mathcal{M}_{t-}^{*,X}$ and $\mathcal{M}_{t-}^{*,Y}$, so by Theorem 14 their nonequivalence implies that the postdiction processes cannot be nonlinearly equivalent for all t.

Hence we see that nonlinear equivalence of the predictions is not preserved under interchange of past and future.

REFERENCES

[1]. Blumenthal, R., Getoor, R. K., and McKean, H. P. Jr., *Markov processes with identical hitting distributions*, Illinois J. Math. **6** (1962), 402–420.

[2]. Dellacherie, C. and Meyer, P.-A., *Probabilités et Potentiel, Chap. I–IV* (1980), Hermann, Paris.

[3]. Doob, J. L., *Stochastic Processes* (1953), Wiley, New York,.

[4]. Knight, F. and Orey, S., *Construction of Markov processes with identical hitting probabilities*, J. Math. Mech. **13** (1964), 857–873.

[5]. Knight, F. B., *A remark on Markovian germ fields*, Z. Wahr. verw. Geb. **15** (1970), 291–296.

[6]. Knight, F. B., *Essays on the Prediction Process*, I.M.S. Lecture Notes Series, S. Gupta, Ed., Inst. Math. Statistics **1** (1981).

[7]. Knight, F. B., *Foundations of the Prediction Process* (1991), Oxford University Press, Oxford.

[8]. Loeve, M., *Probability Theory, Third Edition* (1963), D. Van Nostrand, Princeton.

Professor Frank B. Knight
Department of Mathematics
University of Illinois at Urbana–Champaign
1409 West Green Street
Urbana, Illinois 61801
U.S.A.

BROWNIAN INTERPRETATIONS OF AN ELLIPTIC INTEGRAL

by

S.M. KOZLOV, J.W. PITMAN and M. YOR

1. Introduction.

This paper presents some interpretations in terms of Brownian motion of the *Legendre first order elliptic integral* displayed in (1.a) below. We express the probability that a complex valued Brownian motion hits one subinterval of the real line before another in terms of the Legendre elliptic integral. Then we find the asymptotic distribution of the Legendre integral along a Brownian path, and deduce asymptotic laws for looping numbers of the Brownian path on the associated Riemann surface.

The Legendre first order elliptic integral is

$$(1.a) \qquad \int_0^{z_0} \frac{dz}{\sqrt{(1 - z^2)(1 - k^2 z^2)}} ,$$

where $0 < k < 1$, and the integral is defined along a path from 0 to $z_0 \in \mathbb{C}$ which avoids the singularities ± 1 and $\pm 1/k$. Starting with the convention that at $z = 0$ the integrand is $+1$, the integrand and hence the integral are defined continuously along the path. The value of the integral depends on how the path of integration winds around the singularities. Let $\phi(z_0)$, the *principal value* of the integral, be defined for $z_0 \in \mathbb{C} - \{(-\infty, -1] \cup [1, \infty)\}$, by integration along the straight line from 0 to z_0. It is well known that ϕ defines a one-to-one conformal mapping of $\mathbb{C} - \{(-\infty, -1] \cup [1, \infty)\}$ onto the open rectangle

$$(-K, K) \times (-K', K') = \{w : -K < \text{Re}(w) < K, \ -K' < \text{Im}(w) < K'\},$$

where K and K' are the positive reals defined by

$$(1.b) \quad K = \int_0^1 \frac{dt}{\sqrt{(1 - t^2)(1 - k^2 t^2)}} ; \qquad iK' = \int_1^{1/k} \frac{dt}{\sqrt{(1 - t^2)(1 - k^2 t^2)}} .$$

Research supported in part by NSF grant DMS91-07531

See Figure 1.0. The inverse of this mapping extends by reflection to define a mero-morphic function on the whole plane, which is periodic with periods $4K$ and $2iK'$. This is Jacobi's elliptic sine function sn. The set of possible values of the integral (1.a), over all continuous paths from 0 to z_0 which avoid the singularities, is

(1.c) $$\mathrm{sn}^{-1}(z_0) \;=\; \{2mK + 2inK' + (-1)^m \phi(z_0)\}$$

where m and n range over integers. See for instance Henrici (1974, Sec 5.13), Ahlfors (1966, Sec. 6.2.3), Nehari (1952), Lawden (1989, Sec. 8.14).

Figure 1.0. (Inspired by Bowman (1953, Ch. 5, Fig. 4)).

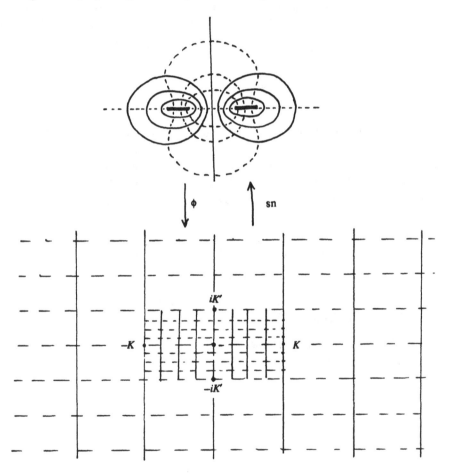

Proposition 1.1. *Each of the functions* Reϕ *and* | Im ϕ | *has a unique continuous extension from* $\mathbb{C} - \{(-\infty, -1] \cup [1, \infty)\}$ *to the whole complex plane* $\mathbb{C}$. *These continuous extensions admit the following interpretations in terms of a planar Brownian motion* $Z = (Z_t, t \geq 0)$ *starting from* $Z_0 = z_0 \in \mathbb{C}$:

(1.d) $$P \ (Z \text{ hits } [1, 1/k] \text{ before } [-1/k, -1]) \ = \ \frac{1}{2} + \frac{\text{Re}\,\phi(z_0)}{2K}$$

(1.e) $$P \ (Z \text{ hits } [1, 1/k] \text{ or } [-1/k, -1] \text{ before the imaginary axis}) \ = \ | \text{Re}\,\phi(z_0) | / K$$

(1.f) $$P \ (Z \text{ hits } (-\infty, -1/k] \text{ or } [1/k, \infty) \text{ before } [-1, 1]) \ = \ | \text{Im}\,\phi(z_0) | / K'.$$

The continuous extensions of Re ϕ and | Im ϕ | to the real axis have values on $[0, \infty)$ as shown in the following table, where all square roots are positive. Values on $(-\infty, 0]$ are obtained from these by symmetry: on the real axis, Re ϕ is an odd function and | Im ϕ | is even.

Table 1.2.

range of $x \in \mathbf{R}$	$[0, 1]$	$[1, 1/k]$	$[1/k, \infty)$
value of Re $\phi(x)$	$\int_0^x \dfrac{dt}{\sqrt{(1 - t^2)(1 - k^2 t^2)}}$	K	$\int_x^\infty \dfrac{dt}{\sqrt{(t^2 - 1)(k^2 t^2 - 1)}}$
value of | Im $\phi(x)$ |	0	$\int_1^x \dfrac{dt}{\sqrt{(t^2 - 1)(1 - k^2 t^2)}}$	K'

Following Spitzer (1958), who discovered the limiting case of (1.d) for
$$k = 0, \quad \text{when } K = \pi/2, \quad \text{and} \quad \phi = \sin^{-1},$$
expressions (1.d) and (1.f) can be interpreted for real x as hitting probabilities for the Cauchy process derived from Z as its trace on the real line. See also Pitman-Yor (1986b) for related results.

Since the real and imaginary parts of the analytic function ϕ are harmonic on $\mathbb{C} - \{(-\infty, -1] \cup [1, \infty)\}$, once details of the boundary behaviour are checked, Proposition 1.1 follows from the well known identification of Brownian hitting probabilities with the solution of a Dirichlet boundary value problem (Doob (1954)). In Section 2 we review the electrostatic analogue of Proposition 1.1, following Smythe (1968). The following proposition is then obtained from the probabilistic interpretation of capacity, reviewed in Burdzy, Pitman and Yor (1988), and the fact that the capacity of the segment $[1, 1/k]$, when the segment $[-1/k, 1]$ is grounded, is equal to $K'/2K$:

Proposition 1.3. *For a Borel subset A of the plane let*

$$\text{time}(A,t) = \int_0^t 1(Z_s \in A) \, ds$$

be the time that the Brownian motion Z spends in A up to time t. Let N_t be the number of crossings of the path of Z from $[-1/k, -1]$ to $[1, 1/k]$ that are completed by time t. Then

(1.g)
$$\frac{N_t}{\text{time}(A,t)} \xrightarrow{a.s.} \frac{1}{2} \frac{K'}{K} \frac{1}{\text{area}(A)}$$

(1.h)
$$\frac{N_t}{\log t} \xrightarrow{d} \frac{1}{4\pi} \frac{K'}{K} e$$

where e has the standard exponential distribution:

(1.i)
$$P(e > x) = e^{-x}, \quad x \geq 0.$$

We now turn to a discussion of the stochastic Legendre integral

(1.j)
$$\Phi_t = \int_0^{Z_t} \frac{dz}{\sqrt{(1 - z^2)(1 - k^2 z^2)}} = \int_0^t \frac{dZ_s}{\sqrt{(1 - Z_s^2)(1 - k^2 Z_s^2)}}.$$

The first integral is the Legendre integral along the Brownian path $(Z_s, 0 \leq s \leq t)$ starting at $z_0 = 0$. The second may be interpreted as either an Itô or a Stratonovich stochastic integral, with integrand defined continuously along the path. With probability one, all three interpretations of the integral give the same process $(\Phi_t, t \geq 0)$. See Yor (1977), or Revuz and Yor (1991, Exercise V.2.19). As a variation of a result of Messulam and Yor (1982), we obtain the following asymptotic law, which will be proved in Section 3.

Proposition 1.4. *As $t \to \infty$*

(1.k)
$$\frac{\Phi_t}{\sqrt{\log t}} \xrightarrow{d} \sqrt{\frac{KK'}{\pi}} \sqrt{2e} (\gamma_1 + i\gamma_2),$$

jointly with the convergence in law (1.h), where e has standard exponential distribution, γ_1 and γ_2 have standard Gaussian distribution, and e, γ_1 and γ_2 are independent.

Standard formulae for the resolvent of the Brownian semigroup, from Erdélyi (1954, (1): 146, (26) and (27)), imply

(1.l)
$$P(\sqrt{2e}\,\gamma_i \in dx) = \frac{1}{2} e^{-|x|} dx,$$

(1.m)
$$P(\sqrt{2e}\,\gamma_1 \in dx, \sqrt{2e}\,\gamma_2 \in dy) = \frac{1}{2\pi} K_0(\sqrt{x^2 + y^2}) \, dx \, dy$$

where K_0 is the usual modified Bessel function of index zero.

Consider now the Riemann surface of the analytic function $\sqrt{(1-z^2)(1-k^2z^2)}$. This two leafed surface consists of two copies of $\mathbb{C}$, each glued to the other along cuts on the intervals $[-1/k, -1]$ and $[1, 1/k]$. It is well known that when compactified by two points at ∞ (one for each sheet), this surface is homeomorphic to a torus. See for instance Klein (1893), Section II.15, for nice pictures. Consequently, every closed curve on the Riemann surface is homologically equivalent to $n\gamma + m\beta$ for some integers n and m, where γ and β are the two non-trivial independent cycles on the surface as in the following figure:

Figure 1.5. Solid parts of the curves are on the upper leaf of the Riemann surface and dotted parts on the lower leaf:

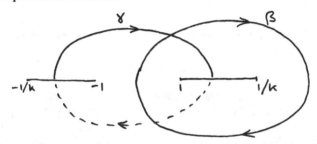

Now lift the planar Brownian motion Z to obtain a Brownian motion $\hat{Z}$ on the Riemann surface. Follow $\hat{Z}$ on the surface up to time t, then return to the origin by some direct route, say by a straight line if $\hat{Z}_t$ is on the same leaf as when it started, and by a single passage between leaves followed by a straight line in case $\hat{Z}_t$ is on the other leaf. The closed path Γ_t so obtained is then homotopic to $n_t \gamma + m_t \beta$ for some random integers n_t and m_t which describe the looping of Γ_t on the torus.

Corollary 1.6. *As $t \to \infty$,*

$$\frac{(n_t, m_t)}{\sqrt{\log t}} \xrightarrow{d} \sqrt{\frac{2e}{\pi}} \left[\frac{1}{4} \sqrt{\frac{K'}{K}} \gamma_1, \frac{1}{2} \sqrt{\frac{K}{K'}} \gamma_2 \right],$$

where this convergence holds jointly with (1.h) *and* (1.k) *for* e, γ_1, γ_2 *as in Propositions* 1.3 *and* 1.4.

Proof. Since the Legendre integrals along γ and β are $4K$ and $2K'i$ respectively, the Legendre integral along Γ_t is

$$4Kn_t + 2K'm_t i = \Phi_t + \xi_t,$$

for Φ_t as in (1.j), where ξ_t, the integral incurred along the return path, is such that

$| \operatorname{Re} \xi_t | \leq 4K$ and $| \operatorname{Im} \xi_t | \leq 4K'$. □

2. Electrostatic interpretations.

If in Figure 1.0 the solid contours in the z-plane (level sets of $\operatorname{Re}\phi$) are interpreted as equipotentials and the dotted ones (level sets of $\operatorname{Im}\phi$) as lines of force, the contour graph of ϕ may be interpreted as the equilibrium electric field in the plane associated with a positive charge on the interval $[1, 1/k]$ and a negative charge on the interval $[-1/k, -1]$. See Smythe (1968, Section 4.29), from which much of the following account is drawn. In the physical model, the two intervals are understood as the intersections with the plane of two parallel conducting strips perpendicular to the plane. We suppress from now on any mention of the third dimension, but all quantities in the plane can be understood as two dimensional sections of corresponding three dimensional quantities.

While ϕ cannot be extended continuously to the whole of $\mathbb{C}$, its real part can, due to the symmetry property

$$\overline{\phi(z)} = \phi(\overline{z}),$$

where for $z = x + iy$, $\overline{z} = x - iy$. The electrostatic interpretation is that $\operatorname{Re}\phi(z)$ is the equilibrium potential at z for the configuration of charge on $[-1/k, -1]$ and $[1, 1/k]$ which achieves potential $-K$ on one interval and $+K$ on the other. Such a configuration of equal and opposite positive and negative charges at equilibrium on the surfaces of two conductors is called a *condenser*.

Since the equipotentials form closed curves which separate the two conductors, the magnitude of the charge on each conductor is the dielectric constant ε times the absolute value of the increment of $\operatorname{Im}\phi$ obtained by going once around a level contour of $\operatorname{Re}\phi$. This increment of $\operatorname{Im}\phi$ is readily computed for any $1/k < x < \infty$ as

(2.a) $\lim_{\delta \to 0} [\operatorname{Im}\phi(x + \delta i) - \operatorname{Im}\phi(x - \delta i)] = 2K'$.

So the total charge on $[1, 1/k]$ is $2K'\varepsilon$. This charge of $2K'\varepsilon$ should be thought of as the total of two identical charge distributions, one for the "top side" of the interval, and one for the "bottom side", each with density per unit length

(2.b) $\varepsilon \left| \dfrac{d\phi}{dz} \right|_{z \to x} = \dfrac{\varepsilon}{\sqrt{(x^2 - 1)(1 - k^2 x^2)}}, \qquad 1 \leq x \leq 1/k$.

The same formula gives half the density of negative charge on $[-1/k, -1]$. Since the absolute charge on each interval is $Q = 2K'\varepsilon$ and the potential difference between the intervals is $V = 2K$, the *capacity* C of the condenser is

(2.c) $$C = Q/V = \varepsilon K'/K .$$

The translation into probabilistic terms of these concepts from classical potential theory is well known. In particular, the probabilistic interpretation of $\mathrm{Re}\,\phi(z)$ is given by (1.d). Formulae (1.e) and (1.f) correspond to similar electrostatic setups.

According to results summarized in Burdzy, Pitman and Yor (1988), if we take $\varepsilon = 1/2$, the capacity of the condenser can be interpreted as an equilibrium rate of crossings of the Brownian path from one conductor to the other. Proposition 1.3 follows immediately from this interpretation of the capacity, the ergodic theorem for additive functionals of planar Brownian motion, and the Kallianpur-Robbins (1953) law

(2.d) $$\frac{1}{\log t} \int_0^t g(Z_s)\,ds \;\overset{d}{\to}\; \frac{e}{2\pi} \iint g(x+iy)\,dx\,dy,$$

valid for any Lebesgue integrable g such that the left side is a.s. finite for some (and hence all) $t > 0$.

When normalized by its total mass $\varepsilon K'$, the equilibrium charge distribution (2.b) is interpreted as the asymptotic distribution as $n \to \infty$ of Z_{T_n}, or of Z_{L_n}, where T_n is the moment at which the nth passage from $[-1/k,-1]$ to $[1,1/k]$ is completed, and L_n is the subsequent time at which the path leaves $[1,1/k]$ on its next journey to $[-1/k,-1]$. The factor of 2 in (2.a) corresponds to the fact that on each passage the BM is equally likely to approach the interval from above or below, and similarly for departures.

Variations of the above argument give the fields, charge distributions and capacities, with similar Brownian interpretations, for a variety of similar problems. For example, the contour graph of ϕ restricted to the right half plane ($Re\ z \geq 0$) can be interpreted as the equilibrium electrical field for a distribution of negative charge $-2K'$ on the imaginary axis with density at iy equal to

$$\frac{1}{\sqrt{(1+y^2)(1+k^2y^2)}} ,$$

and the same distribution of positive charge $+2K'$ on $[1,1/k]$ as before. This condenser has capacity twice that of the previous one. Also, scaling, translation and inversion of the present results yield corresponding formulae for any two subintervals of the line instead of the intervals $\pm[1,1/k]$. Lawden (1989, Section 5.9) gives expressions in terms of elliptic functions for the electric field and capacity associated with a parallel plate capacitor. Given AB and CD two opposite sides of a rectangle in the plane, this field determines the probability that Brownian motion started at z, anywhere in the plane, hits AB before CD, and the capacity gives the equilibrium rate of crossings. Formula 5.8.9 and Exercise 5.9 of the same text give products of elliptic functions, which, when multiplied by $-\dfrac{2}{\pi}$, can be interpreted as the probability that a Brownian motion, started at point (x,y) in the rectangle

$$\{(x,y): 0 \leq x \leq K, 0 \leq y \leq K'\},$$

first hits the boundary of the rectangle on a prescribed union of sides.

For an application of the elliptic function appearing here to analysis of a signed additive functional of one-dimensional Brownian motion see McGill (1989, Example 3 of Section 4).

3. Asymptotics for the stochastic Legendre integral.

We eventually show that Proposition 1.4 follows by a slight variation of the stochastic calculus arguments used by Messulam and Yor (1982), and Pitman and Yor (1989) to establish similar results. But first we indicate a more elementary approach, which is closely connected to the preceding discussion. This is similar to the method used by Lyons and McKean (1984) and Pitman and Yor (1986b) in the study of windings. This approach establishes the convergence in distribution separately for the real and imaginary parts in (1.k). But it is not clear how to complete this argument to establish *joint* convergence in distribution to the limit defined by mutually independent e, γ_1 and γ_2. For this it seems necessary to take the stochastic calculus approach.

For simplicity we choose some point $z_0 \in [1, 1/k]$, as the starting point of the Brownian motion, and make some arbitrary initial choice of sign for the Legendre integrand. It is easy to see that neither choice has any effect on the asymptotics. Let

$$T_1 = \inf\{t > 0: Z_t \in [-1/k, -1]\}, \text{ the first passage time to } [-1/k, -1],$$

$$L_1 = \text{last time } Z \text{ is in } [1, 1/k] \text{ before } T_1,$$

$$T_2 = \inf\{t > T_1: Z_t \in [1, 1/k], \text{ time of the next return to } [1, 1/k],$$

$$L_2 = \text{last time in } [-1/k, -1] \text{ before } T_2, \text{ and so on:}$$

$$0 < L_1 < T_1 < L_2 < T_2 < L_3 < T_3 < \cdots \uparrow \infty \text{ a.s.}$$

Notice that

(3.a) $|.\text{Re}\Phi_t| < 2K \text{ for } 0 \leq t < T_1,$

(3.b) $\text{Re}\Phi_{T_1} = \pm 2K,$

(3.c) $| \text{Re}\Phi_t - \text{Re}\Phi_{T_1}| < 2K \text{ for } T_1 \leq t < T_2,$

$$\mathrm{Re}\Phi_{T_2} = \mathrm{Re}\Phi_{T_1} \pm 2K, \quad \text{and so on. Thus}$$

(3.d) $$\mathrm{Re}\,\Phi_t = \pm 2K \pm 2K \cdots \pm 2K + \varepsilon_t,$$

where the number of terms of type $\pm 2K$ is $M_t = \#\{k : T_k \le t\}$, say, and

(3.e) $$|\varepsilon_t| \le 2K.$$

The value of the nth sign $\pm$ associated with the crossing from one interval to the other over time $[L_n, T_n]$ is the product of one sign determined by the looping of the path around the singularities $\{-1/k, -1, 1, 1/k\}$ strictly before time L_n, and another sign determined by whether the path leaves the real axis at time L_n on the side of $+i$ or $-i$. By symmetry and the last exit decomposition at time L_n, the latter sign is equally likely to be $+$ or $-$, independently of the former, indeed independently of the whole path before time L_n. Similarly, the value of the next passage time T_n is independent of the first n signs. Thus the sequence of signs $\pm$ in (3.d) is determined by a fair coin-tossing process independently of the sequence $(T_1, T_2, ...)$, hence also independent of the counting process $(M_t, t \ge 0)$. Since $|M_t - 2N_t| \le 1$ for N_t as in Proposition 1.3, (1.h) gives

(3.f) $$\frac{M_t}{\log t} \xrightarrow{d} \frac{1}{2\pi} \frac{K'}{K} \mathbf{e}.$$

Combining this with (3.d), (3.e) and de Moivre's normal approximation for fair coin tossing gives the real part of (1.k) as follows. Informally, with $\overset{d}{\cong}$ denoting "approximate equality in distribution", the argument is

$$\frac{\mathrm{Re}\,\Phi_t}{\sqrt{\log t}} \overset{d}{\cong} \frac{2K}{\sqrt{\log t}} [\,\text{Sum of } M_t \pm 1\text{'s}\,]$$

$$\overset{d}{\cong} \frac{2K}{\sqrt{\log t}} \beta(M_t) \quad \text{for a Brownian motion } \beta \text{ independent of } M_t$$

$$\overset{d}{\cong} \frac{2K}{\sqrt{\log t}} \beta\left[\frac{\log t}{2\pi} \frac{K'}{K} \mathbf{e}\right] \quad \text{by (3.f)}$$

$$\overset{d}{\cong} \sqrt{\frac{KK'}{\pi}} \sqrt{2\mathbf{e}}\, \beta(1) \quad \text{by Brownian scaling.}$$

Details of rigor are easily supplied. The imaginary part of (1.k) follows similarly using the dual process of crossings between $[-1, 1]$ and $(-\infty, -1/k] \cup [1/k, \infty]$, related to formula (1.f). But we do not see how to push this sort of argument to establish the joint convergence of the real and imaginary parts. We appeal instead to the following proposition, which is a variation of a result of Messulam and Yor (1982). The proposition is applied to

(3.g)
$$f(z) = \frac{1}{\sqrt{(1 - z^2)(1 - k^2 z^2)}},$$

for some arbitrary Borel measurable choice of the square root, and σ_s the factor of ± 1 determined by the Brownian path up to time s which corrects the choice of square root to make it continuous along the path.

Proposition 3.1. *Let (σ_s) be a process with values in the unit circle, which is progressively measurable with respect to the filtration of a complex Brownian motion (Z_t) starting from $Z_0 = z_0$. Suppose $f : \mathbb{C} \to \mathbb{C}$ is a Borel function, satisfying the following three conditions:*

(3.h) f *is bounded in some neighbourhood of z_0,*

(3.i) f *is Lebesgue square integrable, and*

(3.j) $z \to f(z)/z$ *is Lebesgue integrable.*

Then as $t \to \infty$

(3.h)
$$(\log t)^{-1/2} \int_0^t \sigma_s f(Z_s)\, dZ_s \xrightarrow{d} \sqrt{2e}(\eta(f) + i\chi(f)),$$

where e, η and χ are independent, e is a standard exponential variable, and η and χ are two independent Gaussian measures on $\mathbb{C}$, with intensity $(2\pi)^{-1}$ per unit area.

Proof. The argument used to prove Theorem 6.1 of Pitman and Yor (1989) goes through without change, since the factor σ_s has no effect on the quadratic variation of any of the relevant martingales. The required estimate (6.h) in that paper follows from the Kallianpur-Robbins law (2.d), using the assumptions on f. $\square$

Remark 3.2.. The process $(\Phi_t, t \geq 0)$ is a conformal martingale (see Getoor and Sharpe (1972)), with increasing process

(3.l)
$$U_t = \int_0^t |f(Z_s)|^2\, ds, \quad t \geq 0,$$

for f as in (3.g). It follows that

(3.m) $\Phi_t = \Theta(U_t)$

for some complex Brownian motion Θ. For $c > 0$ define Θ^c by $\Theta^c(u) = \frac{1}{c}\Theta(c^2 u)$, $u \geq 0$, so Θ^c is a complex BM too. Then

$$\Phi_t / \sqrt{\log t} = \Theta^{\sqrt{\log t}}(U_t / \log t).$$

The proof of Proposition 3.1 shows that as $t \to \infty$

(3.n) $(U_t / \log t \; ; \; \Theta^{\sqrt{\log t}}(u), u \geq 0) \xrightarrow{d} (\dfrac{2KK'}{\pi} e \; ; \Theta^\infty(u), u \geq 0),$

where Θ^∞ is a complex BM, starting at 0, independent of the standard exponential variable e which comes from the Kallianpur-Robbins law (2.d). The factor KK' comes from the easily verified identity

(3.o) $\int\limits_{\mathbb{C}} |f|^2 (x + iy) \, dx dy = 4KK'.$

The convergence in (3.n) is convergence of distributions on the product of $\mathbb{R}$ and $C \; (\mathbb{R}_+, \mathbb{C})$ equipped with the topology of uniform convergence on compacts.

On the other hand, the discussion at the beginning of Section 2 implies that

$$Z_t = \text{sn}(\Phi_t),$$

so that

$$d\Phi_t = f(Z_t) dZ_t = f \circ \text{sn}(\Phi_t) dZ_t.$$

Now $\Phi_t = \Theta(U_t)$ can be recovered from Θ alone using the fact (Revuz and Yor (1991, Proposition V.1.10)) that (U_t) is the inverse of the additive functional of (T_u) of Θ defined by

(3.t) $T_u = \int\limits_0^u |f \circ \text{sn}|^{-2}(\Theta_s) \, ds = \int\limits_0^u g(\Theta_s) \, ds \;,$ say, where

(3.u) $g(w) = |1 - \text{sn}^2(w)| \; |1 - k^2 \text{sn}^2(w)|,$

is a doubly periodic function of w with periods $2K$ and $2iK'$ (c.f. the periods $4K$ and $2iK'$ of sn). The Kallianpur-Robbins law for (U_t) translates into the following asymptotic law for the Brownian additive functional (T_u)

(3.v) $\dfrac{\log T_u}{u} \xrightarrow{d} \dfrac{\pi}{2KK'e}$ as $u \to \infty.$

This very crude asymptotic law is dictated by the order of the poles of g, and is similar to results mentioned by Burdzy-Pitman-Yor (1988, page 71) and Mountford (1991). The random factor of $1/e$ is attributable to randomness in the log of the distance of closest approach of the Brownian motion to one of the poles. See Le Gall-Yor (1986, Theorem 7.4) for the same phenomenon on the sphere instead of the torus.

References.

AHLFORS, L. V. (1966) *Complex Analysis.* 2nd. ed. New York, McGraw-Hill.

BOWMAN, F. (1953) *Introduction to Elliptic Functions.* Wiley, New York.

BURDZY, K., PITMAN, J. and YOR, M. (1988). Some asymptotic laws for crossings and excursions. *Colloque Paul Lévy sur les Processus Stochastiques, Société Mathématique de France, Astérisque* **157-158**, 59-74.

BURDZY, K., PITMAN, J. and YOR, M. (1990). Brownian crossings between spheres. *J. of Mathematical Analysis and Applications,* **148**, No 1, 101-120.

COURANT, R. and HURWITZ, A. (1929). *Functionentheorie.* Springer. Berlin.

DOOB, J. L. (1954). *Classical potential theory and its probabilistic counterpart.* Springer-Verlag.

DOOB, J. L. (1954). Semi-martingales and subharmonic functions. *Trans. Am. Math. Soc.***77**, 86-121.

ERDELYI, A. (1954) *Higher transcendental functions.* Bateman Manuscript Project. New York, McGraw-Hill.

GETOOR, R. K. & SHARPE, M. J. (1972). Conformal martingales. *Invent. Math.* **16,**271-308.

HENRICI, P. (1974). *Applied and computational complex analysis. Vol I* Wiley, New York.

HENRICI, P. (1986). *Applied and computational complex analysis. Vol III* Wiley, New York.

KALLIANPUR, G. and ROBBINS, H. (1953). Ergodic property of the Brownian motion process. *Proc. Nat. Acad. Sci. USA,* **39**, p. 525-533.

KLEIN, F. (1893). *On Riemann's theory of algebraic functions and their integrals.* Reprinted by Dover, New York, 1963.

KOZLOV, S. M. (1985). The method of averaging and random walks in inhomogeneous environments. *Russian Math. Surveys* **40,** 2 , 73-145.

LAWDEN, D. F. (1989) *Elliptic functions and applications.* New York. Springer-Verlag.

LE GALL, J. F. and YOR M. (1986). Etude asymptotique de certains mouvements browniens complexes avec drift. *Probab. Th. Rel. Fields,* **71**, 183-229.

LYONS, T.J. and McKEAN, H.P. (1984). Winding of the plane Brownian motion. *Advances in Math.* **51**, 212-225.

MESSULAM, P. and YOR, M (1982). On D. Williams 'pinching' method and some applications. *Journal London Math. Soc (2)*, **26**, 348-364.

MOUNTFORD, T. (1991). The asymptotic distribution of the number of crossings between tangential circles by planar Brownian motion. *Journal London Math. Soc., to appear.*

McGILL, P. (1989). Wiener-Hopf factorisation of Brownian motion. *Probability Theory and Related Fields*, **83**, 355 - 389.

NEHARI, Z. (1952) *Conformal Mapping.* New York, McGraw-Hill.

PITMAN, J. and YOR, M. (1986). Level crossings of a Cauchy process. *Annals of Probab.*, **14,**, p. 780-792.

PITMAN, J. and YOR, M. (1989). Further asymptotic laws of planar Brownian motion. *Annals of Probability,* **17,** No. 3, 965-1011.

REVUZ, D. and YOR, M. (1991). *Continuous Martingales and Brownian motion,* Springer-Verlag.

SMYTHE, W. R. (1968) *Static and dynamic electricity. 3d ed.* New York, McGraw-Hill.

SPITZER, F. (1958). Some theorems concerning 2-dimensional Brownian motion. *Trans. Amer. Math. Soc.* **87**, 187-197.

WHITTAKER, E.T., and WATSON, G.N. (1927). *A course of modern analysis* Vol II. Cambridge University Press.

YOR, M. (1977). Sur quelques approximations d'intégrales stochastiques. *Séminaire de Probabilités XI,* Lecture Notes in Math. 581, Springer, 518-528.

S.M. KOZLOV
Moscow Civil Engineering
Institute.
129337 Moscow
Yaroslavskoe Chausse 26
U.S.S.R.

J.W. PITMAN
Department of Statistics
U.C. Berkeley
Berkeley
Ca 94720
U.S.A.

M. YOR
Lab. de Probabilités
Université P. et M. Curie
4, Place Jussieu - Tour 56
75252 Paris Cedex 05,
France

L-Shapes for the Logarithmic η-Model for DLA in Three Dimensions

GREGORY F. LAWLER

1. INTRODUCTION

There has been a lot of study recently of what can be called *nearest neighbor cluster models*. These are Markov chains A_n, with state space of the set of finite connected subsets of the integer lattice Z^d, $A_1 = \{0\}$, and such that A_{n+1} is obtained from A_n by adding one point from the boundary of A_n. In this paper we discuss a new result for one such model, a variant of diffusion limited aggregation (DLA) first studied by Kesten [2].

We start by defining diffusion limited aggregation. On any finite subset $A \subset Z^d$ there is a well-defined probability measure called harmonic measure which intuitively is the hitting measure by random walk from infinity. To be more precise, let S_j be a simple random walk in Z^d and

$$\tau_A = \inf\{j \geq 1 : S_j \in A\}.$$

Then harmonic measure $H_A(x)$ is defined by

$$H_A(x) = \lim_{|y| \to \infty} P^y\{S(\tau_A) = x \mid \tau_A < \infty\}.$$

(Here we write P^y to indicate probabilities assuming $S(0) = y$. If the y is omitted the assumption will be that $S(0) = 0$.) This limit is known to exist [3, 5] and if $d \geq 3$,

$$H_A(x) = \frac{\mathrm{Es}_A(x)}{\sum_{z \in A} \mathrm{Es}_A(z)},$$

where $\mathrm{Es}_A(x)$ is the escape probability defined by

$$\mathrm{Es}_A(x) = P^x\{\tau_A = \infty\}.$$

We write

$$\partial A = \{y \in Z^d \setminus A : |x - y| = 1 \text{ for some } x \in A\},$$

and $\overline{A} = A \cup \partial A$. Then diffusion limited aggregation (DLA) is the nearest neighbor cluster model such that for each $y \in \partial A_n$,

$$P\{A_{n+1} = A_n \cup \{y\} \mid A_n\} = \frac{H_{\partial A_n}(y)}{\sum_{z \in \partial A_n} H_{\partial A_n}(z)}$$

$$= \frac{H_{\overline{A}_n}(y)}{\sum_{z \in \overline{A}_n} H_{\overline{A}_n}(z)}.$$

This model was first introduced by Witten and Sander [7] and has been studied extensively (from heuristic and numerical viewpoints) in the physics literature, see e.g. [6]. Strong evidence exists that the clusters A_n become fractal-like with a dimension around 1.7 for $d = 2$. Rigorous results have been very sparse for this model, see [1].

There is a version of DLA which is analogous to the η-model for dielectric breakdown where one chooses a sequence $\eta(n)$ and then takes the cluster model with

$$P\{A_{n+1} = A_n \cup \{y\} \mid A_n\} = Z^{-1}[H_{\partial A_n}(y)]^{\eta(n)},$$

where

$$Z = Z(A_n, \eta(n)) = \sum_{z \in \partial A_n} [H_{\partial A_n}(z)]^{\eta(n)}.$$

Kesten [2] considered the logarithmic η-model where $\eta(n) = C \ln n$ for some $C > 0$ (or more generally $\eta(n) \sim C \ln n$), and asked whether the cluster can form an "L-shape" or "generalized plus sign" with positive probability. More precisely, let $e_1, \ldots, e_d$ represent the standard unit vectors and let

$$V_i(k_1, k_2) = \{j e_i : k_1 \leq j \leq k_2\}.$$

We say that the cluster forms an L-shape if

$$A_n \in V_1(0, n) \cup V_2(0, n), \text{ for all } n \geq 0,$$

and the ratio of the length of the longer leg to the shorter leg stays bounded. Kesten showed that if $d = 2$ it is not possible for this model to form an L-shape. However, if $d \geq 4$, he showed that it is possible with positive probability if C is sufficiently large; moreover, any generalized plus sign (i.e., any union of half-line segments) can be formed with positive probability. If $d \geq 4$ and an L-shape forms, the sides of the L must grow with equal proportions, i.e.,

$$A_n = V_1(0, r_1(n)) \cup V_2(0, r_2(n)),$$

where $r_1(n) + r_2(n) = n - 1$ and $r_1(n)/n \to 1/2$. The question of what happens for $d = 3$ was left open. In some sense $d = 3$ is the critical dimension for this problem.

Here we discuss the $d = 3$ case and show that L-shapes can be formed. However, an interesting phase transition develops based on the value of C. If C is large, then the legs of the L grow not with equal proportions but rather with some proportion depending on C. If C is smaller, however, the legs grow with equal proportions. For very small C, as in the $d \geq 4$ case, one can show that one cannot get L-shapes. We expect that similar results can be proved for generalized plus signs in three dimensions, but for ease we restrict ourselves to the case of L-shapes. Also for ease we will assume that $\eta(n) = C \ln n$ but the same results will hold under the assumption that $\eta(n) \sim C \ln n$.

To analyze this problem, it is easier to first consider a cluster which is conditioned to stay in $V = V_1(0, \infty) \cup V_2(0, \infty)$. Let B_n be a nearest neighbor cluster model which always stays in V, i.e.,

$$B_n = V_1(0, r_1(n)) \cup V_2(0, r_2(n)),$$

with $r_1(n) + r_2(n) = n - 1$. Assume that B_n has the transitions of the logarithmic η-model conditioned to stay in V,

$$P\{B_{n+1} = B_n \cup \{(r_i + 1)e_i\} \mid B_n\} =$$

$$[\frac{Es_{\partial B_n}((r_i + 1)e_i)}{Es_{\partial B_n}((r_1 + 1)e_1) + Es_{\partial B_n}((r_2 + 1)e_2)}]^{C \ln n},$$

where $r_i = r_i(n), i = 1, 2$. Let $R(n) = \max\{r_1(n)/r_2(n), r_2(n)/r_1(n)\}$. We prove the following.

Theorem 1.1. *If B_n is the cluster model described above, then with probability one either $R(n) \to \infty$ or $R(n) \to \gamma = \gamma(C)$, where γ is the largest root of the equation*

$$\gamma = [\frac{\gamma + \sqrt{\gamma^2 + 1}}{1 + \sqrt{\gamma^2 + 1}}]^{C/2}.$$

Moreover, the latter possibility occurs with positive probability for every $C > 0$; $\gamma = 1$ if $C \leq 2(1 + \sqrt{2})$ and $\gamma > 1$ if $C > 2(1 + \sqrt{2})$.

In the remainder of this section we will show how to relate this theorem to the logarithmic η-model for DLA. Let U be the subset of Z^3,

$$U = \{(0, 0, 0)\} \cup \{je_1 \pm e_i : j = 1, 2, \ldots, \quad i = 2, 3\}$$

Note that the boundary of B_n looks locally like U (if we translate so that $(r_1(n) + 1)e_1$ is moved to the origin). Let $y_0 = 0$ and for $j > 0$ let $y_j = je_1 + e_2$. Let S_i be a simple random walk in Z^3, let

$$\tau = \tau_U = \inf\{i \geq 1 : S_i \in U\},$$

and let

$$\xi_n = \inf\{i > 0 : |S_i| \geq n\}.$$

It has been proved [2, 4] that there exist constants $a_j > 0$ such that

$$P^{y_j}\{\tau > \xi_n\} \sim a_j(\ln n)^{-1/2}.$$

Also, (see [2, Lemma 5])

$$a_0 > a_1 > a_2 > \cdots,$$

and $a_j \to 0$. From this it is fairly straightforward to get the following lemma (see [2]).

Lemma 1.2. *For every* $0 < \rho_2 < \rho_1 < a_1/a_0$ *there exists a* $k = k(\rho_2)$ *such that for any* $\epsilon > 0$, *if* n *is sufficiently large and* $\epsilon n \leq r_1(n) \leq (1 - \epsilon)n$, $y \in \partial B_n$, $y \notin \{(r_1(n) + 1)e_1, (r_2(n) + 1)e_2\}$,

$$\frac{\mathrm{Es}_{\partial B_n}(y)}{\mathrm{Es}_{\partial B_n}((r_1(n) + 1)e_1)} \leq \rho_1,$$

and if in fact $|y - r_1(n)e_1| \geq k$ *and* $|y - r_2(n)e_2| \geq k$,

$$\frac{\mathrm{Es}_{\partial B_n}(y)}{\mathrm{Es}_{\partial B_n}((r_1(n) + 1)e_1)} \leq \rho_2.$$

Let

$$\mu = [\ln(a_0/a_1)]^{-1},$$

and suppose $C > \mu$. Choose ρ_1 with $\mu < [-\ln\rho_1]^{-1} < C$ and choose ρ_2 with $(-\ln\rho_2)^{-1} \leq C/3$. Let $\epsilon > 0$ and assume $A_n = V_1(0, r_1(n)) \cup V_2(0, r_2(n))$ where $\epsilon n \leq r_1(n) \leq (1 - \epsilon)n$. Then by Lemma 1.2, for any $y \in \partial A_n, y \notin \{(r_1(n) + 1)e_1, (r_2(n) + 1)e_2\}$,

$$P\{A_{n+1} = A_n \cup \{y\} \mid A_n\} \leq [\frac{\mathrm{Es}_{\partial A_n}(y)}{\mathrm{Es}_{\partial A_n}((r_1(n) + 1)e_1)}]^{C\ln n}$$

$$\leq \rho_1^{C\ln n} = n^{C\ln\rho_1}.$$

Also, all except for $K = K(C)$ such boundary points y satisfy

$$P\{A_{n+1} = A_n \cup \{y\} \mid A_n\} \leq [\frac{\mathrm{Es}_{\partial A_n}(y)}{\mathrm{Es}_{\partial A_n}(r_1(n) + 1)e_1)}]^{C\ln n}$$

$$\leq \rho_2^{C\ln n} \leq n^{-3}.$$

Hence,

$$P\{A_{n+1} \subset V \mid A_n\} \geq 1 - n^{-2} - Kn^{C \ln \rho_1}.$$

By Theorem 1.1, there exists an $\epsilon > 0$ such that conditioned on A_n staying in V the probability that $\epsilon n \leq r_1(n) \leq (1 - \epsilon)n$ for all n sufficiently large is positive. Therefore, by the Borel-Cantelli Lemma,

$$P\{A_n \subset V \text{ for all } n\} > 0.$$

We then can use Theorem 1.1 again to deduce the limiting shape and get the following.

Theorem 1.3. *If $C > \mu$, then with positive probability A_n takes the following form:*

$$A_n = V_1(0, r_1(n)) \cup V_2(0, r_2(n)),$$

with

$$\lim_{n \to \infty} \frac{r_1(n)}{r_2(n)} = \gamma,$$

where $\gamma = \gamma(C)$ is as defined in Theorem 1.1

We can also show, but omit the details, that if $C \leq \mu$, then

$$P\{A_n \subset V \text{ for all } n\} = 0.$$

It would be nice to determine the value of μ. We know no way of determining the value exactly; however, it is relatively straightforward to do computer simulations to estimate the value. Such simulations give an estimate of $\mu \simeq 2.6$. Note that this value is significantly smaller than $2(1 + \sqrt{2})$ so we can be fairly confident that there are values of C such that the legs of A_n can grow with asymptotically equal lengths.

2. PROOF OF THEOREM 1.1

Let B_n be the nearest neighbor cluster model with transitions of the logarithmic η-model conditioned to stay in V as defined in the previous section, and let $r_1(n), r_2(n)$ be as previously defined,

$$B_n = V_1(0, r_1(n)) \cup V_2(0, r_2(n)).$$

Let $\Delta(n) = |r_1(n) - r_2(n)|$. Then $\Delta(n)$ is a nearest neighbor, time inhomogeneous Markov chain on the nonnegative integers. Let $p(n, x) = p(n, x, C)$ denote the probability of a jump of $+1$, i.e.,

$$P\{\Delta(n + 1) = \Delta(n) + 1 \mid \Delta(n) = x\} = p(n, x),$$

$$P\{\Delta(n + 1) = \Delta(n) - 1 \mid \Delta(n) = x\} = 1 - p(n, x).$$

Then, the assumptions of the model are that

$$\frac{p(n,x)}{1-p(n,x)} = \left[\frac{\text{Es}_{\partial B_n}((r_i(n)+1)e_i)}{\text{Es}_{\partial B_n}((r_{3-i}(n)+1)e_{3-i})}\right]^{C\ln n},$$

where i is the longer leg of B_n. To analyze this Markov chain, we need sharp estimates of the escape probability of the tip of an L-shape. We will state two such lemmas, Lemmas 2.1 and 2.3 in this section. The proofs of these lemmas will be in the next section. We start with the first lemma. As before, we let

$$R(n) = \max\{\frac{r_1(n)}{r_2(n)}, \frac{r_2(n)}{r_1(n)}\} = \frac{n-1+\Delta(n)}{n-1-\Delta(n)}.$$

Lemma 2.1. *For every $K < \infty$, there exists a $c = c(K) < \infty$ such that if $1 \le R(n) \le K$,*

$$\left|\left[\frac{\text{Es}_{\partial B_n}((r_i+1)e_i)}{\text{Es}_{\partial B_n}((r_{3-i}+1)e_{3-i})}\right]^{\ln n} - \left[\frac{R+\sqrt{R^2+1}}{1+\sqrt{R^2+1}}\right]^{1/2}\right| \le c\frac{(\ln\ln n)^2}{\ln n},$$

where $R = R(n), r_j = r_j(n)$, and i is the longer leg of B_n.

Let Z be the (random) set of cluster points of the sequence $\{R(n)\}$. Clearly $Z \subset [1,\infty)$. Also, since $R(n)$ takes jumps of size $O(n^{-1})$, Z is a connected set. Since for every $t < \infty$,

$$\frac{t+\sqrt{t^2+1}}{1+\sqrt{t^2+1}} < 2,$$

it follows from Lemma 2.1 that for all $\epsilon > 0$, and all n sufficiently large (depending on ϵ),

$$(1) \qquad\qquad p(n,x) \le \frac{2^{C/2}}{2^{C/2}+1}, \quad 0 < x \le (1-\epsilon)n.$$

Let $\epsilon < [2^{C/2}+1]^{-1}$ and choose odd m sufficiently large so that (1) holds for all $n \ge m$. It is standard then to show that

$$P\{\Delta(n) \le [\frac{2^{C/2}}{2^{C/2}+1}]n \text{ for all } n \ge m \mid \Delta(m) = 0\} > 0.$$

Hence, with positive probability, $R(n)$ stays bounded and Z is non-empty. Let

$$T = T_C = \{t : t \ne [\frac{t+\sqrt{t^2+1}}{1+\sqrt{t^2+1}}]^{C/2}\}.$$

We will show that if $t \in T$, then there exists a $\delta > 0$ such that

$$P\{R(n) \in (t-\delta, t+\delta) \text{ i.o.}\} = 0.$$

By covering T with a countable collection of such intervals we see that this implies

$$P\{Z \cap T \neq \emptyset\} = 0.$$

Hence, the only possible cluster points for $R(n)$ are t with

$$t = [\frac{t + \sqrt{t^2 + 1}}{1 + \sqrt{t^2 + 1}}]^{C/2}.$$

The next lemma discusses the roots to this equation. Clearly, $t = 1$ is a root for all C.

Lemma 2.2. *Let $f = f_C : [0, \infty) \to [0, \infty)$ be the function*

$$f(x) = [\frac{x + \sqrt{x^2 + 1}}{1 + \sqrt{x^2 + 1}}]^{C/2}.$$

Then,
(i) if $C \leq 2(\sqrt{2} + 1)$, then $f(x) < x$ for all $x > 1$;
(ii) if $C > 2(\sqrt{2} + 1)$, then there exists a $\gamma = \gamma(C) > 1$ such that: $f(\gamma) = 1; f(x) > x, x \in (1, \gamma); f(x) < x, x \in (\gamma, \infty)$.

Proof. If we differentiate f we get,

$$f'(x) = \frac{Cf(x)h(x)}{2x},$$

where

(2) $$h(x) = \frac{x(x + 1 + \sqrt{x^2 + 1})}{(x + \sqrt{x^2 + 1})(1 + \sqrt{x^2 + 1})\sqrt{x^2 + 1}}.$$

A messy calculation (done easily by Maple) shows that $h'(x) < 0$ for $x > 1$. Note that $f'(1) = C(\sqrt{2} - 1)/2$. Hence, $f'(1) > 1$ if and only if $C > 2(\sqrt{2} + 1)$.

If $C \leq 2(\sqrt{2} + 1)$, then $f(x) < x$ for $x \in (1, 1 + \epsilon)$ for some $\epsilon > 0$ (this can be verified for $C = 2(\sqrt{2} + 1)$ by checking that $f''(1) < 0$). Assume $f(x) = x$ for some $x > 1$ and let x_0 be the minimum such x. Then $f'(x_0) \geq 1$. But

$$f'(x_0) = \frac{Cf(x_0)h(x_0)}{2x_0} = \frac{Ch(x_0)}{2} < \frac{Ch(1)}{2} = f'(1) \leq 1,$$

which is a contradiction. Hence $f(x) < x$ for all $x > 1$.

Similarly, if $C > 2(\sqrt{2} + 1)$, let $\gamma = \gamma(C)$ be the smallest $x > 1$ with $f(x) = x$. Such a γ must exist since $f(x) > x$ for $x \in (1, 1 + \epsilon)$ and f is a bounded function. Then $f'(\gamma) \leq 1$. Also

$$
\begin{aligned}
f''(\gamma) &= (C/2\gamma^2)(f'(\gamma)h(\gamma)\gamma + f(\gamma)h'(\gamma)\gamma - f(\gamma)h(\gamma)) \\
&\leq (C/2\gamma^2)(h(\gamma)\gamma + \gamma^2 h'(\gamma) - \gamma h(\gamma)) < 0.
\end{aligned}
$$

Hence, $f(x) < x$ for $x \in (\gamma, \gamma + \epsilon)$ for some $\epsilon > 0$. Again, let x_0 be the smallest number greater than γ with $f(x_0) = x_0$ (assuming such a number exists). Then $f'(x_0) \geq 1$, but

$$f'(x_0) = \frac{C}{2}h(x_0) < \frac{C}{2}h(\gamma) = f'(\gamma) \leq 1,$$

which is a contradiction. Hence, no such x_0 exists and $f'(x) < x$ for all $x > \gamma$.   Q.E.D.

Let $\Delta(n)$ be any nearest neighbor Markov chain on the nonnegative integers with transitions

$$\begin{aligned} p(n, x) &= P\{\Delta(n+1) = x + 1 \mid \Delta(n) = x\} \\ &= 1 - P\{\Delta(n+1) = x - 1 \mid \Delta(n) = x\}, \end{aligned}$$

with $\Delta(1) = 0$. Let $R(n) = [(n-1) + \Delta(n)]/[(n-1) - \Delta(n)]$ be the ratio of the number of moves to the right to the number of moves to the left. Suppose $u \in (1, \infty)$, $p(n, x) \geq u(1 - p(n, x))$ for all $n \geq N$, and that $R(N) \geq u$. Let $\delta > 0$ and consider the event

$$V = V(N, \delta, u) = \{R(n) \leq u - \delta \text{ for some } n \geq N\}.$$

If $R(n) \leq u - \delta$ for some $n \geq N$, then

$$\begin{aligned} \Delta(n) - &\Delta(N) - (\frac{u-1-\delta}{u+1-\delta})(n - N) \\ &\leq \frac{u-1-\delta}{u+1-\delta}(n-1) - \frac{u-1}{u+1}(N-1) - \frac{u-1-\delta}{u+1-\delta}(n-N) \\ &\leq -v(N-1), \end{aligned}$$

where

$$v = \frac{u-1}{u+1} - \frac{u-1-\delta}{u+1-\delta} > 0.$$

Now since $p(n, x) \geq u(1 - p(n, x))$,

$$Y_j = \Delta(N+j) - \Delta(N) - j(u-1-\delta)/(u+1-\delta)$$

is stochastically bounded below by a sum of j independent, identically distributed random variables, each with positive mean, and bounded by 2. It is then easy to see that there exists an $a = a(u, \delta) > 0$ such that

$$P(V \mid R(N) \geq u) \leq e^{-aN}.$$

Similarly, suppose for some $\epsilon > 0$, $p(n, x) \geq (u+\epsilon)(1 - p(n, x))$ for $n \geq N$ and x satisfying

$$R(n, x) \doteq \frac{n-1+x}{n-1-x} \in (u - 2\epsilon, u + 2\epsilon).$$

Let $\delta \in (0, \epsilon)$ and let

$$\sigma = \sigma(N, \delta, u, \epsilon) = \inf\{j \geq N : R(j) \leq u - \delta \text{ or } R(j) \geq u + \epsilon\}.$$

Then there exists an $a = a(u, \delta, \epsilon) > 0$ such that for any $\tilde{u} \in (u - (\delta/2), u + (\delta/2))$,

$$P\{S(\sigma) \geq u + \epsilon \mid R(N) = \tilde{u}\} \geq 1 - e^{-aN}.$$

Now suppose $f(u) > u$. Find $\epsilon > 0$ such that for all $y \in (u - 2\epsilon, u + 2\epsilon)$, $f(y) > u + 2\epsilon$. Choose $\delta < \epsilon/2$ and assume that

$$R(n) \in (u - \delta, u + \delta) \text{ infinitely often,}$$

Then by the argument above,

$$P\{R(n) \leq (u - 2\delta) \text{ i.o. } \mid R(n) \in (u - \delta, u + \delta) \text{ i.o. }\} = 0,$$

$$P\{R(n) \geq (u + \epsilon) \text{ i.o. } \mid R(n) \in (u - \delta, u + \delta) \text{ i.o. }\} = 1.$$

However, since $f(u + \epsilon) > u + \epsilon$, the same argument can be used to show that

$$P\{R(n) \leq u + \delta \text{ i.o.} \mid R(n) \geq u + \epsilon \text{ i.o. }\} = 0.$$

Hence,

(3) $$P\{R(n) \in (u - \delta, u + \delta) \text{ i.o. }\} = 0.$$

A similar argument shows that if $f(u) < u$ there is a $\delta > 0$ such that (3) holds.

Therefore the only possible cluster points of $R(n)$ are the u with $f(u) = u$. If $C \leq 2(1 + \sqrt{2})$, by Lemma 2.2, there is only one such root, $u = 1$, so with probability one $R(n) \to 1$ or $R(n) \to \infty$. For $C > 2(1 + \sqrt{2})$, there are two roots, $u = 1$ and $u = \gamma$, so more work will be needed to determine which can be a limit. We will show below that if $C > 2(1 + \sqrt{2})$, then there exists an $\epsilon = \epsilon(C) > 0$ such that with probability one, $\Delta(n) \geq \epsilon n$ infinitely often. This implies $R(n) \geq (1 + \epsilon)/(1 - \epsilon)$ infintely often. Since we know $R(n)$ converges either to $1, \gamma,$ or ∞ , this will imply that $R(n) \to \gamma$ if it stays bounded.

Lemma 2.3. *Let $B_n = V_1(0, r_1(n)) \cup V_2(0, r_2(n))$. Then for every $u < \sqrt{2} - 1$ there exists an $\epsilon = \epsilon(u) > 0$ such that for all n sufficiently large: if $r_1(n) - r_2(n) \in (0, \epsilon n)$,*

$$[\frac{Es_{\partial B_n}((r_1(n) + 1)e_1)}{Es_{\partial B_n}((r_2(n) + 1)e_2)}]^{\ln n} \geq 1 + \frac{u(r_1(n) - r_2(n))}{n}.$$

Now fix $C > 2(1 + \sqrt{2})$ and choose $u < \sqrt{2} - 1$ so that $uC > 2$. Let $4\delta = uC - 2$. Then by Lemma 2.3, there exists an $\epsilon = \epsilon(C, u) > 0$ such that for all n sufficiently large and all $x < \epsilon n$,

$$\frac{p(n, x)}{1 - p(n, x)} \geq 1 + (2 + 3\delta)\frac{x}{n},$$

or equivalently, there exists an ϵ (perhaps slightly smaller) such that for all n sufficiently large and all $x < \epsilon n$,

$$p(n, x) \geq \frac{1}{2} + \frac{(1 + \delta)x}{2n}.$$

The theorem then follows from the following proposition which is similar to Proposition 10 of [2].

Proposition 2.4. *Suppose $\Delta(n)$ is a nearest neighbor Markov chain on the nonnegative integers with*

$$\begin{aligned}
p(n, x) &= P\{\Delta(n+1) = x + 1 \mid \Delta(n) = x\} \\
&= 1 - P\{\Delta(n+1) = x - 1 \mid \Delta(n) = x\}.
\end{aligned}$$

Suppose there exists a $\kappa > 1/2$ and $\epsilon \in (0, 1/2)$ such that for all n sufficiently large and all $x < \epsilon n$,

(4)
$$p(n, x) \geq \frac{1}{2} + \frac{\kappa x}{n}.$$

Then with probability one,

$$\Delta(n) \geq \epsilon n, \text{ infinitely often.}$$

Proof. Without loss of generality we may assume that (4) holds for all $x < \epsilon n$ and that $p(n, x) \geq 1/2$ for all n, x. We can then write $\Delta(n)$ as

$$\Delta(n) = \Delta(0) + \sum_{j=1}^{n} Y_j + \sum_{j=1}^{n} Z_j,$$

where the Y_j are independent,

$$P\{Y_j = 1\} = P\{Y_j = -1\} = \frac{1}{2},$$

and the Z_j depend on $\Delta(j-1)$ and Y_j,

$$P\{Z_j = 0 \mid Y_j = 1\} = 1,$$

$$P\{Z_j = 2 \mid Y_j = -1, \Delta(j-1) = x\} =$$
$$1 - P\{Z_j = 0 \mid Y_j = -1, \Delta(j-1) = x\} = 2p(n, x) - 1.$$

One advantage of this characterization is that we immediately see that $\Delta(n) - \Delta(0)$ is stochastically bounded below by a simple random walk $S(n) = \sum_{j=1}^{n} Y_j$. In particular, with probability one $\Delta(n) \geq 2\sqrt{n}$ infinitely often. What we will show below is that for some $c > 0$, if $b \geq 2\sqrt{n}$,

$$(5) \qquad P\{\Delta(j) \geq \epsilon j \text{ for some } j \geq n \mid \Delta(n) = b\} \geq c.$$

The proposition then follows easily. In proving (5), we will assume that

$$p(n, x) \geq \min\{1, \frac{1}{2} + \frac{\kappa x}{n}\}$$

for all n, x. Clearly, this also gives no loss of generality.

For any $\alpha, \beta > 1$, let V_k be the event

$$V_k = V_k(\alpha, \beta, n) = \{\Delta(\beta^k n) < (\beta \alpha)^k \sqrt{n}\}$$

(for ease we write $\Delta(\beta^k n)$ for $\Delta([\beta^k n])$). We will show the following: for every $\kappa > 1/2$ there exist $\alpha, \beta > 1$ and $q_k < 1$ with $\sum_{k=1}^{\infty} q_k < \infty$ such that if $b \geq (\beta \alpha)^{k-1} \sqrt{n}$, then

$$P\{V_k \mid \Delta(\beta^{k-1} n) = b\} \leq q_k,$$

provided that $\alpha^k < \sqrt{n}/2$. For any n, choose K so that

$$\epsilon \sqrt{n} \leq \alpha^K \leq \sqrt{n}/2$$

(such a K can always be found if α is chosen sufficiently close to 1 and one can check that the choice of α made below can be made as close to 1 as desired). Then, if $b \geq 2\sqrt{n}$,

$$
\begin{aligned}
P\{\Delta(j) \geq \epsilon j \text{ for some } j \geq n \mid \Delta(n) = b\} \\
\geq \quad & P\{V_K^c \mid \Delta(n) = b\} \\
\geq \quad & P\{V_1^c \cap \cdots \cap V_K^c \mid \Delta(n) = b\} \\
\geq \quad & \prod_{k=1}^{K}(1 - q_k) \\
\geq \quad & \prod_{k=1}^{\infty}(1 - q_k) \doteq c > 0.
\end{aligned}
$$

Choose $\alpha, \beta > 1, \delta > 0$ such that

$$1 + 2\kappa(\beta - 1)(1 - \delta)^2 = \alpha\beta.$$

Note that this can be done since $2\kappa > 1$. Suppose $\Delta(\beta^{k-1} n) = b_{k-1} \geq (\beta\alpha)^{k-1}$. Let $U = U_k$ be the event

$$U = \{S(j) - S(\beta^{k-1} n) \leq -\delta(\beta\alpha)^{k-1}\sqrt{n} \text{ for some } j \in (\beta^{(k-1)} n, \beta^k n)\},$$

$$W = \{ \sum_{\beta^{k-1}n<j\leq\beta^k n} Z_j \leq (\beta^k - \beta^{k-1})n\kappa(1-\delta)^2(\beta\alpha)^{k-1}n^{-1/2} \}.$$

Then,

$$P(V_k \mid \Delta(\beta^{k-1}n) = b_{k-1}) \leq P(U) + P(W \mid U^c, \Delta(\beta^{k-1}n) = b_{k-1}).$$

By the reflection principle and Chebyshev's inequality,

$$
\begin{aligned}
P(U) &\leq 2P\{S(\beta^k n - \beta^{k-1}n) \leq -\delta(\beta\alpha)^{k-1}\sqrt{n}\} \\
&\leq 2\frac{(\beta^k n - \beta^{k-1}n)}{(-\delta(\beta\alpha)^{k-1}\sqrt{n})^2} \\
&\leq 2(\alpha/\delta)^2\alpha^{-2k}.
\end{aligned}
$$

If $\Delta(\beta^{k-1}n) \geq (\beta\alpha)^{k-1}\sqrt{n}$, then on U^c,

$$\Delta(j) \geq (1-\delta)(\beta\alpha)^{k-1}\sqrt{n},$$

for $\beta^{k-1}n \leq j \leq \beta^k n$. Therefore on U^c,

$$X \doteq \frac{1}{2}\sum_{\beta^{k-1}n<j\leq\beta^k n} Z_j$$

is stochastically bounded below by a binomial random variable with parameters $\beta^k n - \beta^{k-1}n$ and $\kappa(1-\delta)(\beta\alpha)^{k-1}n^{-1/2}$. Standard exponential estimates for binomial random variables say that for every $s, t > 0$, there exists an $a > 0$ such that if Φ is a binomial random variable with parameters n and p with $p \geq tn^{-1/2}$, then

$$P\{\Phi \leq (1-s)pn\} \leq \exp\{-a\sqrt{n}\}.$$

Therefore for some $a > 0$,

$$P(W) \leq \exp\{-a\sqrt{n(\beta^k - \beta^{k-1})}\} \leq \exp\{-a\sqrt{\beta^k - \beta^{k-1}}\}.$$

Therefore, we have the result with

$$q_k = 2(\alpha/\delta)^2\alpha^{-2k} + \exp\{-a\sqrt{\beta^k - \beta^{k-1}}\}. \quad \text{Q.E.D.}$$

3. HARMONIC MEASURE ESTIMATES

In this section we prove the necessary lemmas about harmonic measure which were cited in the last section. Because many of the arguments are similar, we will only prove some of the necessary results allowing the reader to supply the details for other cases. Let e_1, e_2, e_3 be the standard unit vectors in Z^3. We let $\phi_n = (\ln \ln n)^2 / \ln n$. Let

$$U_+ = \{je_1 : j > 0\}, \quad U = \{je_1 : j \in Z\},$$

$$V = \overline{U}_+; \quad \tilde{V} = \overline{U}, \quad W = V \setminus \{0, e_1 \pm e_2, e_1 \pm e_3\}.$$

As before, let

$$\xi_n = \inf\{i : |S_i| \geq n\}.$$

Lemma 3.1. [4, Section 3] *There exist constants* $\alpha_1, \ldots, \alpha_4$ *such that if* $j = 2, 3$,

$$P\{\tau_V > \xi_n\} = \alpha_1 (\ln n)^{-1/2}(1 + O(\phi_n)),$$

$$P^{e_1 \pm e_j}\{\tau_V > \xi_n\} = \alpha_2 (\ln n)^{-1/2}(1 + O(\phi_n)),$$

$$P\{\tau_W > \xi_n\} = \alpha_3 (\ln n)^{-1/2}(1 + O(\phi_n)),$$

$$P^{e_1 \pm e_j}\{\tau_W > \xi_n\} = \alpha_4 (\ln n)^{-1/2}(1 + O(\phi_n)).$$

Lemma 3.2. [4, Lemma 3.3]

$$P\{\tau_U > \xi_n\} = \frac{\pi}{3}(\ln n)^{-1}(1 + O((\ln n)^{-1})).$$

In the next two propositions there will be a constant $K < \infty$. The $O(\cdot)$ terms appearing in the remainder of this section may depend on this number K but they do not depend on anything else.

Proposition 3.3. *Let*

$$A = A(n, a, b) = \{je_2 : 1 \leq j \leq an\} \cup \{je_1 + [an]e_2 : 1 \leq j \leq bn\}.$$

Then for every $K < \infty$ *and every* $1 \leq a, b \leq K$,

$$P\{S_j \in A \text{ for some } j > \xi_n\} =$$

$$\frac{1}{2}[1 + \ln(b + \sqrt{a^2 + b^2})](\ln n)^{-1}(1 + O(\phi_n)).$$

Proof. By a last-exit decomposition (see e.g. [4, Proposition 3.6]),

$$P\{S_j \in A \text{ for some } j \geq \xi_n\} = \sum_{x \in A} [G(x) - G_n(x)] \operatorname{Es}_A(x),$$

where G is the standard Green's function for the random walk and G_n is the Green's function for the walk killed upon leaving the ball of radius n. We have (see [4, Proposition 3.6])

$$\frac{2\pi}{3}(G(n) - G_n(x)) = \begin{cases} n^{-1} + O(n^{-2}), & |x| \leq n, \\ |x|^{-1} + O(n^{-2}), & |x| \geq n. \end{cases}$$

Suppose $z = ke_1 + [an]e_2 \in A$. By Lemma 3.2, if $k \geq n(\ln n)^{-3}$ and $|k - bn| \geq n(\ln n)^{-3}$, then

$$\operatorname{Es}_A(z) = \frac{\pi}{3}(\ln n)^{-1}(1 + O(\phi_n)).$$

(Actually, this estimates uses more than Lemma 3.2. If $\tilde{A} = A \cap \{y : |z - y| < n(\ln n)^{-3}\}$, then we also use the fact that

$$\operatorname{Es}_A(z) = \operatorname{Es}_{\tilde{A}}(z)(1 + O(\phi_n)).$$

This type of estimate is proved in Sections 2 and 3 of [4].) Similarly, if $z = ke_2 \in A$ with $k \geq n(\ln n)^{-3}$ and $|k - an| \geq n(\ln n)^{-3}$,

$$\operatorname{Es}_A(z) = \frac{\pi}{3}(\ln n)^{-1}(1 + O(\phi_n)).$$

If we estimate $\operatorname{Es}_A(z)$ first by 0 and then by 1 for other z, we see

$$P\{S_j \in A \text{ for some } j \geq \xi_n\} =$$

$$(1 + O(\phi_n))\frac{1}{2}(\ln n)^{-1}\sum_{z \in A}(n^{-1} \wedge |z|^{-1}).$$

The lemma then follows easily by estimating the sum, and noting that

$$P\{S(\xi_n) \in A\} \leq cn^{-1}. \quad \text{Q.E.D.}$$

Proposition 3.4. *Let* $A = A(n, a, b)$ *be as in Proposition 3.3. Then for every* $K < \infty$, *if* $1 \leq a, b \leq K$,

$$\operatorname{Es}_{\tilde{A}}(0) = P\{\tau_V > \xi_n\}[1 - \frac{1}{2}[1 + \ln(b + \sqrt{a^2 + b^2})](\ln n)^{-1}(1 + O(\phi_n))].$$

Proof. Note that

$$\text{Es}_{\overline{A}}(0) = P\{\tau_{\overline{A}} = \infty\} = P\{\tau_V > \xi_n\}P\{\tau_{\overline{A}} = \infty \mid \tau_V > \xi_n\}.$$

Hence it suffices to prove that

$$P\{\tau_{\overline{A}} < \infty \mid \tau_V > \xi_n\} = \frac{1}{2}[1 + \ln(b + \sqrt{a^2 + b^2})](\ln n)^{-1}(1 + O(\phi_n)).$$

Let C_n be the discrete ball of radius n,

$$C_n = \{z \in Z^3 : |z| < n\},$$

$\Lambda = \Lambda_n = \partial C_n, H(z) = H_n(z) = P\{S(\xi_n) = z\}$, and

$$\tilde{H}(z) = P\{S(\xi_n) = z \mid \tau_V > \xi_n\}.$$

Then,

$$\begin{aligned}
P\{\tau_{\overline{A}} < \infty \mid \tau_V > \xi_n\} &= \sum_{z \in \Lambda} \tilde{H}(z)P^z\{\tau_{\overline{A}} < \infty\} \\
&= \sum_{z \in \Lambda} H(z)P^z\{\tau_{\overline{A}} < \infty\} \\
&\quad + \sum_{z \in \Lambda}[\tilde{H}(z) - H(z)]P^z\{\tau_{\overline{A}} < \infty\}.
\end{aligned}$$

By [4, Proposition 2.5],

(6)
$$\tilde{H}(z) \le H(z)(1 + O(\frac{\ln \ln n}{\ln n})),$$

and by [3, Lemma 1.7.4]

(7)
$$H(z) \le cn^{-2}.$$

The first inequality implies

$$\sum_{z \in \Lambda}(\tilde{H}(z) - H(z))^+ = \sum_{z \in \Lambda}(H(z) - \tilde{H}(z))^+ = O(\frac{\ln \ln n}{\ln n}).$$

Also [4, Lemma 3.5] , if $\text{dist}(z, \overline{A}) \ge n(\ln n)^{-5}$,

$$P^z\{\tau_{\overline{A}} < \infty\} \le c\frac{\ln \ln n}{\ln n}$$

Hence,

$$\left| \sum_{z \in \Lambda} (\tilde{H}(z) - H(z)) P^z \{\tau_{\overline{A}} < \infty\} \right|$$

$$\leq \quad c(\frac{\ln \ln n}{\ln n}) \sum_{\text{dist}(z,\overline{A}) \geq n(\ln n)^{-5}} |\tilde{H}(z) - H(z)|$$

$$+ \sum_{\text{dist}(z,\overline{A}) \leq n(\ln n)^{-5}} (\tilde{H}(z) + H(z))$$

$$(8) \qquad\qquad \leq \quad c(\frac{\ln \ln n}{\ln n})^2.$$

Let $y = ([bn] + 1)e_1 + [an]e_2$. If $\tilde{A} = \{w \in \overline{A} : |w| \geq n(\ln n)^{-5}$ and $|w - y| \geq n(\ln n)^{-5}\}$, then for $w \in \tilde{A}$, Lemma 3.2 implies

$$P^w \{\tau_A = \infty\} \leq \frac{c}{\ln n}.$$

But for $z \in \partial C_n$,

$$P^z \{\tau_{\overline{A} \backslash \tilde{A}} < \infty\} \leq \sum_{x \in \overline{A} \backslash \tilde{A}} G(z, x) \leq c(\ln n)^{-5}.$$

Therefore, for $z \in \partial C_n$,

$$P^z \{\tau_{\overline{A}} < \infty\} = P^z \{\tau_A < \infty\}(1 + O(\frac{1}{\ln n})).$$

Hence,

$$P\{\tau_{\overline{A}} < \infty \mid \tau_V > \xi_n\} =$$

$$O((\frac{\ln \ln n}{\ln n})^2) + \sum_{z \in \Lambda} H(z) P^z \{\tau_A < \infty\}(1 + O(\phi_n)).$$

The result then follows from Proposition 3.3.   Q.E.D.

By Proposition 3.4,

$$[\text{Es}_{\overline{A}}(0)]^{\ln n} =$$

$$[P\{\tau_V > \xi_n\}]^{\ln n} \exp\{-\frac{1}{2}(1 + \ln(b + \sqrt{a^2 + b^2}))\}(1 + O(\phi_n)),$$

and if $y = ([bn] + 1)e_1 + [an]e_2$,

$$[\text{Es}_{\overline{A}}(y)]^{\ln n} =$$

$$[P\{\tau_V > \xi_n\}]^{\ln n} \exp\{-\frac{1}{2}(1 + \ln(a + \sqrt{a^2 + b^2}))\}(1 + O(\phi_n)).$$

We then get Lemma 2.1 by taking the quotient.

To prove Lemma 2.3 we will need significantly sharper estimates. In the next lemma we illustrate the ideas needed. Let

$$U_n = \{je_1 : 0 \le j \le n\}.$$

Then it is known (see [4, (15)]) that

$$\mathrm{Es}_{U_n}(0) = \alpha_5 (\ln n)^{-1/2}(1 + O(\phi_n)),$$

where α_5 is some positive constant. The next lemma estimates the "derivative" of $\mathrm{Es}_{U_n}(0)$ with respect to n and shows that the answer one would expect by differentiation is correct.

Lemma 3.5.

$$\mathrm{Es}_{U_n}(0) - \mathrm{Es}_{U_{n+1}}(0) = \frac{1}{2}\alpha_5 (\ln n)^{-3/2} n^{-1}(1 + O(\phi_n)).$$

Proof. Let $y = (n+1)e_1; m = [n/3]; C_m(y) = \{w : |w - y| < m\}; \Lambda = \partial C_m; \Phi = \partial C_m(y)$. Let $\xi = \xi_m$,

$$\zeta = \inf\{i \ge 1 : S_i \in \Phi\},$$

and

$$H_1(z) = P\{S(\xi) = z\}, \quad H_2(w) = P^y\{S(\zeta) = w\}.$$

Then using a last-exit decomposition (at the last visit to Φ before visiting y) we can see that

$$
\begin{aligned}
G(y) &= \sum_{w \in \Phi} G(w)H_2(w) \\
&= \sum_{z \in \Lambda} \sum_{w \in \Phi} H_1(z)G(z,w)H_2(w).
\end{aligned}
$$

Hence by [3, Theorem 1.5.4],

$$\sum_{z \in \Lambda} \sum_{w \in \Phi} H_1(z)G(z,w)H_2(w) = \frac{3}{2\pi n}(1 + O(n^{-1})).$$

Let $\tau_{n+1} = \tau_{U_{n+1}}, \tau_n = \tau_{U_n}$, and

$$\tilde{H}_1(z) = P\{S(\xi) = z \mid \tau_{n+1} > \xi\}, \quad \tilde{H}_2(w) = P^y\{S(\zeta) = w \mid \tau_{n+1} > \zeta\}.$$

Since for each $z \in \Lambda$,

$$\sum_{w \in \Phi} G(z,w)H_2(w) = G(z,y) \asymp n^{-1},$$

it is easy to see using an argument similar to that in (8) that

$$\sum_{z \in \Lambda} \sum_{w \in \Phi} \tilde{H}_1(z)G(z,w)H_2(w) = \frac{3}{2\pi n}(1 + O(\frac{\ln \ln n}{\ln n})).$$

Similarly,

$$\sum_{z \in \Lambda} \sum_{w \in \Phi} \tilde{H}_1(z) \tilde{H}_2(w) G(z, w) = \frac{3}{2\pi n}(1 + O(\frac{\ln \ln n}{\ln n})).$$

Let $g(z, w)$ be the Green's function on U_{n+1}^c, i.e., for $z, w \notin U_{n+1}$,

$$g(z, w) = E^z[\sum_{j=0}^{\tau_{n+1}-1} I\{S_j = w\}].$$

We claim that if $z \in \Lambda, w \in \Phi$, $\mathrm{dist}(z, U_{n+1}) \geq n(\ln n)^{-3}$, $\mathrm{dist}(w, U_{n+1}) \geq n(\ln n)^{-3}$, then

(9) $$g_{n+1}(z, w) = G(z, w)[1 - O(\frac{\ln \ln n}{\ln n})].$$

To see (9), note that

(10) $$G(z, w) - g(z, w) = \sum_{x \in U_{n+1}} P^z\{S(\tau_{n+1}) = x\} G(x, w).$$

If $n/2 \leq |x| \leq 3n/4$, let $\eta_x = \inf\{j : |S_j - x| \geq n/16\}$. Then

$$P^z\{S(\tau_{n+1}) = x\} \leq P^x\{\eta_x < \tau_{n+1}\} \sup\{G(s, z) : |s - x| \geq n/8\}$$
(11) $$\leq cn^{-1}(\ln n)^{-1}.$$

Hence if $\mathrm{dist}(w, U_{n+1}) \geq n(\ln n)^{-3}$, the sum over such $x \in U_{n+1}$ in (10) is bounded above by

$$cn^{-1}(\ln n)^{-1} \sum_{x \in U_{n+1}} G(x, w) \leq cn^{-1}\frac{\ln \ln n}{\ln n}.$$

For other $x \in U_{n+1}$, $G(x, w) \leq cn^{-1}$. Hence the sum over these x in (10) is bounded above by

$$cn^{-1}P^z\{\tau_{n+1} < \infty\} \leq cn^{-1}\frac{\ln \ln n}{\ln n},$$

provided that $\mathrm{dist}(z, U_{n+1}) \geq n(\ln n)^{-3}$. Since $G(z, w) \asymp n^{-1}$ for $z \in \Lambda, w \in \Phi$, this gives (9).

From (9) and the estimate $\tilde{H}_1(z) \leq cn^{-2}$ (see (6) and (7)), we see that

$$\sum_{z \in \Lambda} \sum_{w \in \Phi} \tilde{H}_1(z) \tilde{H}_2(w) g(z, w) = \frac{3}{2\pi n}(1 + O(\frac{\ln \ln n}{\ln n})).$$

Now by using another last-exit decomposition we can see that

$$\mathrm{Es}_{U_n}(0) - \mathrm{Es}_{U_{n+1}}(0) = P\{S(\tau_{n+1}) = y\} \mathrm{Es}_{U_n}(y) =$$

$$P\{\xi < \tau_{n+1}\} P^y\{\zeta < \tau_{n+1}\} \sum_{z \in \Lambda} \sum_{w \in \Phi} \tilde{H}_1(z) g(z, w) \tilde{H}_2(z) \, \mathrm{E} s_{U_n}(y).$$

If we define $\alpha_6 > 0$ by

$$\mathrm{E} s_{U_n}(y) \sim \alpha_6 (\ln n)^{-1/2} (1 + O(\phi_n))$$

(such an α_6 exists by [4, Section 3]), we get

$$\mathrm{E} s_{U_n}(0) - \mathrm{E} s_{U_{n+1}}(0) = \alpha_7 (\ln n)^{-3/2} n^{-1} (1 + O(\phi_n)),$$

where $\alpha_7 = \alpha_6 \alpha_5^2 (3/2\pi)$. To evaluate this constant we note that

$$
\begin{aligned}
\alpha_5 (\ln n^2)^{1/2} &(1 + O(\phi_n)) \\
&= \mathrm{E} s_{U_{n^2}}(0) \\
&= \mathrm{E} s_{U_n}(0) - \sum_{j=n}^{n^2-1} [\mathrm{E} s_{U_{j+1}}(0) - \mathrm{E} s_{U_j}(0)] \\
&= (1 + O(\phi_n))\{\alpha_5 (\ln n)^{-1/2} + \sum_{j=n}^{n^2-1} \alpha_7 (\ln j)^{-3/2} j^{-1}\} \\
&= (1 + O(\phi_n))\{2\alpha_7 (\ln n^2)^{-1/2} + (\alpha_5 - 2\alpha_7)(\ln n)^{-1/2}\}.
\end{aligned}
$$

Hence $\alpha_5 = 2\alpha_7$ and the lemma is complete. Q.E.D.

Now let V_n be the boundary of $U_n \setminus \{0\}$,

$$V_n = \partial\{je_1 : 1 \le j \le n\}.$$

The following two lemmas can be proved is the same way as Lemmas 3.2 and 3.5, so we omit the proofs.

Lemma 3.6.

$$P^{e_2}\{\tau_{\tilde{V}} > \xi_n\} = \frac{\pi}{12}(\ln n)^{-1}(1 + O(\frac{1}{\ln n})).$$

Lemma 3.7. *If α_1 is the constant defined in Lemma 3.1,*

$$\mathrm{E} s_{V_n}(0) - \mathrm{E} s_{V_{n+1}}(0) = \frac{1}{2}\alpha_1(\ln n)^{-3/2} n^{-1}(1 + O(\phi_n)).$$

The L-shapes that we are considering consist of two line segments so it is not surprising that the same kind of argument can be used to give a "derivative" form of Proposition 3.4.

Lemma 3.8. *Let A be defined as in Proposition 3.3 with $a = 1$ and let $A_1 = A \cup \{([bn] + 1)e_1 + ne_2\}$. Then if $1 \le b \le K$,*

(i) $\operatorname{Es}_{A \cup \{0\}}(0) - \operatorname{Es}_{A_1 \cup \{0\}}(0) =$

$$P\{\tau_{U_*} > \xi_n\}\frac{1}{2}(\ln n)^{-1}(1 + b^2)^{-1/2}n^{-1}(1 + O(\phi_n)).$$

(ii) $\operatorname{Es}_{\overline{A}}(0) - \operatorname{Es}_{\overline{A_1}}(0) =$

$$P\{\tau_V > \xi_n\}\frac{1}{2}(\ln n)^{-1}(1 + b^2)^{-1/2}n^{-1}(1 + O(\phi_n)).$$

Proof. Since the proof is very similar to that of Lemma 3.5, we only sketch the ideas. We restrict ourselves to (i). For notational ease we write A for $A \cup \{0\}$ and A_1 for $A_1 \cup \{0\}$. Let $\tau = \tau_{A_1}$ and $y = ([bn] + 1)e_1 + ne_2$. Then

$$\operatorname{Es}_A(0) - \operatorname{Es}_{A_1}(0) = P\{S(\tau) = y\} \operatorname{Es}_A(y).$$

Let $\xi = \xi_n$ and $\zeta = \inf\{j : |S(j) - y| \ge n/3\}$. Define $H_1, H_2, \tilde{H}_1, \tilde{H}_2$ as in Lemma 3.5. Let $\Lambda = \partial C_n, \Phi = \partial C_{n/3}(y)$. Then by a last-exit decomposition,

$$G(0, y) = \sum_{z \in \Lambda}\sum_{w \in \Phi} H_1(z)G(z, w)H_2(w).$$

$$P\{S(\tau) = y\} = P\{\tau_{U_*} > \xi_n\}P^y\{\tau > \zeta\}\sum_{z \in \Lambda}\sum_{w \in \Phi} \tilde{H}_1(z)g(z, w)\tilde{H}_2(z),$$

where $g(z, w)$ is the Green's function on A_1^c,

$$g(z, w) = E^z \sum_{j=0}^{\tau-1} I\{S_j = w\}.$$

Then arguing as in Lemma 3.5, we prove

$$P\{S(\tau) = y\} = P\{\tau_U > \xi_n\}P^y\{\tau > \zeta\}G(0, y)(1 + O(\phi_n)).$$

Combining this with the estimates

$$G(0, y) = \frac{3}{2\pi n}(1 + b^2)^{-1/2}(1 + O(n^{-1})),$$

$$P^y\{\tau > \zeta\} = \alpha_5(\ln n)^{-1/2}(1 + O(\phi_n)),$$

$$\operatorname{Es}_A(y) = \alpha_6(\ln n)^{-1/2}(1 + O(\phi_n)),$$

as well as the relation $\alpha_5\alpha_6 = \pi/3$, we get the lemma.   Q.E.D.

Lemma 3.9. *Let A be as in Proposition 3.3 with $b = 1$. Let $A_2 = A \cup \{0, -e_2\}$. Then if $1 \le a \le K$,*

$$\mathrm{Es}_{A \cup \{0\}}(0) - \mathrm{Es}_{A_2}(-e_2) =$$

$$P\{\tau_{U_n} > \xi_n\}(2an \ln n)^{-1}(1 - (1 + a^2)^{-1/2})(1 + O(\phi_n)).$$

Proof. Again we write A for $A \cup \{0\}$. Without loss of generality we may assume that an is an integer. Let $x = ane_2, W = \{je_2 : j \in Z\}, Y = A \cap Z, Y_2 = A_2 \cap Z, Q = A \setminus Y = A_2 \setminus Y_2$. Then

$$
\begin{aligned}
\mathrm{Es}_A(0) - \mathrm{Es}_{A_2}(-e_2) \\
= \quad & P^{-e_2}\{\tau_{A_2} < \infty\} - P\{\tau_A < \infty\} \\
= \quad & P^{-e_2}\{\tau_{Y_2 \setminus \{x\}} < \infty\} + P^{-e_2}\{\tau_{Y_2 \setminus \{x\}} = \infty, \tau_x < \infty\} + \\
& P^{-e_2}\{\tau_{Y_2} = \infty, \tau_Q < \infty\} - P\{\tau_Y < \infty\} \\
& -P\{\tau_Y = \infty, \tau_Q < \infty\}.
\end{aligned}
$$

By translation invariance of simple random walk,

$$P^{-e_2}\{\tau_{Y_2 \setminus \{x\}} < \infty\} = P\{\tau_Y < \infty\}.$$

Also, by Lemma 3.5,

$$P^{-e_2}\{\tau_{Y_2 \setminus \{x\}} = \infty, \tau_x < \infty\} =$$

(12) $$\frac{1}{2}P\{\xi_{n/3} < \tau_A\}(\ln n)^{-1}(an)^{-1}(1 + O(\phi_n)).$$

Let $y_j = je_1 + ane_2$. Then

$$
\begin{aligned}
P^{-e_2}\{\tau_{Y_2} &= \infty, \tau_Q < \infty\} - P\{\tau_Y = \infty, \tau_Q < \infty\} \\
= \quad & \sum_{j=1}^{n}[P^{-e_2}\{S(\tau_{A_2}) = y_j\} \mathrm{Es}_{Y_2}(y_j) - \\
& \qquad P\{S(\tau_A) = y_j\} \mathrm{Es}_Y(y_j)] \\
= \quad & \sum_{j=1}^{n} \mathrm{Es}_Y(y_j)[P^{-e_2}\{S(\tau_{A_2}) = y_j\} - P\{S(\tau_A) = y_j\}] \\
& + \sum_{j=1}^{n}[\mathrm{Es}_{Y_2}(y_j) - \mathrm{Es}_Y(y_j)]P^{-e_2}\{S(\tau_{A_2}) = y_j\}.
\end{aligned}
$$

One can show as in Lemma 3.5,

$$|\mathrm{Es}_{Y_2}(y_j) - \mathrm{Es}_Y(y_j)| \le cn^{-1}(\ln n)^{-1/2}.$$

Hence,

$$\left| \sum_{j=1}^{n} [\text{Es}_{A_2}(y_j) - \text{Es}_A(y_j)] P^{-e_2} \{S(\tau_{A_2}) = y_j\} \right|$$

$$\leq cn^{-1}(\ln n)^{-1/2} \sum_{j=1}^{n} P^{-e_2}\{S(\tau_{A_2}) = y_j\}$$

$$\leq cn^{-1}(\ln n)^{-1/2} P^{-e_2}\{\xi_{n/3} < \tau_{A_2}\} P^{-e_2}\{\tau_Q < \infty \mid \xi_{n/3} < \tau_{A_2}\}$$

$$(13) \quad \leq cn^{-1}(\ln n)^{-2}.$$

We now consider the other term. Fix j and let $y = y_j, m = [n/3], \Lambda = \partial C_m, \Phi = \Phi_j = \partial C_m(y), \xi = \xi_m$, and $\zeta = \zeta_j = \inf\{i \geq 1 : S_i \in \Phi\}$. Also define $H_1, H_2, \tilde{H}_1, \tilde{H}_2$ as before. Then by the last-exit decomposition as before,

$$G(0, y) = \sum_{z \in \Lambda} \sum_{w \in \Phi} H_1(z) H_2(w) G(z, w),$$

$$G(-e_2, y) = \sum_{z \in \Lambda} \sum_{w \in \Phi} H_1(z) H_2(w) G(z - e_2, w).$$

By [3, Theorem 1.5.5],

$$(14) \quad G(0, y) - G(-e_2, y) = \frac{3}{2\pi}[|y|^{-1} - |y + e_2|^{-1}] + O(|y|^{-3})$$

Hence,

$$\sum_{z \in \Lambda} \sum_{w \in \Phi} H_1(z) H_2(w)[G(z, w) - G(z - e_2, w)] =$$

$$\frac{3}{2\pi}\left(\frac{|y + e_2| - |y|}{|y|^2}\right) + O(n^{-3}).$$

Let $g(z, w)$ and $g_2(z, w)$ denote the Green's functions on A^c and A_2^c respectively, i.e., for $z \notin A_2$

$$g(z, w) = E^z \sum_{j=0}^{\tau_A - 1} I\{S_j = w\}, \quad g_2(z, w) = E^z \sum_{j=0}^{\tau_{A_2} - 1} I\{S_j = w\}.$$

Then again by last-exit decompositions we have

$$P\{S(\tau_A) = y\} = P\{\xi < \tau_A\} P^y\{\zeta < \tau_A\} \sum_{z \in \Lambda} \sum_{w \in \Phi} \tilde{H}_1(z) \tilde{H}_2(z) g(z, w),$$

$$P\{S(\tau_{A_2}) = y\} = P\{\xi < \tau_A\} P^y\{\zeta < \tau_A\} \sum_{z \in \Lambda} \sum_{w \in \Phi} \tilde{H}_1(z) \tilde{H}_2(z) g_2(z - e_2, w).$$

For $z \in \Lambda, w \in \Phi$, we can estimate

$$
\begin{aligned}
g(z,w) - g_2(z,w) &\leq cP^z\{S(\tau_{A_2}) = -e_2\}P^{-e_2}\{S(\tau_{A \cup \{w\}}) = w\} \\
&\leq c[n^{-1}(\ln n)^{-1/2}][n^{-1}(\ln n)^{-1/2}] \\
&\leq cn^{-2}(\ln n)^{-1}.
\end{aligned}
$$

Hence,

$$
P^{-e_2}\{S(\tau_{A_2}) = y\} - P\{S(\tau_A) = y\} =
$$

$$
P\{\xi < \tau_A\}P^y\{\zeta < \tau_A\}[O(n^{-2}(\ln n)^{-1}) +
$$

$$
\sum_{z \in \Lambda}\sum_{w \in \Phi} \tilde{H}_1(z)\tilde{H}_2(w)(g(z-e_2,w) - g(z,w)).
$$

For any $z \in \Lambda, w \in \Phi$,

$$
G(z,w) - g(z,w) = G(w,z) - g(w,z) = \sum_{s \in A} P^w\{S(\tau_A) = s\}G(s,z),
$$

and hence

$$
\begin{aligned}
|(G(z-e_2,w) - G(z,w)) &- (g(z-e_2,w) - g(z,w))| \\
&= |\sum_{s \in A} P^w\{S(\tau_A) = s\}(G(s,z-e_2) - G(s,z))| \\
&\leq c\sum_{s \in A} P^w\{S(\tau_A) = s\}(|s-z| \vee 1)^{-2}.
\end{aligned}
$$

From this it is easy to see that if $w \in \Phi$,

$$
\sum_{z \in \Lambda} \tilde{H}_1(z)(g(z-e_2,w) - g(z,w)) =
$$

$$
O(n^{-2})P^w\{\tau_A < \infty\} + \sum_{z \in \Lambda} \tilde{H}_1(z)(G(z-e_2,w) - G(z,w)),
$$

and hence we can also see that

$$
\sum_{z \in \Lambda}\sum_{w \in \Phi} \tilde{H}_1(z)\tilde{H}_2(w)(g(z-e_2,w) - g(z,w))
$$

$$
= O(\frac{\ln \ln n}{n^2 \ln n}) + \sum_{z \in \Lambda}\sum_{w \in \Phi} \tilde{H}_1(z)\tilde{H}_2(w)(G(z-e_2,w) - G(z,w)).
$$

It is not difficult to see that the second term is bounded below by a constant times n^{-2}. By replacing $\tilde{H}_1$ and $\tilde{H}_2$ by H_1 and H_2 as before we get

$$
\begin{aligned}
& P^{-e_2}P\{S(\tau_{A_2}) = y_j\} - P\{S(\tau_A) = y_j\} \\
&= (1 + O(\phi_n))P\{\xi < \tau\}P^{y_j}\{\zeta < \tau_A\} \\
&= \sum_{z \in w}\sum_{w \in \Phi} H_1(z)H_2(w)(G(z - e_2, w) - G(z, w)) \\
&= (1 + O(\phi_n))P\{\xi < \tau\}P^{y_j}\{\zeta < \tau_A\}\frac{3}{2\pi}\frac{|y_j| - |y_j + e_2|}{|y_j|^2}.
\end{aligned}
$$

For $n(\ln n)^{-3} \le j \le n - n(\ln n)^{-3}$,

$$
P^{y_j}\{\zeta < \tau_A\} = \frac{\pi}{3}(\ln n)^{-1}(1 + O(\phi_n)),
$$

and hence

$$
\text{Es}_Y(y_j)P^{y_j}\{\zeta < \tau_A\} = \frac{\pi}{3}(\ln n)^{-1}(1 + O(\phi_n)).
$$

If we estimate this quantity for other j first by 0 and then by 1 we get

$$
\begin{aligned}
& \sum_{j=1}^{n} \text{Es}_Y(y_j)[P^{-e_2}\{S(\tau_{A_2}) = y_j\} - P\{S(\tau_A) = y_j\}] \\
&= -(1 + O(\phi_n))P\{\xi < \tau\}\frac{1}{2}(\ln n)^{-1}\sum_{j=1}^{n}\frac{|y_j + e_2| - y_j}{|y_j|^2} \\
&= -(1 + O(\phi_n))P\{\xi_n < \tau_A\}\frac{1}{2}(\ln n)^{-1}n^{-1}(a\sqrt{1 + a^2})^{-1}.
\end{aligned}
$$

If we combine (12), (13), and the above we get the lemma.   Q.E.D.

The following can then be proven in the same way.

Lemma 3.10. *If A is defined as in Proposition 3.3 with $b = 1$ and $A_2 = A \cup \{0\}$, then for all $1 \le a \le K$,*

$$
\text{Es}_{\overline{A}}(0) - \text{Es}_{\overline{A_2}}(-e_2) = P\{\tau_V > \xi_n\}(2an\ln n)^{-1}(1 - (1 + a^2)^{-1/2})(1 + O(\phi_n)).
$$

Now let A be defined as in Proposition 3.3 with $a = 1$ and let $A_1 = A \cup \{([bn] + 1)e_1 + ne_2\}$ as in Lemma 3.8. Fix K and assume $1 \le b \le K$. By Lemma 3.8,

$$
\frac{\text{Es}_{\overline{A_1}}(0)}{\text{Es}_{\overline{A}}(0)} = 1 - \frac{1}{2}(n\ln n)^{-1}(1 + b^2)^{-1/2}(1 + O(\phi_n)).
$$

By Lemma 3.10, if $y = ([bn] + 1)e_1 + ne_2, y_1 = y + e_1$,

$$
\frac{\text{Es}_{\overline{A_1}}(y_1)}{\text{Es}_{\overline{A}}(y)} = 1 - \frac{1}{2}(n\ln n)^{-1}b^{-1}(1 - (1 + b^2)^{-1/2})(1 + O(\phi_n)).
$$

Therefore,

$$\frac{\text{Es}_{\overline{A_1}}(y_1)}{\text{Es}_{\overline{A_1}}(0)} = \frac{\text{Es}_{\overline{A}}(y)}{\text{Es}_{\overline{A}}(0)}[1 + \frac{1}{2}(n\ln n)^{-1}[(b^{-1} + 1)(1 + b^2)^{-1/2} - 1](1 + O(\phi_n)),$$

and

$$[\frac{\text{Es}_{\overline{A_1}}(y_1)}{\text{Es}_{\overline{A_1}}(0)}]^{\ln n} \geq [\frac{\text{Es}_{\overline{A}}(y)}{\text{Es}_{\overline{A}}(0)}]^{\ln n} + \frac{1}{2n}[(b^{-1} + 1)(1 + b^2)^{-1/2} - 1](1 + O(\phi_n)).$$

Lemma 2.3 can now be derived easily by repeated application of the above. Note that the factor of $1/2$ disappears since $r_1(n)$ and $r_2(n)$ are each approximately $n/2$ rather than n.

4. Acknowledgments

I would like to thank Emily Puckette for the computer simulations leading to the approximate value of μ and Bill Allard for help with Maple in proving Lemma 2.2. This research was partially supported by National Science Foundation.

References

1. H. Kesten (1990). Upper bounds for the growth rate of DLA. Physica A **168** 529-535.
2. H. Kesten (1991). Some caricatures of multiple contact diffusion-limited aggregation and the η-model, to appear in Proceedings of Durham Symposium on Stochastic Analysis.
3. G. Lawler (1991). *Intersections of Random Walks*. Birkäuser Boston.
4. G. Lawler (1991). Escape probabilities for slowly recurrent sets, to appear.
5. F. Spitzer (1976). *Principles of Random Walk*. Springer-Verlag.
6. T. Vicsek (1989). *Fractal Growth Phenomena*. World Scientific.
7. T. Witten and L. Sander (1981). Diffusion-limited aggregation, a kinetic growth phenomenon. Phys. Rev. Lett. **47** 1400-1403.

Department of Mathematics
Duke University
Durham, NC 27706

Remark on the intrinsic local time *

PAUL McGILL

The intrinsic local time is a semimartingale in the excursion filtration. We indicate, inter alia, a new proof of the fact that its martingale part generates an orthogonal martingale measure in the sense of Walsh. The calculations avoid explicit use of excursion theory, relying instead on stochastic calculus in the space variable.

Suppose B_t is a real Brownian motion, started at zero, and write τ_t^x for the right continuous inverse of $t \to A_t^x = \int_0^t 1_{(B_s < x)} ds$. The intrinsic local time $\tilde{L}_x$ is an infinite dimensional process defined by

$$\tilde{L}_x^t = \tfrac{1}{2} L(x, \tau_t^x)$$

where $L(x,t)$ is the local time of B_t — normalisation is $A_t^x = \int_{-\infty}^x L(a,t) da$.

One knows $\tilde{L}_x$ is adapted to the excursion filtration $\mathcal{E}^x = \sigma\{B_{\tau_t^x} : t > 0\}$, which is indexed by $x \in R$ and satisfies the usual conditions [1]. Moreover, it is a simple consequence of the Ray-Knight-Williams theorem — stated at (a) below but see also the picture — that for fixed t the process $x \to \tilde{L}_x^t + x^-$ is an $\mathcal{E}^x$ supermartingale. We write its Doob-Meyer decomposition

$$\tilde{L}_x^t = M_x^t - D_x^t - x^- \tag{1}$$

* Research supported by: Département de Mathématiques, Université Louis Pasteur, 67000 Strasbourg, France.

with $x \to D_x^t$ increasing.

In [1] we noted that for $0 < u \leq s < t$

$$dD_x^t = \frac{d}{dt}\left(\tilde{L}_x^t\right)^2 dx \quad ; \quad d\langle M^t\rangle_x = 2\tilde{L}_x^t dx \quad ; \quad \langle M^u, M^t - M^s\rangle_x = 0$$

Our reasoning was rather involved. We used 'first order' excursion computations to calculate the Laplace transform of $t \to D_x^t$, then invoked second order calculations to check the above properties of M_x^t. From this we deduced that (1) has a bicontinuous version, allowing us to invert the Laplace transform and so identify D_x^t.

Here we prove that (1) has a bicontinuous version without using excursion theory. We will show that

$$\langle M^t - M^s\rangle_y - \langle M^t - M^s\rangle_x = 2\int_x^y \left(\tilde{L}_a^t - \tilde{L}_a^s\right) da \tag{2}$$

by computing the left side from the quadratic variation of $\tilde{L}_x^t - \tilde{L}_x^s$. Bicontinuity of (1) follows easily, as we shall see later.

To simplify notation we give details only for $x > 0$. Also, writing

$$\tilde{L}_y^t - \tilde{L}_x^t = \tfrac{1}{2}L(y, \tau_t^x) - \tfrac{1}{2}L(x, \tau_t^x) - \tfrac{1}{2}L(y, \tau_t^x) + \tilde{L}_y^t,$$

we introduce $U = \int_0^{\tau_t^x} 1_{(x < B_s < y)} ds$ and a 'remainder' term $R_a = \tfrac{1}{2}L(a, \tau_U^x) \circ \theta_{\tau_t^x}$ — the interval $[x, y]$ is fixed temporarily. For keeping track of time at different levels there is a picture

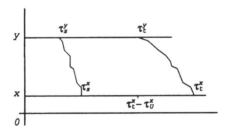

(horizontal axis measures time). So U is the time shift produced in going from level x to level y and

$$\tilde{L}_y^t - \tilde{L}_x^t = \tfrac{1}{2}L(y, \tau_t^x) - \tfrac{1}{2}L(x, \tau_t^x) - R_y \tag{3}$$

Our proof of (2) depends on the following three facts, stated for $y > x > 0$.

($\mathfrak{a}$) The process $y \to L(y, \tau_t^x) \circ \theta_{\tau_t^x}$ is a continuous $\mathcal{E}^y$ martingale with quadratic variation $4 \int_x^y daL(a, \tau_t^x) \circ \theta_{\tau_t^x}$. Hence, it is equivalent in law to a $\mathfrak{bes}_0^2$ process.

($\mathfrak{b}$) If $p \geq 1$ then $||\tilde{L}_y^t||_{2p} \leq c_p t^{1/2}$.

($\mathfrak{c}$) For $p \geq 1$ we have $\mathbf{E}\left[R_y^p\right] \leq c_t |x - y|^{(3p+1)/6}$.

The first result can be found in [3], the second comes from the fact that $t \to \tilde{L}_x^t$ is the supremum of (some) Brownian motion. Proof of the (rather crude) bound ($\mathfrak{c}$) is postponed to the end.

Since $\tau_t^x \leq \tau_t^0$, we find from ($\mathfrak{a}$) that $x \to \tilde{L}_x^t$ is bounded pathwise by a $\mathfrak{bes}_0^2$ process, so its supremum has moments of all orders. This fact will be used below without further comment.

Remark: The picture shows why $\tilde{L}_x^t$ is a supermartingale: from ($\mathfrak{a}$) local time is a martingale on vertical lines and, since the curve $x \to \tau_t^x$ tends monotonically to the y-axis, one imagines D_x^t being produced by time sliding backwards as x moves up. This is the motivation for starting with (3).

Notation: The letter c represents a generic constant, independent of $|x - y|$ and $|t - s|$. Subscripts may help clarify our reasoning.

There is one difficulty we have glossed over up to now. Invoking the Doob-Meyer decomposition presumed existence of a regularised version for $\tilde{L}_x^t$ (cf. [1]). So, for the reader's convenience, we prove

Proposition *The process* $(x, t) \to \tilde{L}_x^t$ *has a bicontinous version.*

Proof: To estimate $\frac{1}{2}L(y, \tau_t^x) - \frac{1}{2}L(x, \tau_t^x)$ in the L_{2p} norm we use ($\mathfrak{a}$) and a BDG inequality

$$\mathbf{E}[|L(y, \tau_t^x) - L(x, \tau_t^x)|^{2p}] \leq c_p \mathbf{E}\left[\left(\int_x^y L(a, \tau_t^x) da\right)^p\right]$$

$$\leq c_p (y - x)^{p-1} \int_x^y \mathbf{E}[L^p(a, \tau_t^x)] da \leq c_{p,t} |y - x|^p$$

(the last part is an estimate with $\mathfrak{bes}_0^2$). From (3), and using ($\mathfrak{c}$), this gives $||\tilde{L}_y^t - \tilde{L}_x^t||_{2p} \leq c_p|y - x|^{1/2}$. Next we use ($\mathfrak{b}$), and the strong Markov property at time τ_s^y, to obtain $||\tilde{L}_y^t - \tilde{L}_y^s||_{2p} \leq c_p|t - s|^{1/2}$. This means

$$||\tilde{L}_y^t - \tilde{L}_x^s||_{2p} \leq ||\tilde{L}_y^t - \tilde{L}_y^s||_{2p} + ||\tilde{L}_y^s - \tilde{L}_x^s||_{2p} \leq c_p(|x - y|^{1/2} + |t - s|^{1/2})$$

Now apply the Kolmogoroff criterion to get bicontinuity.

Accepting (2) for the moment, we can explain why (1) has a bicontinuous version. First remark, from (2) with $s = 0$ and a BDG inequality, that

$$\mathbf{E}\left[(M_y^t - M_x^t)^{4p}\right] \leq c_p \mathbf{E}\left[\left(\int_x^y \tilde{L}_a^t da\right)^{2p}\right],$$

so ($\mathfrak{b}$) gives the bound $c_p|y - x|^{2p}t^p$. However, if we normalise by $M_0^t \equiv 0$, the same estimate with $0 < s < t$ yields

$$||M_y^t - M_x^s||_{4p} \leq ||M_y^t - M_x^t||_{4p} + ||M_x^t - M_x^s||_{4p} \leq c_p(|x - y|^{1/2} + |t - s|^{1/4})$$

By Kolmogoroff M_x^t has a bicontinuous version. Bicontinuity of (1) follows from the proposition.

Turning to the proof of (2), we look first at the case $s = 0$. From (3) and ($\mathfrak{a}$) we have

$$\mathbf{E}\left[(\tilde{L}_y^t - \tilde{L}_x^t)^2|\mathcal{E}^x\right] = \int_x^y \mathbf{E}[L(a, \tau_t^x)|\mathcal{E}^x]da$$

$$+\mathbf{E}\left[R_y L(y, \tau_t^x) - R_y L(x, \tau_t^x) + R_y^2|\mathcal{E}^x\right]$$

where the first term on the right is just $2(y - x)\tilde{L}_x^t$.

Repeating this argument on intervals $x = x_0 < x_1 \ldots < x_n = y$ and taking sums, we find, using dyadic refinement, that the left side converges to $\mathbf{E}[\langle M^t \rangle_y - \langle M^t \rangle_x|\mathcal{E}^x]$. On the right, using continuity of $\tilde{L}_a^t$ in the space variable, the sums of integrals converge almost surely to $\mathbf{E}\left[2\int_x^y \tilde{L}_a^t da|\mathcal{E}^x\right]$, so the limit comes out like

$$\mathbf{E}[\langle M^t \rangle_y - \langle M^t \rangle_x|\mathcal{E}^x] = \mathbf{E}\left[2\int_x^y \tilde{L}_a^t da|\mathcal{E}^x\right] + S(x, y)$$

with $S(x, y)$ the contribution of various 'remainder' terms. However, $S(x, y)$ vanishes because, by (c) with $p = 2$, the terms involving R converge to zero in L_1 (use Hölder and (a) to deal with the cross term). Our equation now says that $\langle M^t \rangle_x$ and $2 \int^x \tilde{L}_a^t da$ give the same random weight to $\mathcal{E}^x$ predictable rectangles — they have the same $\mathcal{E}^x$ dual predictable projection. Each process is continuous and adapted, so they must be equal. This proves (1) when $s = 0$.

To prove (2) for $0 < s < t$ we use the same argument, except that now we have extra terms coming from time s. The details are more complicated but the conclusion is the same — time shifts do not contribute to the quadratic variation.

All that remains is the proof of (c). We use the Ray-Knight theorem: given $z = B_{\tau_t^y}$ and U, the process $R_a - R_x - 2(a \wedge z - x \wedge z)$ is a square integrable martingale with quadratic variation $4 \int_x^a R_b db$. By Ito's formula

$$\mathbf{E}_z[R_y^p] \leq \mathbf{E}_z[R_x^p] + 2p^2 \int_x^y \mathbf{E}_z[R_a^{p-1}] da,$$

so if we grant the estimate $\mathbf{E}[R_x^p] \leq c|x - y|^{(3p+1)/6}$ then, since R_a^p is a submartingale, induction on p proves (c).

The last step is the bound for $\mathbf{E}[R_x^p] = \mathbf{E}\left[\left(\tilde{L}_x^U \circ \theta_{\tau_t^y}\right)^p\right]$. Remark that $U \leq \int_x^y L(a, \tau_t^0) da$, so by (a) and BDG we find $\mathbf{E}[U^p] \leq c|x - y|^p$. Next, taking $\delta = |x - y|^{1/3}$ and defining T_x as the B_t hitting time of x, the strong Markov property applied at times τ_t^y and $T_x \circ \theta_{\tau_t^y}$ shows

$$\mathbf{E}\left[\left(\tilde{L}_x^U \circ \theta_{\tau_t^y}\right)^p; B_{\tau_t^y} \leq x - \delta\right] \leq \mathbf{E}\left[\left(\tilde{L}_x^{U'}\right)^p\right] \mathbf{P}_\delta[T_0 < U']$$

with U' an independent copy of U. Then from (b) we have the bound

$$c\mathbf{E}\left[\int_0^U \frac{\delta}{\sqrt{2\pi s^3}} e^{-\delta^2/2s} ds\right]$$

This decreases faster than any power of δ, thereby implying, again from (b), that

$$\mathbf{E}\left[\left(\tilde{L}_x^U \circ \theta_{\tau_t^y}\right)^p; x - \delta < B_{\tau_t^y} < y\right] \leq \mathbf{E}\left[U^{p/2}; x - \delta < B_{\tau_t^y} < y\right]$$

will dominate the estimate for $\mathbf{E}[R_x^p]$. But, from Hölder and the above estimate on $\mathbf{E}[U^p]$, we bound this by

$$\mathbf{E}^{\frac{1}{2}}[U^p] \mathbf{P}^{\frac{1}{2}}\left[x - \delta < B_{\tau_t^y} < y\right] \leq c|x - y|^{p/2} \delta^{1/2} \frac{1}{t^{1/4}}$$

using the explicit density for the reflecting Brownian law $B_{r_i^y}$. This proves (c), and hence (2).

Remarks: (a) An argument similar to the proof of (2) shows that

$$\langle M^u, M^t - M^s \rangle_x = 0 \qquad\qquad (u \leq s < t)$$

so disjoint time intervals give rise to orthogonal martingales. This echoes the fact, proved in [3], that $\langle L(., \tau_{t-s}^x) \circ \theta_{r_s^x}, L(., \tau_u^x) \rangle = 0$.

(b) Calculating D_x^t is much more difficult. Although (3) gives us $E[R_y | \mathcal{E}^x] = E[D_x^t - D_y^t | \mathcal{E}^x]$, we still lack *a priori* information about the right side. [2] provides an elegant way out of this predicament.

REFERENCES

[1] McGill, P. *Integral representation of martingales in the excursion filtration.* Séminaire de Probabilités XX. Springer Lecture Notes in Mathematics 1204 (1986) 465-502.

[2] Rogers, L.C.G. and Walsh J.B. *The intrinsic local time sheet of a Brownian motion.* Th. Prob. Rel. Fields, Vol. 88, 3 (1991) 363-379.

[3] Walsh, J.B. *Excursions and local time.* In: Temps Locaux, Astérisque Vol. 52/3, Soc. Math. France, 1978.

Department of Mathematics
University of California at Irvine
Irvine, CA 92717, USA

HARMONIC FUNCTIONS ON DENJOY DOMAINS

T.S. MOUNTFORD* AND S.C. PORT

Let D be a non-empty open subset of R^d and let $X(t)$ be a Brownian motion on R^d. Set $T_D = \inf\{t > 0 : X(t) \notin D\}$ and let $P_x(\cdot)$ be the law of $X(t)$ when $X(0) = x$.

Suppose f is a bounded harmonic function on D. A fundamental result due to Doob shows that with $P_x(\cdot)$ probability one, for every $x \in D$, $\lim_{t \to T_D} f(X(t)) = Y$ exists and $f(x) = E_x Y$. Moreover, there is a unique bounded harmonic function g such that $g(x) = O(P_x(T_D < \infty))$ and a unique constant α such that $f(x) = g(x) + \alpha P_x(T_D = \infty)$.

Let $H_D(x, A) = P_x(X(T_D) \in A; T_D < \infty)$. In MP[1] we called f representable iff there was a bounded measurable φ on ∂D such that $g(x) = H_D\varphi(x)$. We also called D Poissonian iff every bounded f was representable.

Suppose H is a hyperplane of R^d and C is a closed subset of H. The open set $D = R^d \backslash C$ is called a Denjoy domain. A problem of some interest in analysis was to give necessary and sufficient conditions on C for D to be Poissonian, and when D is not Poissonian, to give necessary and sufficient conditions on f for it to be representable.

In this paper we will establish the following results.

THEOREM 1. *Let f be a bounded harmonic function on the Denjoy domain $D = R^d \backslash C$. Then f is representable iff it is symmetric with respect*

---

*Research supported in part by NSF grant DMS 86-01800

to reflection across the hyperplane.

THEOREM 2. *In order that every bounded harmonic function on the Denjoy domain $D = R^d \backslash C$ be representable it is necessary and sufficient that C have hyperplane Lebesgue measure 0.*

The study of Poissonian sets was initiated by us in MP[1]. In that paper we characterized Poissonian sets in terms of their Martin boundaries and gave various properties that such sets possess. Although our results were purely analytical the methods we used were purely probabilistic. Theorems 1 and 2 were first proved by such probabilistic methods.

It was of some interest to see if the results we obtained could be done by analytic methods. In Bi[1] Bishop accomplished that task. He also obtained an elegant characterization of Poissonian domains that did not involve the Martin boundary. As it turns out this characterization can also be very easily obtained by probabilistic methods (see MP[1]). Bishop's analytical methods yield purely analytical proofs of Theorems 1 and 2. However, our original probabilistic methods are of some interest in their own right and the techniques used may prove useful in other contexts. For these reasons we will give these proofs here.

Our first result is on general open sets. The result of this proposition is the key to Bishop's characterization of Poissonian domains. See MP[1] for details.

PROPOSITION 1. *Suppose D is a non-Poissonian open set. There is then a bounded harmonic function f on D with values in $[0, 1]$ and an initial distribution μ concentrated on D such that for Brownian motion with initial point distributed on D according to μ,*

(i) *$P(\lim_{t \uparrow T_D} f(X(t)) \in \{0, 1\}) = 1$ and both limiting values occur with positive probability,*

(ii) *$P[\lim_{t \uparrow T_D} f(X(t)) = 1 \mid X(T_D) = y] > 0$,*

for all y except those in a set having $P(X(T_D) \in \cdot)$ measure 0.

PROOF. Let δ_x be the unit mass at x. Let $\{q(j)\}$ be dense in D and let $\mu = \sum_j 2^{-j} \delta_{q(j)}$. By assumption there is a non-representable bounded harmonic function g on D. We know that $P(\lim_{t \uparrow T_D} f(X_t) = Y) = 1$ for some random variable Y. Let $K(x, dy)$ be a regular conditional distribution of Y given $X(T_D)$. By assumption $K(x, \cdot)$ is not the unit mass at a point for $P(X(T) \in \cdot)$ a.e. x. Thus there is a set B having $P(X(T_D) \in B) > 0$ such that for all $x \in B$, $K(x, \cdot)$ is not the unit mass at a point. Let $B(y) = \{x \in B : 0 < K(x, (-\infty, y]) < 1\}$. Then for Q the rationals, $\cup_y B(y) = B$ and since $P(X(T_D) \in B) > 0$ it follows that for some $y \in Q$,

$P(X(T_D) \in B(y)) > 0$. Hence there is a set F with $P(X(T_D) \in F) > 0$ and an a such that $0 < K(x, (-\infty, a]) < 1$ for all $x \in F$.

We now take as our harmonic function f the function $P_x(\Gamma)$ where $\Gamma = [\lim_{t \uparrow T_D} X(t) \notin F$ or $Y > a]$.

A simple martingale argument (see the proof of Lemma 3.2 in MP[1] for details) shows that a.s. the limiting value of $f(X(t))$ as t tends to T_D is equal to I_Γ, so condition (i) is satisfied. By our choice of a and F for $x \in F$ $Q(x, \Gamma) > 0$ and $Q(x, \Gamma^c) > 0$, where $Q(,)$ is a regular conditional probability on the space of paths given X_T.   □

To prove Theorems 1 and 2 it suffices to consider C a closed subset of the hyperplane, $H = \{x : x_1 = 0\}$. For a point $x \in R^d$ we will write $x = (x_1, y)$ where y is on the hyperplane. The proof of Theorem 1 will be given first. This will be accomplished via a sequence of lemmas and propositions.

We say that a function f on R^d is reflexive if $f(x_1, y) = f(-x_1, y)$.

PROPOSITION 2. *In order for a bounded harmonic function f on D to be representable it is necessary for f to be reflexive. If C has zero $(d-1)$ dimensional Lebesgue measure then every bounded harmonic function is reflexive. If C has positive $(d-1)$ dimensional measure then there is a bounded harmonic function that is not representable.*

PROOF. Let $T = \inf\{t > 0 : X(t) \in H\}$ then

$$P_x(X(T) \in dz) = \frac{\Gamma(d/2)}{\pi^{d/2}} \frac{|x_1|}{(|y - z|^2 + x_1^2)^{d/2}} dz .$$

Suppose that f is representable. Then

$$f(x) = E_x H_D\varphi(X_T) + \alpha P_x(T_D = \infty)$$
$$= \int_H \frac{\Gamma(d/2)}{\pi^{d/2}} \frac{|x_1|}{(|y - z|^2 + x_1^2)^{d/2}} H_D\varphi(z)dz + \alpha P_x(T_D = \infty) .$$

Hence $f(x) = f(x')$. Suppose that C has zero $(d-1)$ dimensional measure. Then $P_x(X(T) \in C) = 0$ and so $f(x) = E_x f(X(T))$. Thus f is reflexive. Finally suppose C has positive $(d-1)$ dimensional measure. Let

$$f(x) = \int_C \frac{x_1}{(|y - z|^2 + x_1^2)^{d/2}} dz .$$

Then f is bounded by $\dfrac{\Gamma(d/2)}{\pi^{d/2}}$. It is obviously harmonic on $R_+ \times R^{d-1}$ and on $R_- \times R^{d-1}$. It is harmonic on $H\backslash C$ by the averaging property. However $f(x) \neq f(x')$ so it is not representable.   □

To finish the proof of Theorem 1 we must show that if C has zero $d-1$ dimensional Lebesgue measure then every bounded f is representable.

The essential idea behind the proof of this fact is as follows. For D to be non-Poissonian it must be possible for two Brownians paths to hit C at the same point, but for the limit of f (as produced in Proposition 1) to be one along one path and to be zero along the other. But Brownian motion is of such variability that values of f along two such paths must be of comparable magnitude infinitely often as the paths tend toward the hitting point of C. This will provide a contradiction.

The details are much simpler in the case $d = 2$, so we will first establish the theorem in this case.

We will write a point in R^2 as a complex number z. We require some simple lemmas.

LEMMA 1.   *Suppose that $\{c(r) : r \geq 0\}$ is a curve in $R_+ \times R^1$ which satisfies*

(a) $c(0) = 0$.

(b) $\lim_{r \to \infty} |c(r)| = \infty$.

Denote by K the totality of points on $\{c(r) : r \geq 0\}$ and by E the domain $(R_+ \times R^1)$. For each $z \in E$ with $arg(z) \in \{\pi/6, 0, -\pi/6\}$

$$H_E(z, K) \geq \delta = \frac{1}{\pi}(\pi/2 - \tan^{-1}(\sqrt{3})).$$

PROOF.   Let z be as prescribed. Then by the Cauchy distribution for the hitting point of $\{0\} \times R^1$

$$H_{R_+ \times R^1}(z, \{0\} \times R_+) \quad \text{and} \quad H_{R_+ \times R^1}(z, \{0\} \times R_-)$$

are both greater than δ. Now the curve c must separate z from either $\{0\} \times R_+$ or $\{0\} \times R_-$. Without loss of generality assume the former. Then

$$H_E(x, K) \geq H_{R_+ \times R^1}(x, \{0\} \times R_+) .   \qquad \square$$

COROLLARY 1.   *Let $\{f(r) : r \in [0, \delta)\}$ be a curve in $R_+ \times R^1$ which satisfies*

(a) $f(r) \neq 0$ *for $r \in (0, \delta)$.*

(b) $f(0) = 0$.

Then (with K and E defined as in Lemma 1)

$$\lim_{r \to 0} H_E(re^{i\theta}, K) \geq \delta$$

for $\theta \in \{\pi/6,\ \pi/2,\ 5\pi/6\}$.

PROOF.   Let $\{c(r) : r \geq 0\}$ be any continuous extension of f to $(0, \infty)$ so that

(a) $\lim_{r \to \infty} |f(r)| = \infty$.

(b)   For all $r > \delta$, $|f(r)| \geq |f(\delta)|$.

Let K' be the totality of points in c. From Lemma 1 we know that $H_E(re^{i\theta}, K') \geq \delta$. Let $A = \{(R_+ \times R^1) \cap \{z : |z| = |f(\delta)|\}$ and $B = \{z : |z| = |f(\delta)|\}$. Then

$$H_E(re^{i\theta}, K) \geq \delta - H_A(re^{i\theta}, B) .$$

But the last term goes to zero as r goes to zero and the corollary follows.□

PROOF OF THEOREM 1 IN TWO DIMENSIONS.   Suppose that D is non-Poissonian. Let f be a bounded harmonic function and μ the measure as described in Proposition 1. A time r is a cone point of angle α for a planar Brownian motion $\{X(t) : t \geq 0\}$ if there exists an $h > 0$ and a cone C of angle α such that for $s \in [0, h]$ $X(r + s) - X(t) \in C$. It is known (see B[1] or S[1]) that planar Brownian motion does not possess cone points of angle $\pi/3$. Accordingly for $H_D(\mu, .)$ a.e.x

$$P[A_x | X(T_D) = x] = 1$$

where A_x is the set of paths with the property that $\arg(X(t) - x) \in \{\pi/6,\ \pi/2,\ 5\pi6,\ 7\pi/6,\ 3\pi/6\}$ infinitely often as t tends to T_D.

By Proposition 1 we can find an $x \in K$ and paths ω_1 and ω_2 such that

A. $X(T_D, \omega_i) = x$.

B. $\lim_{t \to T_D} f(X(t, \omega_1)) = 1$ and $\lim_{t \to T_D} f(X(t, \omega_2)) = 0$.

C. $\omega_2 \in A_x$.

Now the function f is preserved by reflection about the y axis. So $\lim_{t \to T_D} f(X(t, \overline{\omega_1})) = 1$ where $\overline{\omega}$ is the reflection of ω about the y axis. Given the path ω_1 let the path $|\omega_1|$ be defined

$$X(t, |\omega_1|) = (|X_1(t, \omega_1)|, X_2(t, \omega_1)) .$$

Let $|\overline{\omega_1}|$ be the reflection of $|\omega_1|$ about the y axis. Let $\theta \in \{\pi/6, \pi/2, 5\pi/6\}$, K be the set of points in $\{X(t, |\omega_1|) : t \in [0, T_D\}$ and E be $R_+ \times R^1$, then by Corollary 1 as r tends to zero

$$f(x + re^{i\theta}) \geq \lim_{r \to \infty} \int_K f(y) H_E(x + r^{i\theta}, dy) \geq \delta/2 .$$

Similarly, let $\theta \in \{7\pi/6, 3\pi/2, 11\pi/6\}$, K be the set of points in $\{X(t, |\overline{\omega_1}|) : t \in [0, T_D\}$ and E be $(R_+ \times R^1)\backslash K$. Then as r tends to zero

$$f(x + re^{i\theta}) \geq \lim_{r \to \infty} \int_K f(y) H_E(x + r^{i\theta}, dy) \geq \delta/2.$$

But by C this implies that

$$\varliminf_{t \to T_D} f(X(t, \omega_2)) \geq \delta/2 .$$

This contradiction establishes the Theorem in two dimensions. $\square$

The proof of Theorem 1 for $d > 2$ will be carried out via a sequence of propositions. In the following argument we will need to use the skew-normal decomposition of d-dimensional Brownian motion. Details can be found in IM[1].

A d-dimensional Brownian motion $\{X(t) : t \geq 0\}$ can be written as

$$X(t) = R(t)\theta_{\tau(t)}$$

where $\{R(t) : t \geq 0\}$ is a d-dimensional Bessel process, $\tau(t) = \int_0^t \frac{1}{R(u)^2} du$ and $\{\theta_r : r \geq 0\}$ is an independent S^{d-1} valued Brownian motion. This is the skew-normal decomposition.

It follows from the skew-normal decomposition and some simple Ito calculus that the process

$$\{\theta^1(r) = \frac{X_1(r)}{R(r)} : r \geq 0\}$$

is the diffusion $\{Y(r) : r \geq 0\}$ on $[-1, 1]$ run with clock $\tau(r)$, where Y has generator

$$\frac{\sqrt{1 - x^2}}{2} \frac{d^2}{dx^2} - (d-1)x \frac{d}{dx} .$$

Consider the process $\{X(r) : r \geq 0\}$ started at the origin. For each integer m define the random (but not stopping) times

$$S_m = \sup\{r : R(r) = 2^{m-1}\} \quad T_m = \inf\{r > S_m : R(r) = 2^m\} \ .$$

PROPOSITION 3. *The random variables* $\{\tau(T_m) - \tau(S_m)\}$ *are independent, identically distributed and strictly positive. The processes*

$$\{\theta(r) : r \in (\tau(S_m),\ \tau(T_m))\}$$

are identically distributed. They are not independent but are conditionally independent given the random variables

$$\{\theta(\tau(S_m)), \theta(\tau(T_m))\ m \in Z\} \ .$$

The proof is split into a number of lemmas.

LEMMA 2. *The processes* $\{R(r + T_m) : r \geq 0\}$ *and* $\{R(r \wedge T_m) : r \geq 0\}$ *are independent.*

Remark. Since for $i \leq m$, S_i, T_i are measurable with respect to $\sigma(R(r \wedge T_m) : r \geq 0$ we see that

$$(\tau(T_i)) - (\tau(S_i)) = \int_{S_i}^{T_i} \frac{1}{R(r)^2} dr \ , \quad i = 1, 2, \ldots,$$

are independent.

PROOF.   Observe first that if $E_i\ i \geq 1$ are independent identically distributed random elements of some measurable space E and $N = \inf\{i : E_i \in A \subset E\}$ then E_N and the sigma-field generated by N and E_i for $i < N$ are independent.

For the process $\{R(t) : t \geq 0\}$ define the stopping times

$$U_1 = \inf\{t : R(t) = 2^m\}, \ V_i = \inf\{t > U_i : R(t) = 2^{m-1}\}, \quad \text{and}$$
$$U_i = \inf\{t > V_{i-1} : R(t) = 2^m\}.$$

By the strong Markov property all the processes $\{R(t + U_i) : t \in [0, U_{i+1} - U_i)\}$ are independent. In addition the process $\{R(t) : t \in [0, U_1)\}$ is independent of these processes. Let the random variable N be defined as the first i such that V_i is infinite. By the opening observation the process $\{R(t + U_N) : t \geq 0\}$ is independent of the sigma-field generated by $\{R(t) : t \in [0, U_1)\}$, $\{R(t + U_i) : t \in [0, U_{i+1} - U_i)\}(i < N)$ and $\{N\}$. This gives the result since $T_m = U_N$ and the latter sigma-field generates $\{R(r \wedge T_m) : r \geq 0\}$. □

LEMMA 3.   *The random variables $\tau(T_m) - \tau(S_m)$ are identically distributed.*

PROOF.   Define the process $\{Z(t) : t \geq 0\}$ by

$$Z(t) = 2X(t/4) .$$

Then Z is also a Brownian motion. Let S'_m, T'_m be the random times for the process Z, corresponding to S_m and T_m. It follows from the definition of Z that $S'_{m+1} = 4S_m$ and $T'_{m+1} = 4T_m$ and

$$\int_{S'_{m+1}}^{T'_{m+1}} dt/|Z(t)|^2 = \int_{4S_m}^{4T_m} dt/|2X(t/4)|^2$$

$$= \int_{S_m}^{T_m} 4dt/|2X(t)|^2$$

$$= \int_{S_m}^{T_m} dt/|X(t)|^2 . \qquad \square$$

PROOF OF PROPOSITION 3.   Lemmas 2 and 3 show that the times S_m and T_m are such that $\tau(T_m) - \tau(S_m)$ are identically distributed. Given that these random times are independent of $\{\theta_t : t \geq 0\}$ and that $\theta_{\tau(S_m)}$ is uniformly distributed we see that the processes

$$\{\theta_r : r \in (\tau(S_m), \tau(T_m))\}$$

are identically distributed. That they are conditionally independent follows from the simple Markov property and the fact that the intervals (S_m, T_m) do not overlap. □

Define p_ϵ as $\sup_{x_1, x_2} P(\sup_{r \in (S_m, T_m)} |\theta^1(r)| < \epsilon \big| \theta^1(S_m) = x_1, \ \theta^1(T_m) = x_2)$.

PROPOSITION 4.   *The quantities p_ϵ satisfy*

$$\lim_{\epsilon \to 0} p_\epsilon = 0 .$$

PROOF. The diffusion $\{\theta^1(r) : r \geq 0\}$ has a density $p(t, x, y)$ continuous in $(0, \infty) \times [-1, 1] \times [-1, 1]$. Fix $\delta > 0$ and let m and M be such that

$$m < p(t, x, y) < M \quad \text{for} \quad (x, y) \in [-1/2, 1/2]^2, \ t \in [\delta/2, 1/\delta] .$$

Then we see for $\varepsilon < 1/2$

$$P(\sup_{r \in (S_m, T_m)} |\theta^1(r)| < \varepsilon | \theta^1(S_m) = x_1, \theta^1(T_m) = x_2)$$
$$< \varepsilon M^2/m + P[T_m - S_m \notin [\delta, 1/\delta]] .$$

Letting δ and ε tend to zero yields the result. $\square$

Let $\{Y(r) : r \geq 0\}$ be a process in $R^d(d > 1)$ with $Y(0) = 0$. For $n \geq 1$ define the random variables

$$I_n^\varepsilon = 1 \text{ if } \sup_{|Y(r)| \in [2^{-(n+1)} 2^{-n})} \frac{|Y^1(r)|}{|Y(r)|} < \varepsilon$$
$$= 0 \text{ otherwise .}$$

$$K_n^\varepsilon = 1 \text{ if } \sup_{|Y(r)| \in [2^{-(n+2)}, 2^{-n+1}]} \frac{|Y^1(r)|}{|Y(r)|} < \varepsilon$$
$$= 0 \text{ otherwise .}$$

PROPOSITION 5. *Let $\{Y(r) : r \geq 0\}$ be a d-dimensional Brownian motion. Let*

$$Z_n = \sum_{i=1}^n I_i^\varepsilon .$$

There is an ε such that if n is sufficiently large then

$$P[Z_n \geq n/5] \leq 2^{-4n} .$$

PROOF. For $t \in (\tau(S_{-m}), \tau(T_{-m})]$, by definition $|Y(t)| \in (2^{-(m+1)}, 2^{-m}]$. So if for some $t \in (\tau(S_{-m}), \tau(T_{-m})]$, $|Y^1(t)|/|Y(t)| \geq \varepsilon$, then $I^{\varepsilon m} = 0$. Thus motivated let us define

$$J_n^\varepsilon = 1 \text{ if } \sup_{r \in [S_{-m}, T_{-m})} \frac{|Y^1(r)|}{|Y(r)|} < \varepsilon$$
$$= 0 \text{ otherwise .}$$

Obviously $\sum_{m=1}^n I_m^\epsilon \le \sum_{m=1}^n J_m^\epsilon$. Now by the definition of p_ϵ, $\sum_{m=1}^n J_m^\epsilon$ is stochastically less than $B(n, p_\epsilon)$. Here $B(n, p)$ denotes a Binomial random variable with parameters n and p.

Chernoff's inequality (see e.g. B[1], page 148) states that

$$P[B(n, p) \ge k] \le \exp\{nH(k/n)\} \quad \text{where}$$
$$H(x) = x \log(p/x) + (1 - x) + (1 - x) \log(q/(1 - x)).$$

By Chernoff's inequality $P\left[\sum_{m=1}^n I_m^\epsilon \ge n/16\right]$ has probability majorized by

$$\exp(n[1/4 \log(4p_\epsilon) + 3/4 \log(4(1 - p_\epsilon)/3)]).$$

If ϵ (and hence p_ϵ) is sufficiently small, this is majorized by $1/2^{4n}$. $\square$

From now on we will fix an ϵ_0 sufficiently small to satisfy the conclusion of Proposition 5.

COROLLARY 2.   *Consider $\{X(t) : t \ge 0\}$, a Brownian motion stopped when it hits $\{x : |x| = 1\}$. For this process $P\left[\sum_{m=1}^n I_m^{\epsilon_0} \ge n/5\right] \le 2^{-4n+2}$.*

PROOF.   This follows since with probability greater than or equal to $1/2$ d-dimensional Brownian motion never returns to $\{x : |x| = 1/2\}$ after it hits $\{x : |x| = 1\}$. $\square$

Given a Brownian motion in d dimensions $\{X(r) : r \ge 0\}$, for time r let $\{Y^r(s) : s \ge 0\}$ be the process $X(r + s) - X(r)$ run until it first has magnitude 2. For such a process indexed by r let the random variables K_m^ϵ be denoted by $K_m^{r,\epsilon}$.

Given a Brownian motion in d-dimensions $\{X(r) : r \ge 0\}$, for time r let $\{Z^r(s) : s \ge 0\}$ be the process $X(r + s) - X(r)$ run until it first has magnitude 1. For such a process indexed by r let the random variables I_m^ϵ be denoted by $I_m^{r,\epsilon}$.

PROPOSITION 6.   *Let $\{X(s); s \ge 0\}$ be a d-dimensional Brownian motion and let ϵ be less than $\epsilon_0/2$. With probability one there does not exist an $r \in R_+^1$ such that $\overline{\lim}_{n \to \infty} \frac{1}{n} \sum_{\nu=1}^n K_\nu^{r,\epsilon} \ge 1/3$.*

PROOF.   Let B_n be the event that such an r occurs in the time interval $[n, n + 1)$. Clearly the probability of such an r occuring is majorized by $\sum_{n=0}^\infty P(B_n)$. However by the stationary independent increments property of Brownian motion $P(B_n) = P(B_0)$ n. Therefore it is sufficient to show that such an r occurs on $[0, 1]$ with probability zero. This will be accomplished via few lemmas.

LEMMA 5. *Suppose vectors x, y and z satisfy*

1. $|x - y| < 3\sqrt{2^{-n} \log(2^n)}$.

2. $|x - z| \in [2^{-(m+1)}, 2^{-m})$ *for $m \leq n/3$.*

and

3. $\frac{|x_1 - z_1|}{|x - z|} > 2\varepsilon$.

Then for n sufficiently large

4. $\frac{|y_1 - z_1|}{|y - z|} > \varepsilon$ *and* $|y - z| \in [2^{-(m+2)}, 2^{-m+1})$.

PROOF. These follow simply from the triangle inequality. □

It is well known (see e.g. IM[1]) that

$$\lim_{h \to 0} \sup_{t \in [0,1]} \frac{\|X(t+h) - X(t)\|}{\sqrt{2h \log(1/h)}} = 1 .$$

This result is due to Levy. Therefore for n sufficiently large

(1)
$$\sup_{t \in [0,1]} \sup_{h \leq 2^{-n}} \frac{|X(t+h) - X(t)|}{\sqrt{h \log(1/h)}} \leq 3.$$

LEMMA 6. *Let $m \in [n/4, n/3]$. If n is so large that (1) holds and if there exists an r such that*

$$\sum_{\nu=1}^{m} K_\nu^{r,\varepsilon} \geq m/3 ,$$

then for some j in $[1, 2^n]$ we have

$$\sum_{\nu=1}^{n/3} I_\nu^{j/2^n, 2\varepsilon} \geq n/12 .$$

PROOF. First observe that if such an r and an m exist then

$$\sum_{\nu=1}^{n/3} K_\nu^{r,\varepsilon} \geq \sum_{\nu=1}^{m} K_\nu^{r,\varepsilon} \geq m/3 \geq n/12 .$$

Now observe that by (1) if $|r - j/2^n| < 2^{-n}$, then $|X(r) - X(j/2^n)| \leq 3\sqrt{2^{-n}\log(2^n)}$. By Lemma 5, $K_\nu^{r,\epsilon} = 1$ must imply that $I_\nu^{j/2^n,2\epsilon} = 1$. $\quad\Box$

PROOF OF PROPOSITION 6. Let B_n be the event that there exist $r \in [0, 1]$ and $m \in [n/4, n/3]$ with $\sum_{\nu=1}^m K_\nu^{r,\epsilon} \geq m/3$ and let $A_{j/2^n}^n$ be the event

$$\sum_{\nu=1}^{n/3} I_\nu^{r,2\epsilon} \geq n/12 .$$

The above discussion shows that

$$B_n \subset \bigcup_{j=1}^{2^n} A_{j/2^n}^n \bigcup \left[\sup_{t\in[0,1]} \sup_{h\leq 2^{-n}} \frac{|X(t+h) - X(t)|}{\sqrt{h\log(1/h)}} \geq 3 \right] .$$

And so, up to a set of probability zero

$$\limsup_n B_n \subset \limsup \bigcup_{j=1}^{2^n} A_{j/2^n}^n .$$

But by Corollary 1, $P\left[\bigcup_{j=1}^{2^n} A_{j/2^n}^n\right] \leq \sum_{j=1}^n P(A_{j/2^n}^n) \leq 2^{-n/3}$. Hence the result follows from the Borel-Cantelli Lemma. $\quad\Box$

PROPOSITION 7. *Let f be a bounded harmonic function on $R^d \backslash C$ taking values in $(0,1)$. Suppose that x and y are vectors in this domain satisfying*

(i) $|x|/8 < |y| < 8|x|$

(ii) $|x_1|/|x|, |y_1|/|y| > \delta$

then there is a constant $c(\delta) > 0$ such that $f(x) > 3/4$ implies $f(y) > c(\delta)$.

PROOF. C has zero $(d-1)$ dimensional measure so by Proposition 1 f is reflexive. Therefore we may assume that x_1 and y_1 are both positive. From simple geometric considerations we can see that it is possible to choose an n not depending on x and y, a chain of intersecting balls B_i $i = 1, 2, \ldots, n$ such that $x \in B_1, y \in B_n$ and such that the dilation by a factor of two of each B_i about its centre is still contained in $R_+ \times R^1$. Then by Harnack's inequality there exists a constant c independent of B_i such u, $v \in B_i$ implies

$$f(u) \leq cf(v). \quad \text{So by iteration} \quad c^{-n}f(y) \leq f(x) \leq c^n f(y). \quad\Box$$

PROOF OF THEOREM 1 FOR $d > 2$. Suppose C has zero measure and that D is not Poissonian; by Proposition 1 it follows that there is a harmonic function f on D satisfying the conclusions of Proposition 1. Consider the conditional probability distribution on the space of Brownian paths given $X(T_D)$, $Q(x, d\omega)$. Let A be the set of paths ω such that for all $r \in R_+^1$

$$\{Y^r(s, \omega) : s \geq 0\} \text{ has}$$

$$\overline{\lim} \frac{\sum_{\nu=1}^n K_\nu^{r,\varepsilon}}{n} < 1/3 \, .$$

By Proposition 6 we may assume that $\forall x, Q(x, A) = 1$. By assumption there is an $x \in \partial D$, ω_1 and ω_2 such that

(i) $X(T_D(\omega_i)) = x$.

and

(ii) A contains w_i.

(iii) $\lim_{t \to T_D} f(X(t(\omega_1))) = 1$ and $\lim_{t \to T_D} f(X(t(\omega_2))) = 0$.

Property (i) and (ii) implies that there are sequences r_n increasing to $T_D(\omega_1)$ and s_n increasing to $T_D(\omega_2)$ such that

$$|X(r_n, \omega_1) - x|/8 < |X(s_n, \omega_2) - x| < 8|X(r_n, \omega_1) - x|$$

and

$$\frac{|X_1(r_n, \omega_1)|}{|X(r_n, \omega_1)|} \, , \, \frac{|X_1(S_n, \omega_2)|}{|X(s_n, \omega_2)|} > \varepsilon \, .$$

But property (iii) and Proposition 7 give a contradiction which completes the proof.   □

PROOF OF THEOREM 2. By Proposition 2 if f is representable then it must be preserved under reflection so it remains to show that conversely this property ensures representability. Let f be a bounded harmonic function on D, preserved under reflection. In the following we will assume, as we may, that f is bounded by a multiple of the probability of exiting D. By Fatou's Theorem we know that f has non-tangential limits a.e. on the hyperplane from above and below. Since f is preserved by reflection both non-tangential limits must be equal. Partition C into $B \bigcup G$ where B is the set of points of C for which f has non-tangential limit. G necessarily has zero Lebesgue measure. Let $\varphi(x)$ for $x \in B$ be the non-tangential limit. By Proposition 6 the event $X(T_D) \in B$; $\lim_{t \to T_D} f(X(t)) \neq \varphi(X(T_D))$ has zero probability. Accordingly by the argument used to establish Proposition 1 if f is not representable we can find a compact subset F of G with positive harmonic measure and an $a \in R^1$ such that

$$x \in F, \ P\left[\lim_{t \to T_D} f(X(t)) > a \mid X(T_D) = x\right] > 0 \qquad \text{and}$$

$$P\left[\lim_{t \to T_D} f(X(t)) < a \mid X(T_D) = x\right] > 0.$$

This implies that F^c is non-Poissonian. However this contradicts Theorem 1 since F is a subset of G and therefore has zero $(d-1)$-dimensional measure. $\square$

REFERENCES

B Billingsley, P. *Probability and Measure Theory*, 2nd Edition. Wiley, New York 1986.

Bi Bishop C. *A characterization of Poissonian Domains*, Arkiv Mat. 29, 1-24, 1991.

Bu Burdzy, K. *Brownian Paths and Cones*, Anals of Prob. **13** 3, 1006-1010, 1985.

IM Ito, K. and McKean, H.P. *Diffusion Processes and Their Sample Paths*, Springer, New York. 1965.

MP Mountford, T.S, and Port, S.C. *Representations of Bounded Harmonic Functions*, Arkiv Mat. 29, 107-126. 1991.

S Shimura, M. *Excursions in a cone for two dimensional Brownian motion*, Journal of Mathematics of Kyoto University **25** 3, 433-443, 1985.

T. S. Mountford
Department of Mathematics
University of California
Los Angeles, CA   90024-1555

S. C. Port
Department of Mathematics
University of California
Los Angeles, CA   90024-1555

Conditional Dawson–Watanabe Processes and Fleming–Viot Processes

by EDWIN A. PERKINS

There has been interest recently in establishing connections between the Dawson-Watanabe and Fleming-Viot superprocesses (eg. Konno-Shiga (1988), Etheridge-March (1991)). Sometimes results are more readily derived for one class of processes but one would like to be able to infer them for the other with minimal effort. Tribe (1989) used the Konno-Shiga (1988) results to analyze the Dawson-Watanabe superprocess near extinction.

Etheridge and March (1991) showed that the Fleming-Viot superprocesses is the Dawson-Watanabe superprocess, conditioned to have total mass one. Our goal is to generalize this pretty result via a different approach.

Let E be a locally compact space with a countable base and one-point compactification $E_\infty = E \cup \{\infty\}$. Its Borel σ-field is $\mathcal{E}$ and $b\mathcal{E}$ denotes the class of bounded $\mathcal{E}$-measurable functions from E to $\mathbf{R}$. $M_F(E)$ and $M_1(E)$ denote the spaces of finite measures, and probability measures, respectively, with the topologies of weak convergence. If $\phi \in b\mathcal{E}$ and $\nu \in M_F(E)$, $\nu(\phi)$ denotes $\int \phi(x) d\nu(x)$.

Let $\Omega = C([0,\infty), M_F(E))$ and $\hat{\Omega} = C([0,\infty), M_1(E))$ (compact-open topologies) and let $\mathcal{F}$ and $\hat{\mathcal{F}}$ be their respective Borel σ-fields. $X_t(w) = w(t)$ and $\hat{X}_t(\hat{w}) = \hat{w}(t)$ denote the coordinate mappings on Ω and $\hat{\Omega}$, respectively, and $\mathcal{F}_t^0 = \sigma(X_s : s \le t)$, $\hat{\mathcal{F}}_t^0 = \sigma(\hat{X}_s : s \le t)$, $\mathcal{F}_t = \mathcal{F}_{t+}^0$, $\hat{\mathcal{F}}_t = \hat{\mathcal{F}}_{t+}^0$.

Let (Y_t, P_y) be an E-valued conservative Feller process with generator A defined on $D(A) \subset C_o(E)$ (continuous functions on E, vanishing at ∞). Recall that this means T_t, the semigroup of Y, is strongly continuous on $C_o(E)$. For

each $\sigma^2 > 0$ and $m \in M_F(E)$ there is a unique probability $\mathbf{P}_m$ on $(\Omega, \mathcal{F})$ such
that

$$\forall \phi \in D(A) \qquad M_t(\phi) = X_t(\phi) - m(\phi) - \int_0^t X_s(A\phi)ds \quad \text{is an}$$

(DW_m) $(\mathcal{F}_t)$ – martingale starting at 0 and such that

$$< M(\phi) >_t = \sigma^2 \int_0^t X_s(\phi^2)ds.$$

For each $\sigma^2 > 0$ and $m \in M_1(E)$ there is a unique probability on $(\hat{\Omega}, \hat{\mathcal{F}})$ such
that

$$\forall \phi \in D(A) \qquad \hat{M}_t(\phi) = \hat{X}_t(\phi) - m(\phi) - \int_0^t \hat{X}_s(A\phi)ds \quad \text{is an}$$

(FV_m) $(\hat{\mathcal{F}}_t)$ – martingale starting at 0 and such that

$$< \hat{M}(\phi) >_t = \sigma^2 \int_0^t \hat{X}_s(\phi^2) - \hat{X}_s(\phi)^2 ds.$$

See Ethier-Kurtz (1986, Ch. 9.4, 10.4) or Roelly-Coppoletta (1986) for the above
results. $\mathbf{P}_m$ and $\hat{\mathbf{P}}_m$ are the laws of the A-Dawson-Watanabe and A-Fleming-Viot
superprocesses, respectively. σ^2 is usually assumed to be one, unless otherwise
indicated, and hence is suppressed in our notation..

Remark 1. (DW_m) and (FV_m) extend to $\phi \equiv 1$ by taking limits through a
sequence $\{\phi_n\} \subset D(A)$ such that $\phi_n \rightarrow 1$ and $A\phi_n \rightarrow 0$, both in the bounded
pointwise sense. For example, let $\phi_n(x) = \int_0^1 T_t f_n(x)dt$ where $f_n(x) = 1 -$
$e^{-nd(x,\infty)}$ and d is a bounded metric on E_∞. The extension will also hold for
$(FV_{m,f})$ described below and will be used without further comment.

If $T > 0$, let $(\hat{\Omega}_{T-}, \hat{\mathcal{F}}_{T-}) = (C([0,T), M_1(E))$, Borel sets) and let $(\hat{\Omega}_T, \hat{\mathcal{F}}_T)$
denote the same space with $[0,T]$ in place of $[0,T)$. $(\Omega_T, \mathcal{F}_T)$ and $(\Omega_{T-}, \mathcal{F}_{T-})$
denote the same spaces with $M_F(E)$ in place of $M_1(E)$. (We are abusing the $\mathcal{F}_T$
notation slightly here.) Each of these spaces is given the compact-open topology.
If $\mathbf{P}$ is a probability on $(\Omega, \mathcal{F})$ (or $(\hat{\Omega}, \hat{\mathcal{F}})$), $\mathbf{P}|_{T-}$ is defined on $(\Omega_{T-}, \mathcal{F}_{T-})$ (or on
$(\hat{\Omega}_{T-}, \hat{\mathcal{F}}_{T-})$) by $\mathbf{P}|_{T-}(A) = \mathbf{P}(X|_{[0,T)} \in A)$ (or use $\hat{X}$ in place of X). Similarly
one defines $\mathbf{P}|_T$.

Here then is a slight restatement of the result of Etheridge and March (1991).

Theorem A. (Etheridge-March (1991)). Assume $m_n \to m$ in $M_F(E)$ where $m(E) = 1$. Let $\epsilon_n \downarrow 0$ and $T_n \to T$, where $T_n \in (0, \infty)$ and $T \in (0, \infty]$. Then

$$\mathbf{P}_{m_n}|_{T-}(\ \cdot\ | \sup_{t \leq T_n} |X_t(1) - 1| < \epsilon_n) \overset{w}{\to} \hat{\mathbf{P}}_m|_{T-} \quad \text{on } (\Omega_{T-}, \mathcal{F}_{T-}).$$

The best way to understand this result is to recall the "particle pictures" of these two processes.

Consider a system of K_N particles which follow independent copies of Y on $[0, 1/N]$ and then at $t = 1/N$ independently produce offspring according to a law ν with mean one and variance one. The offspring then follow independent copies of Y on $[1/N, 2/N]$ and this pattern of alternating branching and spatial motions continues. If $X_N(t)(A)$ is N^{-1} times the number of particles in A at time t and $\mathbf{P}_N$ is the law of X_N on $D([0, \infty), M_F(E))$ then $X_N(0) \to m$ in $M_F(E)$ implies $\mathbf{P}_N \overset{w}{\to} \mathbf{P}_m$.

Now consider a system of N particles which follow independent copies of Y on $[0, 1/N]$. At $t = 1/N$ these N particles produce a vector of offspring in Z_+^N distributed as a multinomial random vector with N trials and $p_1 = ... = p_N = 1/N$. This pattern of alternating spatial motions and "multinomial branching" continues. $\hat{X}_N(t)$ denotes the empirical probability distribution of the N particles at time t and $\hat{\mathbf{P}}_N$ is the law of $\hat{X}_N$ on $D([0, \infty), M_1(E))$. If $\hat{X}_N(0) \to m$ in $M_1(E)$ then $\hat{\mathbf{P}}_N \overset{w}{\to} \hat{\mathbf{P}}_m$.

These results are minor modifications of results in Ethier-Kurtz (1986, Ch. 9.4, 10.4).

If $\{X_i : i \leq N\}$ are Poisson (1) and $S_N = \sum_{i=1}^{N} X_i$, then an easy calculation shows that

$$P((X_1...X_N) \in \cdot | S_N = N) \text{ is multinomial with } N \text{ trials and } p_1 = ... = p_N = \frac{1}{N}.$$

This shows that if we take ν to be Poisson (1) in the above construction of $\mathbf{P}_m$, then

$$\mathbf{P}_N(\ \cdot\ | X_N(t) = 1 \text{ for } t \leq T)|_T = \hat{\mathbf{P}}_N|_T.$$

Letting $N \to \infty$ suggests (but does not prove) the result of Etheridge and March.
Our original proof of our main result (Theorem 3 below) used this particle picture.
The proof given below has sacrificed intuition for brevity.

Let

$$C_+ = \{f : [0, \infty) \to [0, \infty) : f \text{ continuous, } \exists t_f \in (0, \infty] \text{ such that}$$

$$f(t) > 0 \text{ if } t \in [0, t_f) \text{ and } f(t) = 0 \text{ if } t \geq t_f\}$$

with the compact-open topology. If $A \subset C_+$ and $T > 0$ let $A|_{T^-} = \{f|_{[0,T)} : f \in A\}$ and $A|_T = \{f|_{[0,T]} : f \in A\}$.

Theorem 2.

(a) If $f \in C_+$ and $m \in M_1(E)$, there is a unique probability $\hat{\mathbf{P}}_{m,f}$ on $(\hat{\Omega}, \hat{\mathcal{F}})$ such that under $\hat{\mathbf{P}}_{m,f}$:

$$\forall \phi \in D(A) \quad \hat{M}_t(\phi) = \hat{X}_t(\phi) - m(\phi) - \int_0^t \hat{X}_s(A\phi)ds, \quad t < t_f,$$

$(FV_{m,f})$ is an $(\hat{\mathcal{F}}_t)$ – martingale starting at 0 and such that

$$< \hat{M}(\phi) >_t = \int_0^t (\hat{X}_s(\phi^2) - \hat{X}_s(\phi)^2) f(s)^{-1} ds \quad \forall t < t_f.$$

$$\hat{X}_t = \hat{X}_{t_f} \text{ for all } t \geq t_f.$$

(b) If $(m_n, f_n|_{[0,T)}) \to (m, f|_{[0,T)})$ in $M_1(E) \times C_+|_{T^-}$ where $T \leq t_f$, then $\hat{\mathbf{P}}_{m_n,f_n}|_{T^-} \xrightarrow{w} \hat{\mathbf{P}}_{m,f}|_{T^-}$ on $(\hat{\Omega}_{T^-}, \hat{\mathcal{F}}_{T^-})$. In particular if $t_f = \infty$, $\hat{\mathbf{P}}_{m_n,f_n} \xrightarrow{w} \hat{\mathbf{P}}_{m,f}$.

Remark. If $f(s) = \sigma^{-2}$ is constant clearly $\hat{\mathbf{P}}_{m,f}$ is just the unique solution $\hat{\mathbf{P}}_m$ of (FV_m).

The proof of Theorem 2 is easy (although some tedious calculations make it a little long), and given at the end of this work. $\hat{\mathbf{P}}_{m,f}$ will be constructed by making a deterministic time change of a Fleming-Viot process whose underlying Markov process, Y, is time-inhomogeneous.

If $m \in M_F(E) - \{0\}$, let $\bar{m}(A) = m(A)/m(E)$. If $t_X(w) = \inf\{u : X_u(E) = 0\}(w \in \Omega)$, then Tribe (1991) showed that $\lim_{t \uparrow t_X} \bar{X}_t$ exists $\mathbf{P}_m$–a.s. Hence we may $\mathbf{P}_m$-a.s. extend $\{\bar{X}_t : t < t_X\}$ to a continuous $M_1(E)$-valued process on $[0, \infty)$ by setting $\bar{X}_t = \bar{X}_{t_X-}$ for $t \geq t_X$. In fact Tribe's result will follow from our arguments but this is not surprising as we will borrow some of his methods.

Let $Q_y \in M_1(C_+)$ denote the law of the unique solution of

$$Z_t = y + \int_0^t \sqrt{Z_s} dB_s$$

(B a standard Brownian motion). If follows from (DW_m) with $\phi = 1$ that $\mathbf{P}_m(X.(1) \in A) = Q_{m(1)}(A)$.

Theorem 3. If $m \in M_F(E) - \{0\}$, then

$$\mathbf{P}_m(\bar{X} \in A | X.(1) = f) = \hat{\mathbf{P}}_{\bar{m},f}(A) \quad Q_{m(1)} - \text{a.a.f } \forall A \in \hat{\mathcal{F}}.$$

Hence $\hat{\mathbf{P}}_{\bar{m},f}(\cdot)$ is a regular conditional distribution for $\bar{X}$ on $(\Omega, \mathcal{F}, \mathbf{P}_m)$ given $X.(1) = f$.

Proof. If $M_t(\phi)$ is as in (DW_m), $\phi \in D(A)$, $T_n = \inf\{t : X_t(1) \leq n^{-1}\}$, and

(1)
$$\bar{M}_t^n(\phi) = \int_0^t 1(s \leq T_n) X_s(1)^{-1} dM_s(\phi) - \int_0^t 1(s \leq T_n) X_s(\phi) X_s(1)^{-2} dM_s(1),$$

then Ito's Lemma implies

(2)
$$\bar{X}_{t \wedge T_n}(\phi) = \bar{m}(\phi) + \int_0^t 1(s \leq T_n) \bar{X}_s(A\phi) ds + \bar{M}_t^n(\phi).$$

(2) implies that

(3)
$$\sup_{t \leq K, n \in \mathbb{N}} |\bar{M}_t^n(\phi)| \leq 2\|\phi\|_\infty + K\|A\phi\|_\infty.$$

Since $\{\bar{M}_t^n(\phi) : n \in \mathbb{N}\}$ is a martingale in n (t fixed) by (1), it converges a.s. as $n \to \infty$ for each $t \geq 0$ by the Martingale Convergence Theorem and (3). A simple application of the L^2-maximal inequality shows that the convergence is uniform for t in compacts a.s. (by perhaps passing to a subsequence). Hence the limit, $\bar{M}_t(\phi)$, is a continuous martingale which clearly satisfies

(4)
$$\bar{M}_t^n(\phi) = \bar{M}_{t \wedge T_n}(\phi) \quad \forall t \geq 0 \text{ a.s.}$$

(5) $$\sup_{t \leq K} |\bar{M}_t(\phi)| \leq 2\|\phi\|_\infty + K\|A\phi\|_\infty \ a.s.$$

We now may let $n \to \infty$ in (2) to see

(6) $\quad \bar{X}_t(\phi) = \bar{m}(\phi) + \int_0^t 1(s < t_X)\bar{X}_s(A\phi)ds + \bar{M}_t(\phi) \ \forall t \geq 0 \ a.s. \ \forall \phi \in D(A).$

Let $\mathcal{G}_t = \mathcal{F}_t \vee \sigma(X_s(1) : s \geq 0)$. We claim $\bar{M}_t(\phi)$ is a $(\mathcal{G}_t)$-martingale ($\phi \in D(A)$ fixed). Let $s < t$ and let F be a bounded $\sigma(X.(1))$-measurable random variable. The predictable representation theorem of Jacod and Yor (see Yor (1978, Thm. 3) and recall $Q^{m(1)}$ is the law of $X.(1)$) shows that

(7) $$F = \mathbb{P}_m(F) + \int_0^\infty f(s,w)dX_s(1)$$

for some $\sigma(X_s(1) : s \leq t)$-predictable f. Therefore

$\mathbb{P}_m((\bar{M}_{t \wedge T_n}(\phi) - \bar{M}_{s \wedge T_n}(\phi))F|\mathcal{F}_s)$

$= \mathbb{P}_m((\bar{M}_t^n(\phi) - \bar{M}_s^n(\phi))\int_0^\infty f(u)dM_u(1)|\mathcal{F}_s)$ (by (4) and (7))

$= \mathbb{P}_m((\int_s^t 1(u \leq T_n)X_u(1)^{-1}dM_u(\phi) - \int_s^t 1(u \leq T_n)X_u(\phi)X_u(1)^{-2}dM_u(1))\int_s^t f(u)dM_u(1)|\mathcal{F}_s)$

$= \mathbb{P}_m(\int_s^t 1(u \leq T_n)(X_u(\phi)X_u(1)^{-1} - X_u(\phi)X_u(1)^{-1})f(u)du|\mathcal{F}_s)$

$= 0.$

Let $n \to \infty$ in the above and use (4,5) to see that $\mathbb{P}_m((\bar{M}_t(\phi) - \bar{M}_s(\phi))F|\mathcal{F}_s) = 0$ and hence $\bar{M}_t(\phi)$ is a $(\mathcal{G}_t)$-martingale.

It follows from (4) and $\bar{M}_{t \wedge t_X}(\phi) = \bar{M}_t(\phi)$ that

(8) $\quad < \bar{M}(\phi) >_t = \int_0^t 1(s < t_X)(\bar{X}_s(\phi^2) - \bar{X}_s(\phi)^2)X_s(1)^{-1}ds \ \mathbb{P}_m - a.s.$

Let $\{\mathbb{P}(A|f) : A \in \hat{\mathcal{F}}, f \in C_+\}$ be a regular conditional probability for $\bar{X}$ given $X.(1) = f(\cdot)$ (under $\mathbb{P}_m$). If $\phi \in D(A)$ and $f \in C_+$, define $\hat{M}_t^f(\phi) = \hat{M}_t^f(\phi)(\hat{X})$ on $(\hat{\Omega}, \hat{\mathcal{F}})$ by the first equation in $(FV_{\bar{m},f})$ for $t < t_f$ and set $\hat{M}_t^f(\phi) = \hat{M}_{t_f-}^f(\phi)$ for $t \geq t_f$. Note that (6) and $\bar{M}_t(\phi) = \bar{M}_{t \wedge t_X}(\phi)$ imply

(9) $$\bar{M}_t(\phi) = \hat{M}_t^{X.(1)}(\phi)(\bar{X}) \ \forall t \geq 0 \ \mathbb{P}_m - a.s. \ \forall \phi \in D(A).$$

If $G \in b\hat{\mathcal{F}}_s^0$ and $s < t$, then the $(\mathcal{G}_t)$-martingale property of $\bar{M}_t(\phi)$ shows that

$$\mathbf{P}_m((\bar{M}_t(\phi) - \bar{M}_s(\phi))G(\bar{X})|X_.(1)) = 0 \quad \mathbf{P}_m - \text{a.s.}$$

and hence, by (9),

$$\mathbf{P}((\hat{M}_t^f(\phi) - \hat{M}_s^f(\phi))G|f) = 0 \quad Q^{m(1)} - \text{a.a.f.}$$

Consider the null set, Λ, of f's off which the above holds for all rational $s < t$ and all G in C_s, a countable set in $b\hat{\mathcal{F}}_s^0$ whose bounded pointwise closure is $b\hat{\mathcal{F}}_s^0$. By working on Λ^c and taking limits in both s and G one easily shows that

(10) $\{\hat{M}_t^f(u) : t \geq 0\}$ is an $(\hat{\mathcal{F}}_t)$ − martingale under $\mathbf{P}(\cdot|f)$ for $Q^{m(1)}$ − a.a.f.

If $t_n^f = \inf\{u : f(u) \leq 1/n\}$, then (8) implies that

(11) $\hat{M}_{t \wedge t_n^f}^f(\phi)^2 - \int_0^t 1(s < t_n^f)(\hat{X}_s(\phi^2) - \hat{X}_s(\phi)^2)f(s)^{-1}ds$ is an $(\hat{\mathcal{F}}_t)$-martingale under $\mathbf{P}(\cdot|f)$ $\forall n \in \mathbb{N}$ for $Q^{m(1)}$−a.a.f.

Now consider a countable core, D_0, for A and fix f outside a $Q^{m(1)}$-null set so that (10) and (11) hold for all $\phi \in D_0$. Take uniform limits in $(\phi, A\phi)$ (recall the definition of $\hat{M}_t^f$) to see that (10) and (11) hold for all ϕ in $D(A)$ for $Q^{m(1)}$- a.a.f. Therefore $\mathbf{P}(\cdot|f)$ solves $(FV_{\bar{m},f})$ for $Q^{m(1)}$-a.a.f. and so $\mathbf{P}(\cdot|f) = \hat{\mathbf{P}}_{\bar{m},f}(\cdot)$ for $Q^{m(1)}$-a.a.f. by Theorem 2. $\blacksquare$

Corollary 4. Let $\{m_n\} \subset M_F(E) - \{0\}$ satisfy $\bar{m}_n \to m$ in $M_1(E)$. Assume $\{A_n\}$ is a sequence of Borel subsets of C_+, $f \in C_+$ and $T \in (0, t_f]$ satisfy:

(12) $Q^{m_n(1)}(A_n) > 0$ $\forall n$

(13) $\sup\{|g(t) - f(t)| : g \in A_n, t \leq S\} \to 0$ as $n \to \infty$ $\forall S < T$.

Then $\mathbf{P}_{m_n}(\bar{X} \in \cdot|X_.(1) \in A_n)|_{T-} \xrightarrow{w} \hat{\mathbf{P}}_{m,f}|_{T-}$ on $(\hat{\Omega}_{T-}, \hat{\mathcal{F}}_{T-})$.

Proof. Let $\phi : \hat{\Omega}_{T-} \to \mathbb{R}$ be bounded and continuous. Then

$$\mathbf{P}_{m_n}(\phi(\bar{X})|X_.(1) \in A_n) - \hat{\mathbf{P}}_{m,f}(\phi)|$$

$$= |\int_{A_n} \hat{\mathbf{P}}_{\bar{m}_n,g}(\phi) - \hat{\mathbf{P}}_{m,f}(\phi)dQ^{m_n(1)}Q^{m_n(1)}(A_n)^{-1}| \quad \text{(Theorem 3)}$$

$$\leq \sup_{g \in A_n} |\hat{\mathbf{P}}_{\bar{m}_n,g}(\phi) - \hat{\mathbf{P}}_{m,f}(\phi)|$$

which approaches 0 as $n \to \infty$ by (13) and Theorem 2(b). ∎

Corollary 5. Let $\{m_n\}$ and m be as in Corollary 4. Let $f \in C_+$, $T \in (0, t_f]$, $T_n \to T$ $(T_n < \infty)$, $\epsilon_n \downarrow 0$ and assume $|m_n(1) - f(0)| < \epsilon_n$. Then

(a) $\mathbf{P}_{m_n}(\bar{X} \in \cdot \mid \sup_{t \leq T_n} |X_t(1) - f(t)| < \epsilon_n)|_{T-} \xrightarrow{w} \hat{P}_{m,f}|_{T-}$ on $(\hat{\Omega}_{T-}, \hat{\mathcal{F}}_{T-})$

(b) $\mathbf{P}_{m_n}(X/f \in \cdot \mid \sup_{t \leq T_n} |X_t(1) - f(t)| < \epsilon_n)|_{T-} \xrightarrow{w} \hat{\mathbf{P}}_{m,f}|_{T-}$ on $(\Omega_{T-}, \mathcal{F}_{T-})$.

Proof.

(a) follows from Corollary 4 with

$$A_n = \{g \in C_+ : \sup_{t \leq T_n} |g(t) - f(t)| < \epsilon_n\}.$$

(b) Note that for $S < T$ and large n, if $\delta_n = \epsilon_n(\inf_{t \leq S} f(t)(f(t) - \epsilon_n))^{-1}$, then

$$\mathbf{P}_{m_n}(\sup_{t \leq S} |X_t(1)^{-1} - f(t)^{-1}| < \delta_n | A_n) = 1$$

(A_n as above). Since $\delta_n \to 0$, (b) is immediate from (a). ∎

Clearly Theorem A is just (b) of the above result with $f(t) \equiv 1$.

Let $\phi(r) = r^2 \log \log 1/r$ $(r < e^{-1})$ and let $\phi - m(A)$ denote the Hausdorff ϕ-measure of $A \subset \mathbf{R}^d$. Let $S(\nu)$ denote the closed support of $\nu \in M_F(\mathbf{R}^d)$. If $\mathbf{P}_m$ denotes the law of super-Brownian motion then for $d \geq 3$ there is a universal constant $K_d \in (0, \infty)$ such that

(14) $X_t(A) = K_d \phi - m(A \cap S(X_t))$ $\forall A \in B(\mathbf{R}^d)$ $\mathbf{P}_m$ − a.s. $\forall t > 0$

(see Dawson-Perkins (1991, Theorem 5.2)). Theorem 3 therefore shows that for all $t > 0$ and $Q^{m(1)}$-a.a.f,

(15) $\hat{X}_t(A) = f(t)^{-1} K_d \phi - m(A \cap S(\hat{X}_t))$ $\forall A \in B(\mathbf{R}^d)$ $\hat{\mathbf{P}}_{m,f}$ − a.s.

It would of course be nice to know if (15) holds for a particular f such as 1. It seems likely that (14) together with the result of Konno-Shiga (1988) and a version of the 0-1 law used to prove (14) will give (14) for the $\Delta/2$-Fleming-Viot superprocess but with another unknown constant K_d'. I have no idea on how one could prove or disprove $K_d = K_d'$.

We now return to the proof of Theorem 2.

If $g : [0, \infty) \to (0, \infty)$ is continuous, let $G(u) = \int_0^u g(s)ds$. For $(t_o, y) \in [0, \infty) \times E \equiv \vec{E}$ define $P^g_{(t_o,y)}$ on the Borel sets of $D([0, \infty), \vec{E})$ by

$$P^g_{(t_o,y)}(A) = P_y(\vec{Y}^g \in A)$$

where $\vec{Y}^g(t) = (t_o + t, Y(G(t_o + t) - G(t_o)))$. Let $P_{(t_o,y)} = P^1_{(t_o,y)}$ and let $\vec{Y}(t)$ denote the canonical process on $D([0, \infty), \vec{E}))$.

Proposition 6. $(\vec{Y}, P^g_{(t_o,y)})$ is an $\vec{E}$-valued Feller process, that is, it is a strong Markov process with a strongly continuous semigroup, $\vec{T}^g_t$, on $C_o(\vec{E})$.

Proof. This is routine. For example, for the strong continuity of $\vec{T}^g_t f$ use the uniform continuity of G on compacts and the fact that $\lim_{t_o \to \infty} \|f(t_o, \cdot)\|_\infty = 0$ for $f \in C_o(\vec{E})$. ∎

Notation. $C_K(A, B)$ denotes the space of continuous functions from A to B with compact support and $C^1_K(A, B)$ denotes those functions in $C_K(A, B)$ which are continuously differentiable.

$$C_o = \{\psi \in C_o(\vec{E}) : \psi(t, x) = \sum_{i=1}^n \psi_i(t)\phi_i(x), \quad \phi_i \in D(A), \quad \psi_i \in C^1_K([0, \infty), \mathbf{R})\}$$

$$C = \{\psi \in C_o(\vec{E}) : \quad t \mapsto \frac{\partial \psi}{\partial t}t, \ A\psi_t \text{ and } \psi_t \text{ are all functions in } C_K([0, \infty), C_o(E))\}$$

Let $\vec{A}^g$ denote the generator of $\vec{Y}^g$ and write $\vec{A}$ for $\vec{A}^1$.

Proposition 7.

(a) C is a core for $\vec{A}^g$ and if $\psi \in C$, then

$$\vec{A}^g \psi(t, x) = \frac{\partial \psi}{\partial t}(t, x) + g(t)A\psi_t(x).$$

(b) C_o is a core for $\vec{A}$.

Proof. The closure of C_o in $C_o(\vec{E})$ contains all functions $\psi(t, x)$ as in the definition of C_o but with $\phi_i \in C_o(E)$. Apply Stone-Weierstrass to this latter class to see that C_o (and hence also $C \supset C_o$) is dense in $C_o(\vec{E})$. Note that if $\psi(t, x) = \sum_{i=1}^n \psi_i(t)\phi_i(x) \in C_o$, then

$$\vec{T}^g_t \psi(t_o, x) = \sum_{i=1}^n \psi_i(t + t_o)T_{G(t+t_o)-G(t_o)}\phi_i(x).$$

It is now easy to check that $\vec{T}_t^g : C_o \to C$. A theorem of S. Watanabe (Ethier-Kurtz (1986, Chapter 1, Prop. 3.3)) will show C is a core for $\vec{A}^g$ providing we prove $C \subset D(\vec{A}^g)$. Moreover since $\vec{T}_t^1 : C_o \to C_o$, the same argument will also prove (b). A direct calculation shows that if $\psi \in C$ then $\psi \in D(\vec{A}^g)$ and $\vec{A}^g \psi$ is given by the formula in (a) above. We omit the details.∎

Proof of Theorem 2. (a) Consider first the uniqueness of $\hat{\mathbf{P}}_{m,f}$. Claim it suffices to prove the result when $t_f = \infty$ and $T_f \equiv \int_0^{t_f} \frac{1}{f(s)} ds = \infty$. To prove this claim let $T_n \uparrow t_f$ $(T_n < t_f)$ and let $f_n(t) = f(t \wedge T_n)$. Note that $t_{f_n} = T_{f_n} = \infty$. Let $\hat{\mathbf{P}}$ be a solution of $(FV_{m,f})$. A solution, $\hat{\mathbf{P}}_n$, of (FV_{m,f_n}) may be constructed by letting $\hat{\mathbf{P}}_n = \hat{\mathbf{P}}$ on $\hat{\mathcal{F}}_{T_n}^0$ and setting the conditional distribution of $\{\hat{X}_{t+T_n}, t \geq 0\}$ given $\hat{\mathcal{F}}_{T_n}^0$ equal to $\hat{\mathbf{P}}_{\hat{X}_{T_n}}$, the unique solution of $(FV_{\hat{X}_{T_n}})$ with $\sigma^2 = f(T_n)^{-1}$. The assumed uniqueness of $\hat{\mathbf{P}}_n$ establishes the uniqueness of $\hat{\mathbf{P}}|_{\hat{\mathcal{F}}_{T_n}^0}$. Since $\hat{X}_t = \hat{X}_{t_f}$ for all $t \geq t_f$ (if $t_f < \infty$), the uniqueness of $\hat{\mathbf{P}}$ on $(\hat{\Omega}, \hat{\mathcal{F}})$ is clear, and the claim is proved.

Assume now that $t_f = T_f = \infty$. Let $\hat{\mathbf{P}}$ be a solution of $(FV_{m,f})$. It is easy to extend $\{\hat{M}_t(\phi) : \phi \in D(A)\}$ to a martingale measure $\{\hat{M}_t(h) : h \in b\mathcal{E}\}$ such that $< \hat{M}(h) >_t = \int_0^t (\hat{X}_s(h^2) - \hat{X}_s(h)^2) f(s)^{-1} ds$ (all under $\hat{\mathbf{P}}$). $\hat{M}_t$ is then a worthy martingale measure (in the sense of Walsh (1986, Ch. 2)) and so we can extend $\hat{M}_t$ to integrands $h(t,x)$ which are bounded measurable functions on $\hat{E} = [0,\infty) \times E$. For such an h we have

(16) $< \hat{M}(h) >_t = \int_0^t (\hat{X}_s(h_s^2) - \hat{X}_s(h_s)^2) f(s)^{-1} ds \quad \hat{\mathbf{P}} - \text{a.s.}$

An elementary stochastic calculus argument now shows that

(17) $\hat{X}(\psi_t) = m(\psi_o) + \int_0^t \hat{X}_s(\vec{A}\psi_s) ds + \hat{M}_t(\psi) \quad \forall t \geq 0 \quad \hat{\mathbf{P}} - \text{a.s.}$

first for $\psi \in C_o$ and then for all $\psi \in D(\vec{A})$ by Proposition 7(b). Let $C_t = \int_0^t f(s)^{-1} ds$. The conditions $t_f = T_f = \infty$ show C is a continuous function from $[0,\infty)$ onto $[0,\infty)$ with a continuous inverse τ. Let $g(u) = f(\tau_u)$, $\tilde{\mathcal{F}}_u = \hat{\mathcal{F}}_{\tau_u}$ and $\tilde{X}(u) = \hat{X}(\tau_u)$. Let $\phi \in C$ and set $\psi(t,x) = \phi(C_t, x)$. It is easy to check that

$\psi \in C$ and hence (17) holds (since $C \subset D(\vec{A})$). Therefore if $\tilde{M}_u(\phi) = \hat{M}_{T_u}(\psi)$, then, using the expression for $\vec{A}\psi$ from Proposition 7, we have

(18)
$$\tilde{X}_t(\phi_t) = \hat{X}_{T_t}(\psi_{T_t}) = m(\phi_o) + \int_0^{T_t} \hat{X}_s(\frac{\partial \psi}{\partial s} + A\psi_s)ds + \hat{M}_{T(t)}(\psi)$$
$$= m(\phi_o) + \int_0^{T_t} \hat{X}_s(\frac{\partial \phi}{\partial s}(C_s, \cdot)f(s)^{-1} + A\phi_{C_s})ds + \tilde{M}_t(\phi)$$
$$= m(\phi_o) + \int_0^t \hat{X}_u(\vec{A}^g \phi_u)du + \tilde{M}_t(\phi),$$

where we have changed variables and used Proposition 7(a). $\tilde{M}_t(\phi)$ is an $(\tilde{\mathcal{F}}_t)$-martingale with

(19)
$$< \tilde{M}(\phi) >_t = \int_0^{T_t} (\hat{X}_s(\psi_s^2) - \hat{X}_s(\psi_s)^2)f(s)^{-1}ds$$
$$= \int_0^t \tilde{X}_u(\phi_u^2) - \tilde{X}_u(\phi_u)^2 du.$$

Since C is a core for $\vec{A}^g$ by Proposition 7, (18) and (19) hold for all $\psi \in D(\vec{A}^g)$. This proves that $\hat{\mathbf{P}}(\delta. \times \tilde{X}. \in \cdot)$ solves $(FV_{\delta_o \times m})$ for $\vec{A}^g$ and hence $V_t^g = \delta_t \times \tilde{X}_t$ is the $\vec{A}^g - FV$ process starting at $\delta_o \times m$ (under $\hat{\mathbf{P}}$). Since

(20)
$$\hat{X}_t(\phi) = V_{C_t}^g(\bar{\phi}), \quad \bar{\phi}(t, x) = \phi(x)$$

we see that $\hat{\mathbf{P}}$ is unique.

Consider now the problem of existence of $\hat{\mathbf{P}}_{m,f}$. If $t_f = T_f = \infty$, then use (20) to define $\hat{\mathbf{P}}_{m,f}$ in terms of an $\vec{A}^g - FV$ process. It is then straightforward to check that $\hat{\mathbf{P}}_{m,f}$ satisfies $(FV_{m,f})$. Assume now that $t_f \wedge T_f < \infty$. Choose $T_n \uparrow t_f, T_n < t_f$. If $f_n(t) = f(t \wedge T_n)$ then clearly $t_{f_n} = T_{f_n} = \infty$ and so by the above there is a unique $\hat{\mathbf{P}}_{m,f_n}$ on $(\hat{\Omega}, \hat{\mathcal{F}})$ solving (FV_{m,f_n}). It is easy to use uniqueness to see that if $k \geq n$ then $\dot{\mathbf{P}}_{m,f_n} = \hat{\mathbf{P}}_{m,f_k}$ on $\hat{\mathcal{F}}_{T_n}^0$. Taking a projective limit of $\{\hat{\mathbf{P}}_{m,f_n} : n \in \mathbb{N}\}$, we may construct a probability $\hat{\mathbf{P}}_{m,f}$ on $(\hat{\Omega}, \hat{\mathcal{F}}_{t_f-})(\hat{\mathcal{F}}_{\infty-} = \hat{\mathcal{F}})$ such that all the conditions of $(FV_{m,f})$ hold except the last ($\hat{X}_t = \hat{X}_{t_f}$ for $t \geq t_f$). We may, and shall, assume that $t_f < \infty$ (or we are done). Then for $\phi \in D(A) \cup \{1\}$

$$\sup_{t<t_f} |\hat{M}_t(\phi)| \leq 2\|\phi\|_\infty + t_f\|A\phi\|_\infty$$

and so by the Martingale Convergence Theorem, $\lim_{t \uparrow t_f} \hat{M}_t(\phi)$ exists $\hat{\mathbb{P}}_{m,f}$–a.s. It now follows trivially from $(FV_{m,f})$ (on $[0, t_f)$) that $\lim_{t \uparrow t_f} \hat{X}_t(\phi)$ exists $\hat{\mathbb{P}}_{m,f}$-a.s. first for all $\phi \in D(A) \cup \{1\}$ and hence for all $\phi \in C(E_\infty)$ therefore $\lim_{t \uparrow t_f} \hat{X}_t = \hat{X}_{t_f-}$ exists in $M_1(E_\infty)$. It follows easily from $(FV_{m,f})$ that for $\phi \in C(E_\infty)$

$$\hat{\mathbb{P}}_{m,f}(\hat{X}_{t_f-}(\phi)) = \lim_{t \uparrow t_f} \hat{\mathbb{P}}_{m,f}(\hat{X}_t(\phi))$$

$$(21) \qquad\qquad = \lim_{t \uparrow t_f} P_{m(1)}(\phi(Y_t))$$

$$= P_{m(1)}(\phi(Y_{t_f-})).$$

This shows $\hat{X}_{t_f-}(\{\infty\}) = 0$ a.s. and hence $\hat{X}_{t_f-} \in M_1(E)$. Now extend $\hat{\mathbb{P}}_{m,f}$ to $\hat{\mathcal{F}}$ by requiring that $\hat{X}_t = \hat{X}_{t_f-}$ $\forall t \geq t_f$ $\hat{\mathbb{P}}_{m,f}$-a.s. Clearly $\hat{\mathbb{P}}_{m,f}$ solves $(FV_{m,f})$ and the proof of (a) is complete.

(b) Fix $S < T$. Since Ω_{T-} is equipped with the compact-open topology it suffices to prove

$$(22) \qquad\qquad \hat{\mathbb{P}}_{m_n,f_n}|_S \to \hat{\mathbb{P}}_{m,f}|_S \text{ on } \hat{\Omega}_S.$$

If $\epsilon = \inf_{t \leq S} f(t) (> 0)$ (recall $S < T \leq t_f$) then $\inf_{t \leq S} f_n(t) \geq \epsilon/2$ for $n \geq N$ and so for $\phi \in D(A)$ and $n \geq N$,

$$| < \hat{M}(\phi) >_t - < \hat{M}(\phi) >_s | \leq \|\phi\|_\infty^2 \int_s^t f_n(u) du \leq \|\phi\|_\infty^2 2\epsilon^{-1}|t-s| \ \forall s, t \leq S \ \hat{\mathbb{P}}_{m_n,f_n}-\text{a.s.}$$

Standard arguments now give the tightness of $\{\hat{\mathbb{P}}_{m_n,f_n}|_S : n \in \mathbb{N}\}$ viewed as probabilities on $\hat{\Omega}_S^\infty = C([0, S], M_1(E_\infty))$ (see for example Thm. 2.3 of Roelly-Coppoletta (1986) but note she is implicitly working with the vague topology and hence we only get tightness on $\hat{\Omega}_S^\infty$, not $\hat{\Omega}_S$). To obtain tightness in $\hat{\Omega}_S$ introduce $h_p(x) = e^{-pd(x,\infty)}$ (d a bounded metric on E_∞), $g_p = \int_0^1 T_t h_p(\cdot) dt \in D(A)$ and note that $A g_p = T_1 g_p - g_p$. Then by (FV_{m_n,f_n}) we have

$$\sup_{t \leq S} \hat{X}_t(g_p) \leq m_n(g_p) + \sup_{t \leq S} |\hat{M}_t(g_p)| + \int_0^S \hat{X}_u(T_1 g_p) du \ \hat{\mathbb{P}}_{m_n,f_n} - \text{a.s.}$$

Now it is easy to use $< \hat{M}(g_p) >_S \leq \int_0^S \hat{X}_u(g_p^2) f_n(u)^{-1} du \ \hat{\mathbb{P}}_{m_n,f_n} - \text{a.s.}$ and the super-process property (see (21)) to conclude that

$$\lim_{p \to \infty} \sup_n \hat{\mathbb{P}}_{m_n,f_n} (\sup_{t \leq S} \hat{X}_t(g_p)) = 0.$$

Since $\lim_{x \to \infty} g_p(x) = 1$, this proves the compact containment property needed in order to conclude $\{\hat{\mathbb{P}}_{m_n, f_n} |_S : n \in \mathbb{N}\}$ are tight in $\hat{\Omega}_S$.

Let $\mathbb{P}$ be a limit point of the above sequence ($\mathbb{P} \in M_1(\hat{\Omega}_S)$). Since everything in sight is uniformly bounded it is clear that the two equations in $(FV_{m,f})$ are satisfied under $\mathbb{P}$ for $t \leq S$. Extend $\mathbb{P}$ to $(\hat{\Omega}, \hat{\mathcal{F}})$ by setting the conditional distribution of $\{\hat{X}_{t+S} : t \geq 0\}$ given $\hat{\mathcal{F}}_S^0$ equal to $\hat{\mathbb{P}}_{\hat{X}_{S}, g}$ where $g(t) = f(S + t)$. Then $\mathbb{P} = \hat{\mathbb{P}}_{m,f}$ and so $\mathbb{P}|_S = \hat{\mathbb{P}}_{m,f}|_S$. (22) follows and the proof is complete.∎

List of References

Dawson, D.A. and Perkins, E.A. (1991). Historical Processes, *Memoirs of the A.M.S.*no. 454.

Etheridge, A. and March, P. (1991). A note on superprocesses, *Probab. Theory Rel. Fields* 89, 141-1481.

Ethier, S.N. and Kurtz, T.G. (1986). *Markov Processes: Characterization and Convergence*, Wiley, New York.

Konno, N. and Shiga, T. (1988). Stochastic differential equations for some measure-valued diffusions, *Probab. Theory Rel. Fields* 79, 201-225.

Roelly-Coppoletta, S. (1986). A criterion of convergence of measure-valued processes; application to measure branching processes, *Stochastics 17*, 43-65.

Tribe, R. (1991). The behaviour of superprocesses near extinction, to appear in Ann. Prob.

Walsh, J.B. (1986). An introduction to stochastic partial differential equations. *Lecture notes in Math. 1180*, Springer-Verlag, Berlin.

Yor, Marc (1978). Remarques sur la representation des martingales comme inte-
grales stochastiques. *Seminaire de Probabilités XII, Lecture notes in Math. 649,*
502-517, Springer-Verlag, Berlin.

Edwin Perkins

Mathematics Department

U.B.C.

Vancouver, B.C.

Canada

V6T 1Z2

p-VARIATION OF THE LOCAL TIMES OF STABLE PROCESSES AND INTERSECTION LOCAL TIME

by

JAY ROSEN[1]

1 Introduction

Let L_t^x denote the local time of the symmetric stable process of order $\beta > 1$ in $\Re^1$. L_t^x is known to be jointly continuous (Boylan [1964]). We will study the p-variation of L_t^x in x, and generalize results concerning Brownian local time of Bouleau and Yor [1981] and Perkins [1982].

Fix $a, b < \infty$ and let $Q(a, b)$ denote the set of partitions $\pi = \{x_0 = a < x_1 \cdots < x_n = b\}$ of $[a, b]$. We use
$$m(\pi) = \sup_i (x_i - x_{i-1})$$
to denote the mesh size of π.

Theorem 1.1 *Let* $\beta = 1 + \frac{1}{k}$, $k = 1, 2, \ldots$ *then*
$$\sum_{x_i \in \pi} (L_t^{x_i} - L_t^{x_{i-1}})^{2k} \longrightarrow \bar{c} \int_a^b (L_t^x)^k \, dx \tag{1.1}$$
in L^2, *uniformly both in* $t \in [0, T]$ *and* $\pi \in Q(a, b)$ *as* $m(\pi) \to 0$.

Here
$$\bar{c} = (2k)!!(4c)^k, \quad c = \int_0^\infty p_t(0) - p_t(1) dt \tag{1.2}$$
and $p_t(x)$ *is the transition density for our stable process.*

[1]Supported in part by NSF DMS 88 022 88, PSC-CUNY Award, and through US-Israel BSF 86-00285

For $k = 1$, i.e., Brownian motion, we recover the result of Bouleau and Yor [1981] and Perkins [1982]:

$$\sum_{x_i \in \pi} (L_t^{x_i} - L_t^{x_{i-1}})^2 \longrightarrow 4 \int_a^b L_t^x dx.$$

This quadratic variation allows one to develop stochastic integrals with respect to the space parameter of Brownian local time, see also Walsh [1983].

We note that the right-hand side of (1.1) is a k-fold intersection local time for the self-intersections of our stable process in $[a, b]$.

The methods of this paper only allow us to compute p-variations when p is of the form $p = 2k$, which limits results of the form (1.1) to $\beta = 1 + \frac{1}{k}$. In Marcus and Rosen [1990], we obtain analogues of (1.1) for arbitrary $\beta > 1$, in the sense of a.s. convergence. The convergence, however, is not uniform in $Q(a, b)$. If we want to obtain results for arbitrary $\beta > 1$ by the methods of this paper, we will have to be satisfied with the following:

Theorem 1.2 *Let $\beta > 1$, then*

$$\sum_{x_i \in \pi} \left(\frac{L_t^{x_i} - L_t^{x_{i-1}}}{(x_i - x_{i-1})^\gamma} \right)^{2k} \longrightarrow \bar{c} \int_a^b (L_t^x)^k dx \qquad (1.3)$$

in L^2, uniformly in both $t \in [0, T]$ and $\pi \in Q(a, b)$ as $m(\pi) \to 0$, where

$$\gamma = \frac{\beta - 1}{2} - \frac{1}{2k}$$

and $\bar{c}$ is given by (1.2).

The methods of this paper were a natural outgrowth of our second order limit laws for the local times of stable processes, Rosen [1990].

It is a pleasure to thank M. Yor for drawing my attention to the problem of p-variation of stable local times.

2   Proofs

Proof of Theorem 1:   We write, for $\tau \in Q(a, b)$

$$E \left(\left\{ \bar{c} \int_a^b (L_t^x)^k dx - \sum_{x_i \in \tau} (L_t^{x_i} - L_t^{x_{i-1}})^{2k} \right\}^2 \right)$$

$$
\begin{aligned}
&= \bar{c}^2 \int_a^b \int_a^b E\left\{(L_t^x)^k (L_t^y)^k\right\} dx\,dy \\
&- 2\bar{c} \int_a^b \sum_i E\left\{(L_t^{x_i} - L_t^{x_{i-1}})^{2k} (L_t^y)^k\right\} dy \\
&+ \sum_{i,j} E\left\{(L_t^{x_i} - L_t^{x_{i-1}})^{2k} \left(L_t^{x_j} - L_t^{x_{j-1}}\right)^{2k}\right\} \\
&\doteq A - 2B_\epsilon + C_\epsilon, \quad \text{where } \epsilon \doteq m(\tau)
\end{aligned}
\tag{2.1}
$$

We will show that as $\epsilon \to 0$, each of $A, B_\epsilon, C_\epsilon$ converges to

$$
[(2k)!(2c)^k]^2 \sum_{\tilde{\pi}} \int_a^b dx \int_a^b dy \int\cdots\int_{0 \le t_1 \le \cdots \le t_{2k} \le t} \prod_{i=1}^{2k} p_{\Delta t_i}(\tilde{\pi}_i, \tilde{\pi}_{i-1}) dt_i
\tag{2.2}
$$

where the sum runs over all paths $\tilde{\pi} : \{1,\ldots,2k\} \to \{x,y\}$ which visit x, y an equal number of times (i.e. k times each).

The fact that A equals (2.2) is straightforward, so we turn to B_ϵ. We have

$$
\begin{aligned}
&E\left\{(L_t^{x_i} - L_t^{x_{i-1}})^{2k} (L_t^y)^k\right\} \\
&= E\left\{\prod_{\ell=1}^{2k} \int_0^t dL_{s_\ell}^{x_i} - dL_{s_\ell}^{x_{i-1}} \prod_{j=1}^k \int_0^t dL_{s_j}^y\right\} \\
&= (2k)! k! \sum_\pi E\left(\int\cdots\int_{0 \le t_1 \le \cdots \le t_{3k} \le t} \prod_{\ell=1}^{3k} d\mathcal{L}_t^{\pi_\ell}\right)
\end{aligned}
\tag{2.3}
$$

where the sum runs over all paths $\pi : \{1,\ldots,3k\} \longrightarrow \{x_i, y\}$ which visit y exactly k times, and

$$
\begin{aligned}
\mathcal{L}_t^{x_i} &\doteq L_t^{x_i} - L_t^{x_{i-1}} \\
\mathcal{L}_t^y &\doteq L_t^y
\end{aligned}
\tag{2.4}
$$

We will say that a path π is even if its visits to x_i occur in even runs. A path will be called odd if it is not even.

Assume that π is even. Then we can evaluate its contribution to (2.3) by successive application of the Markov property. We use the following observations, where [ ] will be used generically to denote an expression depending only on the path up to the earliest times which are exhibited.

$$
\begin{aligned}
&E\left([\quad] \int_{s_{j-2}}^t \left(dL_{s_{j-1}}^{x_i} - dL_{s_{j-1}}^{x_{i-1}}\right) \int_{s_{j-1}}^t dL_{s_j}^{x_i} - dL_{s_j}^{x_{i-1}}\right) \\
&= E\left([\quad] \int_{s_{j-2}}^t \left(dL_{s_{j-1}}^{x_i} + dL_{s_{j-1}}^{x_{i-1}}\right) \int_{s_{j-1}}^t p_{\Delta s_j}(0) - p_{\Delta s_j}(\Delta x_i) ds_j\right)
\end{aligned}
\tag{2.5}
$$

$$E\left(\begin{bmatrix} & 1 \end{bmatrix}\int_{s_{j-2}}^{t}\left(dL_{s_{j-1}}^{x_i} - dL_{s_{j-1}}^{x_{i-1}}\right)\int_{s_{j-1}}^{t} dL_{s_j}^{x_i} + dL_{s_j}^{x_{i-1}}\right)$$

$$= E\left(\begin{bmatrix} & 1 \end{bmatrix}\int_{s_{j-2}}^{t}\left(dL_{s_{j-1}}^{x_i} - dL_{s_{j-1}}^{x_{i-1}}\right)\int_{s_{j-1}}^{t} p_{\Delta s_j}(0) + p_{\Delta s_j}(\Delta x_i) ds_j\right) \quad (2.6)$$

$$E\left(\begin{bmatrix} & 1 \end{bmatrix}\int_{s_{j-2}}^{t} dL_{s_{j-1}}^{y}\int_{s_{j-1}}^{t} dL_{s_j}^{x_i} + dL_{s_j}^{x_{i-1}}\right)$$

$$= E\left(\begin{bmatrix} & 1 \end{bmatrix}\int_{s_{j-2}}^{t} dL_{s_{j-1}}^{y}\int_{s_{j-1}}^{t} p_{\Delta s_j}(y - x_i) + p_{\Delta s_j}(y - x_{i-1}) ds_j\right) \quad (2.7)$$

$$E\left(\begin{bmatrix} & 1 \end{bmatrix}\int_{s_{j-2}}^{t} dL_{s_{j-1}}^{y}\int_{s_{j-1}}^{t} dL_{s_j}^{y}\right)$$

$$= E\left(\begin{bmatrix} & 1 \end{bmatrix}\int_{s_{j-2}}^{t} dL_{s_{j-1}}^{y}\int_{s_{j-1}}^{t} p_{\Delta s_j}(0) ds_j\right) \quad (2.8)$$

$$E\left(\begin{bmatrix} & 1 \end{bmatrix}\int_{s_{j-2}}^{t}\left(dL_{s_{j-1}}^{x_i} - dL_{s_{j-1}}^{x_{i-1}}\right)\int_{s_{j-1}}^{t} dL_{s_j}^{y}\right)$$

$$= E\left(\begin{bmatrix} & 1 \end{bmatrix}\int_{s_{j-2}}^{t}\left(dL_{s_{j-1}}^{x_i} - dL_{s_{j-1}}^{x_{i-1}}\right)\int_{s_{j-1}}^{t} p_{\Delta s_j}(x_i - y) ds_j\right)$$

$$+ E\left(\begin{bmatrix} & 1 \end{bmatrix}\int_{s_{j-2}}^{t} dL_{s_{j-1}}^{x_{i-1}}\int_{s_{j-1}}^{t} p_{\Delta s_j}(x_i - y) - p_{\Delta s_j}(x_{i-1} - y) ds_j\right) \quad (2.9)$$

$$E\left(\begin{bmatrix} & 1 \end{bmatrix}\int_{s_{j-2}}^{t} dL_{s_{j-1}}^{y}\int_{s_{j-1}}^{t} dL_{s_j}^{x_i} - dL_{s_j}^{x_{i-1}}\right)$$

$$= E\left(\begin{bmatrix} & 1 \end{bmatrix}\int_{s_{j-2}}^{t} dL_{s_{j-1}}^{y}\int_{s_{j-1}}^{t} p_{\Delta s_j}(x_i - y) - p_{\Delta s_j}(x_{i-1} - y) ds_j\right) \quad (2.10)$$

As we see, (2.5), (2.9) and (2.10) give rise to 'difference factors', i.e., factors of the form

$$\int p_s(y) - p_s(y - \Delta x_i) ds \quad (2.11)$$

We will see below in Lemma 1 that such factors give a contribution

$$O(\Delta x_i)^{\beta-1} = O(\Delta x_i)^{\frac{1}{k}},$$

hence whenever we have $> k$ difference factors, the contribution to $\lim_{\epsilon \to 0} B_\epsilon$ will be zero. We can see by using the above formulae recursively that all terms arising from the evaluation of the expectation associated to an even path π have $> k$ difference factors, except for a contribution which can be written as

$$2^k \int\cdots\int_{\sum_{\ell=1}^{2k}\Delta s_\ell + \sum_{j=1}^{k}\Delta t_j \leq t} \prod_{\ell=1}^{2k} p_{\Delta s_\ell}(\tilde{\pi}_\ell, \tilde{\pi}_{\ell-1}) \prod_{j=1}^{k} p_{\Delta t_j}(0) - p_{\Delta t_j}(\Delta x_i) \quad (2.12)$$

where π induces the path $\tilde{\pi} : \{1, \ldots, 2k\} \longrightarrow \{x_i, y\}$, (visiting both x_i and y k times) as follows: since visits of π to x_i occur in pairs, we simply suppress one visit from each pair. Note that in getting (2.12), we e.g. rewrote the factor

$$\int p_{\Delta_s}(y - x_i) + p_{\Delta_s}(y - x_{i-1})ds$$

of (2.7) as

$$2 \int p_{\Delta_s}(y - x_i)ds$$

+ a 'difference factor', and similarly for (2.6) and analogous factors.

We will show below, in Lemma 3, that as $\epsilon \to 0$, the integral in (2.10) summed over i converges to c^k times the integral in (2.2). Furthermore, any given $\tilde{\pi}$ will be induced from precisely one even π which will show that the contribution of even paths to $B_\epsilon \longrightarrow$ (2.2).

To see that odd paths π give zero contribution in the limit, we use (2.5)–(2.10) recursively to see that every term in the expansion of an odd path π has $> k$ 'difference factors'.

We now turn to C_ϵ:

$$E\left\{ (L_t^{x_i} - L_t^{x_{i-1}})^{2k} \left(L_t^{x_j} - L_t^{x_{j-1}} \right)^{2k} \right\}$$

$$= (2k!)^2 \sum_\pi E \left(\int \cdots \int_{0 \le t_1 \le \cdots \le t_{4k} \le t} \prod_{\ell=1}^{4k} dL_{t_\ell}^{x_{\pi_\ell}} - dL_{t_\ell}^{x_{\pi_{\ell-1}}} \right) \tag{2.13}$$

where the sum runs over all paths $\pi : \{1, \ldots, 4k\} \longrightarrow \{i, j\}$ which visit i, j an equal number of times, i.e. $2k$ times each.

We will evaluate

$$E \left(\int \cdots \int_{0 \le t_1 \le \cdots \le t_{4k} \le t} \prod_{\ell=1}^{4k} dL_{t_\ell}^{x_{\pi_\ell}} - dL_{t_\ell}^{x_{\pi_{\ell-1}}} \right) \tag{2.14}$$

by using (2.5)–(2.10) together with

$$E \left([\quad] \int_{s_{\ell-2}}^{t} \left(dL_{s_{\ell-1}}^{x_i} - dL_{s_{\ell-1}}^{x_{i-1}} \right) \int_{s_{\ell-1}}^{t} dL_{s_\ell}^{x_j} + dL_{s_\ell}^{x_{j-1}} \right)$$

$$= E \left([\quad] \int_{s_{\ell-2}}^{t} \cdot \left(dL_{s_{\ell-1}}^{x_i} - dL_{s_{\ell-1}}^{x_{i-1}} \right) \int_{s_{\ell-1}}^{t} p_{\Delta s_\ell}(x_i - x_j) + p_{\Delta s_\ell}(x_i - x_{j-1})ds_\ell \right)$$

$$+ E \left([\quad] \int_{s_{\ell-2}}^{t} dL_{s_{\ell-1}}^{x_{i-1}} \int_{s_{\ell-1}}^{t} \{ p_{\Delta s_\ell}(x_i - x_j) - p_{\Delta s_\ell}(x_{i-1} - x_j) \} \right.$$

$$\left. + \{ p_{\Delta s_\ell}(x_i - x_{j-1}) - p_{\Delta s_\ell}(x_{i-1} - x_{j-1}) \} ds_\ell \right) \tag{2.15}$$

and

$$
\begin{aligned}
E & \left([\quad] \int_{s_{\ell-2}}^{t} \left(dL_{s_{\ell-1}}^{x_i} - dL_{s_{\ell-1}}^{x_{i-1}} \right) \int_{s_{\ell-1}}^{t} dL_{s_\ell}^{x_j} - dL_{s_\ell}^{x_{j-1}} \right) \\
= & \; E \left([\quad] \int_{s_{\ell-2}}^{t} \left(dL_{s_{\ell-1}}^{x_i} - dL_{s_{\ell-1}}^{x_{i-1}} \right) \int_{s_{\ell-1}}^{t} p_{\Delta s_\ell}(x_j - x_i) - p_{\Delta s_\ell}(x_{j-1} - x_i) ds_\ell \right) \\
+ & \; E \left([\quad] \int_{s_{\ell-2}}^{t} dL_{s_{\ell-1}}^{x_{i-1}} \int_{s_{\ell-1}}^{t} \{ p_{\Delta s_\ell}(x_j - x_i) - p_{\Delta s_\ell}(x_{j-1} - x_i) \right. \\
& \left. \quad - p_{\Delta s_\ell}(x_j - x_{i-1}) + p_{\Delta s_\ell}(x_{j-1} - x_{i-1}) \} ds_\ell \right)
\end{aligned}
\tag{2.16}
$$

We now call a path π even if both its visits to i and to j occur in even runs. Such a path uniquely induces a path $\tilde{\pi} : \{1, \ldots, 2k\} \longrightarrow \{i, j\}$ by

$$
\tilde{\pi}(\ell) := \pi(2\ell - 1) = \pi(2\ell)
$$

We refer to a 'difference factor' of the form (2.11) as an 'x_i- difference factor', and note that the terms generated by (2.14) will give zero contribution to (2.13), in the limit, if such a term has k_1 'x_i-difference factors' and k_2 'x_j-difference factors'—and

$$
k \leq k_1 \wedge k_2, \quad k_1 \vee k_2 > k.
$$

We can see using the above formulae recursively that if π is even, the only term giving a non-zero limit will be

$$
\begin{aligned}
2^{2k}(2k!)^2 & \int \cdots \int_{\sum_{\ell=1}^{2k} \Delta t_\ell + \sum_{m=1}^{k} s_m + \sum_{n=1}^{k} \Delta r_n \leq t} \prod_{\ell=1}^{2k} p_{\Delta t_\ell}(x_{\tilde{\pi}_i}, x_{\tilde{\pi}_{i-1}}) \\
& \prod_{m=1}^{k} (p_{\Delta s_m}(0) - p_{\Delta s_m}(\Delta x_i)) \prod_{n=1}^{k} p_{\Delta r_n}(0) - p_{\Delta r_n}(\Delta x_j)
\end{aligned}
\tag{2.17}
$$

and we show below, in Lemma 3, that this summed over i, j converges to the integral in (2.2).

Finally, we turn to odd paths π and show that they contribute 0 in the limit. The only new wrinkle comes from the second term in (2.16), which a-priori generates only one 'difference factor' for the two local time integrals. However, if we fix $\delta > 0$, and if

$$
|u| \doteq |x_{i-1} - x_{j-1}| \geq \delta, \quad m(\tau) < \frac{\delta}{4},
$$

then we will show below in Lemma 2 that

$$\left| \int_0^r p_s(u - \Delta x_i + \Delta x_j) - p_s(u - \Delta x_i) \right.$$
$$\left. - p_s(u + \Delta x_j) + p_s(u) \, ds \right|$$
$$\leq \frac{c}{\delta^2} \Delta x_i \, \Delta x_j, \qquad (2.18)$$

while if $|u| \leq \delta$, we can bound (2.18) by breaking it up into pairs—either with Δx_i or Δx_j as the difference, to get via Lemma 1 a bound

$$c(\Delta x_i)^{\beta-1} \wedge (\Delta x_j)^{\beta-1}. \qquad (2.19)$$

The contribution of (2.18) and (2.19) will then be bounded by

$$c \sum_{\substack{i,j \\ |u| \leq \delta}} \Delta x_i \, \Delta x_j + \frac{c}{\delta^2} \epsilon^\alpha$$

for some $\alpha > 0$, $\epsilon = m(\tau) < \frac{\delta}{4}$, and we now take first $\epsilon \to 0$ then $\delta \to 0$ to see that such terms don't contribute in the limit. This completes the proof of Theorem 1, and that of Theorem 2 is basically the same.

3 Lemmas

Lemma 1

$$\int_0^T |p_t(x) - p_t(y)| \, dt \leq c \left| |x|^{\beta-1} - |y|^{\beta-1} \right| \leq c|x - y|^{\beta-1} \qquad (3.1)$$

and

$$\int_0^T p_t(0) - p_t(x) dt = c|x|^{\beta-1} + O\left(\frac{|x|^2}{T^{3/\beta-1}}\right) \qquad (3.2)$$

where

$$c = \int_0^\infty (p_t(0) - p_t(1)) dt < \infty \qquad (3.3)$$

Proof: $p_t(x)$ is monotone in $|x|$, hence if $|x| \leq |y|$,

$$\int_0^T |p_t(x) - p_t(y)| \, dt = \int_0^T p_t(x) - p_t(y) \, dt$$
$$= \int_0^T (p_t(0) - p_t(y)) - (p_t(0) - p_t(x)) \, dt$$
$$\leq \int_0^\infty (p_t(0) - p_t(y)) - (p_t(0) - p_t(x)) \, dt$$
$$= \int_0^\infty (p_t(0) - p_t(y)) \, dt - \int_0^\infty (p_t(0) - p_t(x)) \, dt \qquad (3.4)$$

since $p_t(0) - p_t(x) \geq 0$ and we will now show it is integrable in t.

For this we use the scaling:

$$p_t(x) = \frac{1}{t^{1/\beta}} \, p_1 \left(\frac{x}{t^{1/\beta}} \right) \tag{3.5}$$

so that

$$
\begin{aligned}
& \int_0^\infty p_t(0) - p_t(x) \, dt \\
&= \int_0^\infty \left(p_1(0) - p_1 \left(\frac{x}{t^{1/\beta}} \right) \right) \frac{dt}{t^{1/\beta}} \\
&= |x|^{\beta-1} \int_0^\infty \left(p_1(0) - p_1 \left(\frac{1}{t^{1/\beta}} \right) \right) \frac{dt}{t^{1/\beta}}
\end{aligned} \tag{3.6}
$$

and the last integral is finite since, for t small we have $|p_1(y)| \leq p_1(0)$ and $\beta > 1$, while for large t, we have from symmetry that

$$\left| p_1(0) - p_1 \left(\frac{1}{t^{1/\beta}} \right) \right| \leq \frac{\tilde{c}}{t^{2/\beta}}. \tag{3.7}$$

It is now easy to see that (3.3) is the integral on the r.h.s. of (3.6). This proves (3.1).

For (3.2) we write

$$
\begin{aligned}
& \int_0^T p_t(0) - p_t(x) \, dt \\
&= \int_0^\infty p_t(0) - p_t(x) dt - \int_t^\infty p_t(0) - p_t(x) dt,
\end{aligned} \tag{3.8}
$$

and use (3.6), together with the bound from (3.5), (3.7)

$$
\begin{aligned}
\int_T^\infty p_t(0) - p_t(x) dt
&\leq \tilde{c}|x|^{\beta-1} \int_{T/x^\beta}^\infty \frac{1}{t^{3/\beta}} \, dt \\
&= \tilde{c}|x|^{\beta-1} \left(\frac{T}{|x|^\beta} \right)^{1-3/\beta} \\
&= \tilde{c} \frac{|x|^2}{T^{3/\beta-1}}
\end{aligned} \tag{3.9}
$$

Lemma 2

$$\int_0^T |\, p_t(x+a+b) - p_t(x+a) - p_t(x+b) + p_t(x) \,| \, dt$$

$$\leq c \frac{|a| \, |b|}{|x|^2}, \tag{3.10}$$

for $|x| \geq 4(a \vee b)$.

Proof: We integrate by parts:

$$\frac{d^2}{dx^2} p_t(x) = \frac{d^2}{dx^2} \int e^{ipx} e^{-tp^\beta} dp$$

$$= -\int e^{ipx} p^2 e^{-tp^\beta} dp$$

$$= \frac{1}{ix} \int e^{ipx} \frac{d}{dp} \left(p^2 e^{-tp^\beta} \right) dp$$

$$= \frac{-1}{x^2} \int e^{ipx} \frac{d^2}{dp^2} \left(p^2 e^{-tp^\beta} \right) dp \qquad (3.11)$$

so that

$$\left| \frac{d^2}{dx^2} p_t(x) \right| \leq \frac{\tilde{c}}{|x|^2} \int \left| \frac{d^2}{dp^2} \left(p^2 e^{-tp^\beta} \right) \right| dp$$

$$\leq \frac{\tilde{c}}{|x|^2} \int e^{-tp^\beta} + tp^\beta e^{-tp^\beta} + \left(tp^\beta \right)^2 e^{-tp^\beta} dp$$

$$\leq \frac{\tilde{c}}{|x|^2} \frac{1}{t^{1/\beta}}$$

Now, the mean value theorem, our assumption that $|x| \geq 4(a \vee b)$, and the integrability of $\frac{1}{t^{1/\beta}}$ on $[0, T]$ finishes the proof of Lemma 2.

Lemma 3 *Let $f \in L^1([0,T]^j)$ and set*

$$F(s) \doteq \int \cdots \int_{\sum_{i=1}^j t_i \leq s} f(t) \, dt_1 \ldots dt_j$$

then

$$\int \cdots \int_{\sum_{i=1}^j r_i + \sum_{\ell=1}^k s_\ell \leq t} f(r_1, \ldots, r_j) \prod_{\ell=1}^k p_{s_\ell}(0) - p_{s_\ell}(x_\ell) \, dr \, ds$$

$$\leq c^k \, |x_1 \ldots x_k|^{\beta-1} \, F(t) \qquad (3.12)$$

and for any $\delta > 0$, we have

$$\int \cdots \int_{\sum_{i=1}^j r_i + \sum_{\ell=1}^k s_\ell \leq t} f(r_1, \ldots, r_j) \prod_{\ell=1}^k p_{s_\ell}(0) - p_{s_\ell}(x_\ell) \, dr \, ds$$

$$= c^k \, |x_1 \ldots x_k|^{\beta-1} \left(F(t) + o(1_\delta) + O\left(\sup_\ell \frac{|x_\ell|^{3-\beta}}{\delta^{3/\beta-1}} \right) \right) \qquad (3.13)$$

where $o(1_\delta)$ means a term which goes to zero when $\delta \to 0$.

Proof: (3.12) is immediate from Lemma 1. To see (3.13), fix $\delta > 0$, and define

$$C = \{(r,s) \mid \sum_{i=1}^{j} r_i + \sum_{\ell}^{k} s_\ell \le t\}$$
$$D_\delta = \{(r,s) \mid s_\ell \le \delta, \quad \text{for all } \ell\}$$

Note that

$$C \cap (D_\delta^c) \subseteq \left\{ (r,s) \;\middle|\; \sum_{i=1}^{j} r_i \le t, \text{ and } s_\ell > \delta \text{ for some } \ell \right\}$$

so that

$$\int \cdots \int_{C \cap (D_\delta^c)} f(r) \prod_{\ell=1}^{k} p_{s_\ell}(0) - p_{s_\ell}(x_\ell) \, dr ds$$

$$\le \tilde{c} F(t) \, |x_1 \ldots x_k|^{\beta-1} \sup_{\ell} \frac{|x_\ell|^{3-\beta}}{\delta^{3/\beta-1}} \tag{3.14}$$

from Lemma 1, and (3.9).

Now set

$$H_\delta = \{(r,s) \mid \sum_{i=1}^{j} r_i \le t - k\delta\}$$

and note that, by Lemma 1,

$$\int \cdots \int_{C \cap D_\delta \cap (H_\delta^c)} f(r) \prod_{\ell=1}^{k} p_{s_\ell}(0) - p_{s_\ell}(x_\ell) \, dr ds$$

$$\le \tilde{c} \, |x_1 \ldots x_k|^{\beta-1} \left(F(t) - F(t - k\delta) \right) \tag{3.15}$$

Finally, note that

$$C \cap D_\delta \cap H_\delta = \left\{ (r,s) \;\middle|\; \sum_{i=1}^{j} r_i \le t - k\delta, \text{ and } s_\ell \le \delta, \quad \text{for all } \ell \right\}$$

so that, from Lemma 1,

$$\int \cdots \int_{C \cap D_\delta \cap H_\delta} f(r) \prod_{\ell=1}^{k} p_{s_\ell}(0) - p_{s_\ell}(x_\ell) \, dr ds$$

$$= F(t - k\delta) \prod_{\ell=1}^{k} \int_0^\delta p_s(0) - p_s(x_\ell) \, ds$$

$$= F(t - k\delta) \left(c^k \, |x_1 \ldots x_k|^{\beta-1} + \tilde{c} |x_1, \ldots x_k|^{\beta-1} \sup_{\ell} \frac{|x_\ell|^{3-\beta}}{\delta^{3/\beta-1}} \right) \tag{3.16}$$

(3.14), (3.15) and (3.16) now complete the proof of (3.13).

References

[1] Bouleau, N. and Yor, M. [1981], "Sur La Variation Quadratique de Temps Locaux de Certaines Semi-Martingales", *C. R. Acad. Sci. Paris, Série I* **292**, pp. 491–492.

[2] Marcus, M. and Rosen, J. [1990], "*p*-Variation of the Local Times of Symmetric Markov Processes Via Gaussian Processes". In preparation.

[3] Perkins, E. [1982], "Local Time is a Semi-Martingale", *Z. Warsch. verw. Geb.* **60**, pp. 79–117.

[4] Rosen, J. [1990], "Second Order Limit Laws for the Local Times of Stable Processes", preprint.

[5] Walsh, J. [1983], "Stochastic Integration with Respect to Local Time", *Seminar on Stochastic Processes 1982*, pp. 237–302, Birkhäuser, Boston.

Jay Rosen
Faculty of Industrial Engineering and Management
Technion—Israel Institute of Technology
Haifa 32000
ISRAEL

Permanent Address:

Department of Mathematics
College of Staten Island, CUNY
Staten Island, New York 10301
U.S.A.

CLOSING VALUES OF MARTINGALES
WITH FINITE LIFETIMES

MICHAEL J. SHARPE[1]

1. Introduction

Throughout this paper, ζ denotes a fixed stopping time that will be considered
to be the time of occurrence of some physical event which signals the end of the
lifetime of a process. We imagine that processes with lifetime ζ are observable
only strictly prior to ζ, so that all meaningful constructions associated with such
processes should be based only on their values strictly prior to ζ. Of particular
interest is the case in which ζ is the lifetime of a right process X. In this
situation, X is supposed to be sent to a dead state after its lifetime, and no
further information is generated by it after that time.

The first general study of martingales with finite lifetime was undertaken by
Maisonneuve [Ma77], who was interested principally in continuous local martingales with finite lifetime ζ, which was allowed to be an arbitrary stopping time.
His definition of local martingale in that case was broader than that used by
later workers such as Yan [Ya82] and Zheng [Zh82]. Zheng studied limits of
martingales with finite *predictable* lifetimes. Yan's treatment of the foundations
of the theory of local martingales and semimartingales with finite lifetime and
the associated stochastic integral were aimed principally at at even more general
theory of local martingales on an optional set. His ideas relied partly on the work
of Jacod [Ja79, Ch. V], especially the latter's notion of local martingales up to

[1]Research supported in part by NSF Grant DMS 8721347

a random time. Yan's definition of local martingale on $[\![0, \zeta[\![$ seems to be more appropriate than Maisonneuve's in the case of discontinuous processes. See also [Sh88, §A6 and §53] for some further developments along these lines.

The aim of this paper is to examine the notion of martingale on $[\![0, \zeta[\![$ from the viewpoint of processes observable only at times strictly prior to ζ, so that the question becomes one of characterizing processes on $[\![0, \zeta[\![$ which have a martingale extension, and to give a procedure for computing that extension.

2. Generalities concerning the lifetime interval

We follow the terminology and conventions laid down in [DM75] and [DM80]. In the formulas written below, it is important to keep in mind the convention $\inf \emptyset := \infty$, and to recall [DM75, IV.60], that stochastic intervals of the form $[\![U, V[\![, \ ]\!]U, V]\!], \ [\![\zeta]\!]$ are subsets of $[0, \infty[\times \Omega$, and not of $[0, \infty] \times \Omega$. Let $\mathcal{P}$ denote the predictable σ-field on $[0, \infty[\times \Omega$.

For the rest of the paper, we shall avoid some trivialities by making the blanket assumption that a.s., $\zeta > 0$. (For the general case, replace $\mathbf{P}$ by the conditional probability $\mathbf{P}\{\cdot | \zeta > 0\}$ where necessary.)

Let $\rho := {}^p(1_{[\![0,\zeta[\![})$ and $\lambda := {}^p(1_{[\![\zeta]\!]})$, the predictable projections of the lifetime interval and the graph of the lifetime respectively. It was shown in [Sh88,§A6] that

$$(2.1) \qquad\qquad [\![0, \zeta[\![\ \subset \{\rho > 0\} \subset [\![0, \zeta]\!].$$

Because $[\![0, \zeta]\!]$ is predictable, as is $\{\rho > 0\}$, the random set $[\![0, \zeta]\!] \setminus \{\rho > 0\}$ is predictable and, by (2.1), its sections contain at most one point. It follows then that $[\![0, \zeta]\!] \setminus \{\rho > 0\}$ is the graph of a predictable stopping time that we denote by ζ_p. Note too that the trivial identity $1_{[\![0,\zeta[\![} + 1_{[\![\zeta]\!]} = 1_{[\![0,\zeta]\!]}$ implies, after taking predictable projections, that $\rho + \lambda = 1_{[\![0,\zeta]\!]}$. As ρ and λ are positive, up to evanescence, it follows that, up to evanescence, $0 \le \rho \le 1_{[\![0,\zeta]\!]}$ and $0 \le \lambda \le 1_{[\![0,\zeta]\!]}$. In particular, by (2.1), $\{\lambda = 1\} \subset [\![\zeta]\!]$, and so

$$(2.2) \qquad\qquad [\![\zeta_p]\!] = \{\lambda = 1\} \subset [\![\zeta]\!].$$

Meyer's well known classification of stopping times into totally inaccessible and accessible parts is closely related to the properties of ρ and λ cited above. Indeed, the totally inaccessible part of ζ is the stopping time ζ_i defined by $[\![\zeta_i]\!] = [\![\zeta]\!] \cap \{\lambda = 0\}$ and the accessible part ζ_a of ζ is the stopping time defined by $[\![\zeta_a]\!] = [\![\zeta]\!] \cap \{\lambda > 0\}$. For, given any predictable time T, $[\![T]\!] \subset [\![\zeta_i]\!]$ implies, by definition of predictable projection, that $\mathbf{P}\, 1_{[\![\zeta_i]\!]}(T) = \mathbf{P}\, 1_{[\![\zeta]\!]} 1_{\{\lambda=0\}}(T) = \mathbf{P}\, \lambda(T) 1_{\{\lambda=0\}}(T) = 0$. That is, T is totally inaccessible. On the other hand, since the predictable projection of an optional process differs from it only on a set with countable sections, the set $\{\lambda > 0\}$ a.s. has countable sections, and thus may be expressed as a countable union of graphs of predictable times T_n. That is, $[\![\zeta_a]\!] \subset \cup_n [\![T_n]\!]$, so ζ_a is an accessible time. The random set $\{\lambda > 0\}$ is evidently the smallest (up to evanescence) predictable set containing the graph of the accessible part of ζ.

The following elementary result is a small complement to the Meyer classification.

(2.3) Lemma. ζ_p is the maximal predictable time with graph contained in $[\![\zeta]\!]$. More precisely, given any predictable time T with $[\![T]\!] \subset [\![\zeta]\!]$, one has $[\![T]\!] \subset [\![\zeta_p]\!]$, up to evanescence.

Proof. Given such a T, by definition of predictable projection,

$$\mathbf{P}\, \rho_T 1_{\{T<\infty\}} = \mathbf{P}\, 1_{[\![0,\zeta[\![} 1_{\{T<\infty\}} = \mathbf{P}\, \{T < \zeta\} = 0.$$

Therefore, $[\![T]\!] \subset \{\lambda = 1\}$, and consequently $[\![T]\!] \subset [\![\zeta]\!] \cap \{\lambda = 1\} = [\![\zeta_p]\!]$.

We call ζ_p the *predictable part* of ζ. It is an obvious consequence of the lemma and the preceding discussion that if we define a stopping time ζ_s by $[\![\zeta_s]\!] = [\![\zeta]\!] \cap \{0 < \rho < 1\} = [\![\zeta]\!] \cap \{0 < \lambda < 1\}$, then ζ_s is an accessible time containing the graph of no predictable time. Note the following obvious consequence of the definition of ζ_p and ρ.

$$(2.4) \qquad\qquad \{\rho > 0\} = [\![0,\zeta]\!] \setminus [\![\zeta_p]\!].$$

We may characterize ζ_p in terms of approximation from strictly below in the following manner.

(2.5) Lemma. *Let ζ be a stopping time with predictable part ζ_p. Then there is an increasing sequence (T_n) of stopping times with (a) $T_n \leq \zeta$; (b) $\lim_n T_n = \zeta$ a.s.; (c) $T_n < \zeta$ a.s. on $\{\zeta = \zeta_p\}$. Conversely, given an increasing sequence (T_n) of stopping times satisfying (a) and (b), the event $\{T_n < \zeta$ for all $n\}$ is contained in $\{\zeta = \zeta_p\}$.*

Proof. For the first assertion, choose an announcing sequence (T_n') for ζ_p, and let $T_n := T_n' \wedge \zeta$, so that the sequence (T_n) clearly satisfies (a), (b) and (c). For the converse, let $\Gamma := \{T_n < \zeta$ for all $n\}$ and note that the stochastic intervals $]\!]T_n, \infty[\![\cap [\![0, \zeta]\!] \in \mathcal{P}$ decrease to $[\![\zeta_\Gamma]\!]$ as $n \to \infty$. (Recall that $\zeta_\Gamma(\omega) := \zeta(\omega)$ if $\omega \in \Gamma$, $= \infty$ otherwise.) Thus ζ_Γ is a predictable time with graph contained in $[\![\zeta]\!]$, and so by Lemma 2.3, $\Gamma \subset \{\zeta = \zeta_p\}$.

3. Martingales with finite lifetime

As in [Sh88, A6], we define a uniformly integrable martingale on $[\![0, \zeta[\![$ to be a process $M_t(\omega)$ defined for $(t, \omega) \in [\![0, \zeta[\![$, such that there exists a (globally defined) uniformly integrable martingale $\bar{M}_t'$ $(0 \leq t \leq \infty)$ whose restriction to $[\![0, \zeta[\![$ is M. See also the discussion in [Ya82]. The following result was proved in [Sh88, A6].

(3.1) Proposition. *Let M be a uniformly integrable martingale on $[\![0, \zeta[\![$. There is then a unique extension $\bar{M}$ of M satisfying the two following conditions:*

(3.2-i)   *$\bar{M}$ stops at ζ: that is, a.s., $\bar{M}_t = \bar{M}_\zeta$ for all $t \geq \zeta$;*

(3.2-ii)   *$\bar{M}_\zeta \in \mathcal{F}_{\zeta-}$.*

We call the uniquely determined $\bar{M}$ satisfying (3.2-i) and (3.2-ii) the *minimal extension* of M. The random variable $\bar{M}_\zeta$ will be called the *closing value* of M. In the simplest case, in which ζ is identically ∞, the definition above reduces to

the usual definition of a uniformly integrable martingale, and the closing value
is $\bar{M}_\infty := \lim_{t\to\infty} M_t$. More generally, if ζ is a predictable time, it is easy to
see that the closing value is given by $\bar{M}_\zeta := \lim_{t\uparrow\uparrow\zeta} M_t$, where the $\uparrow\uparrow$ is meant
to indicate a limit taken strictly from the left. However, by considering the first
jump time ζ of a Poisson process with rate 1, and $M_t = -t$ for $t < \zeta$, which is
the restriction to $[\![0,\zeta[\![$ of the martingale $\bar{M}_t = -t \wedge \zeta + 1_{[\![\zeta,\infty[\![}(t)$, it is clear
that when ζ is not predictable, the closing value is not given by the left limit.

The definition of martingale on $[\![0,\zeta[\![$ given above is mathematically convenient but unsatisfying in that it violates the caveat that it should depend only
on observations of M strictly prior to ζ. It is one of the aims of this paper to
find a formulation that is intrinsic to $[\![0,\zeta[\![$.

An example of considerable importance is the case where ζ is the lifetime of a
right process X with state space E, h a function on E extended, as customary, to
vanish at the dead point Δ, and $M_t := h(X_t)$. Under certain strong conditions
such as X a diffusion killed at a boundary, M is a martingale on $[\![0,\zeta[\![$ if and
only if h is harmonic relative to the generator of X. Under broader conditions
on X, under which the lifetime may not be predictable, the characterization of
the h for which M is a martingale on $[\![0,\zeta[\![$ is not so straightforward, and does
not seem to reduce to identities involving hitting operators relative to X.

The proof of existence and uniqueness of $\bar{M}$ given in [Sh88] is correct but
unnecessarily complicated, seeming to depend on an argument involving the relative density of two dual predictable projections. The following is a simple, direct
proof.

(3.3) Lemma. *Fix a globally defined uniformly integrable martingale $\bar{M}'$ stop-
ping at ζ, let M denote the restriction of $\bar{M}'$ to $[\![0,\zeta[\![$, and define $F := \mathbf{P}\{\bar{M}'_\zeta \mid
\mathcal{F}_{\zeta-}\}$. Then the process $\bar{M}$ defined by $\bar{M}_t := M_t 1_{\{t<\zeta\}} + F 1_{\{t\geq\zeta\}}$ is the (unique)
minimal extension of M.*

Proof. For uniqueness, it suffices to prove that if $\bar{M}$ satisfies (3.2) and if $\bar{M}$

vanishes on $[\![0, \zeta[\![$, then $\bar{M}$ is evanescent. Because $\bar{M}$ is a martingale of integrable total variation, its compensator $\bar{M}$ is evanescent. By (3.2), we may write $d\bar{M} = Z\, d1_{[\![\zeta, \infty[\![}$ for some $Z \in \mathcal{P}$, from which it follows that $0 = d\bar{M}^\sim = Z\, dC$, where C is the compensator of $1_{[\![\zeta, \infty[\![}$. Write $Z = Z^+ - Z^-$ to see that $Z^+\, dC = Z^-\, dC$, from which we find $|Z|\, dC = 0$, hence $|Z|\, d1_{[\![\zeta, \infty[\![} = 0$, so that $\bar{M}$ is evanescent. Turning to existence, we see clearly that $\bar{M}$ is uniformly integrable and satisfies (3.2-i) and (3.2-ii), so it suffices to prove that $\bar{M}$ is a martingale. That is equivalent to proving that $\bar{M}' - \bar{M}$ is a martingale. But $N_t := \bar{M}'_t - \bar{M}_t = (\bar{M}'_\zeta - F)1_{\{t \geq \zeta\}}$, hence for an arbitrary $H \in b\mathcal{F}_t$, $\mathbf{P}\, N_\infty H = \mathbf{P}(N_\infty H 1_{\{t < \zeta\}} + N_\infty H 1_{\{t \geq \zeta\}})$, and since $H 1_{\{t < \zeta\}} \in b\mathcal{F}_{\zeta-}$, the first summand reduces to $\mathbf{P}\left(\mathbf{P}\left\{N_\infty \mid \mathcal{F}_{\zeta-}\right\} H 1_{\{t < \zeta\}}\right) = 0$. On the other hand, $N_\infty = N_t$ on $\{t \geq \zeta\}$, and as $N_t = 0$ for $t < \zeta$, we have $\mathbf{P}\left(N_\infty H 1_{\{t \geq \zeta\}}\right) = \mathbf{P}\left(N_t H 1_{\{t \geq \zeta\}}\right) = \mathbf{P}\left(N_t H\right)$. This proves that N is a martingale.

It is going to turn out that the character of a martingale M on $[\![0, \zeta[\![$ depends on the character of the minimal extension $\bar{M}$ in an essential way. This is already evident in the work of Maisonneuve, who was concerned principally with *continuous* martingales and local martingales on $[\![0, \zeta[\![$. He observed that the correct definition of continuity of M involves continuity of some extension of M on $[\![0, \zeta]\!]$ and not just that of M on $[\![0, \zeta[\![$.

The two central questions we shall focus on in this paper are:

(3.4-i)   Is there a formula by which $\bar{M}$ may be computed from M?

(3.4-ii) Given a process M on $[\![0, \zeta[\![$, how may we tell just by looking at M and ζ whether M is a uniformly integrable martingale on $[\![0, \zeta[\![$? That is, how and under what conditions should we try to construct an extension $\bar{M}$ of M that will be a global martingale?

It is going to turn out that the answer to the second question includes the answer to the first. Note first that the extension $\bar{M}$ of M is easy to calculate on the predictable part of ζ. The closing value of M on that predictable part is just the left limit of M at ζ, just as in the classical case in which $\zeta \equiv \infty$.

(3.5) **Proposition.** *Let M be a uniformly integrable martingale on $[\![0, \zeta[\![$ with minimal extension $\bar{M}$. Then*

$$(3.6) \qquad\qquad \bar{M}_{\zeta_p} = M_{\zeta_p-} \quad \text{a.s. on } \{\zeta_p < \infty\}.$$

Proof. Because ζ_p is predictable, $M_{\zeta_p-} = \bar{M}_{\zeta_p-} = \mathbf{P}\{\bar{M}_\zeta \mid \mathcal{F}_{\zeta_p-}\}$, and since $\bar{M}_\zeta \in \mathcal{F}_{\zeta-}$ by hypothesis, $\bar{M}_{\zeta_p} = \bar{M}_\zeta 1_{\zeta=\zeta_p} \in \mathcal{F}_{\zeta-} \subset \mathcal{F}_{\zeta_p-}$. The result follows.

4. P-measures carried by the graph of ζ

Let $\mathcal{A}^p$ (resp., $\mathcal{A}^m$) denote the class of predictable (resp., measurable) processes A with integrable total variation. More precisely, $\mathcal{A}^p$ (resp., $\mathcal{A}^m$) consists of the predictable (resp., measurable) A such that

(4.1-i) $A_0 = 0$, $A_\infty = \lim_{t \to \infty} A_t$;

(4.1-ii) $t \to A_t$ is a.s. right continuous with left limits, and of finite variation on

   finite intervals;

(4.1-iii) $\mathbf{P} \int_0^\infty |dA_t| < \infty$.

Recall [DM80] that a signed measure ν on $([\![0, \infty[\![, \mathcal{P})$ is a P-measure provided $|\nu|(\Gamma) = 0$ whenever $\Gamma \in \mathcal{P}$ is evanescent with respect to $\mathbf{P}$. It is shown there that the class of (signed) P-measures is in 1-1 correspondence with the class $\mathcal{A}^p$ under the correspondence given by

$$(4.2) \qquad A \leftrightarrow \nu_A \quad \Longleftrightarrow \quad \nu_A(H) = \mathbf{P} \int_0^\infty H_t \, dA_t \qquad \forall H \in b\mathcal{P}.$$

More generally, given $A, B \in \mathcal{A}^m$, the P-measures ν_A, ν_B are equal if and only if $A^p = B^p$, where A^p denotes the dual predictable projection, or compensator, of A. Let $\mathcal{A}^p_{\text{loc}}$ denote the predictable processes of *locally* integrable variation, and similarly for $\mathcal{A}^m_{\text{loc}}$. Thus, for example, $A \in \mathcal{A}^m_{\text{loc}}$ if and only if there exists an increasing sequence T_n of stopping times with limit $+\infty$ a.s. so that for every n, A^{T_n} (A stopped at T_n) is in $\mathcal{A}^m$. The right side of (4.2) is then not in general defined for every $H \in b\mathcal{P}$, but it is defined as a signed measure on $[\![0, T_n]\!]$ for all n.

(4.3) Definition. *(i) A process $A \in \mathcal{A}^m$ is carried by $[\![\zeta]\!]$, the graph of ζ, provided $\mathbf{P} \int_0^\infty H_t \, dA_t = 0$ for all $H \in b\mathcal{P}$ satisfying $H_\zeta = 0$ a.s. on $\{\zeta < \infty\}$. (ii) A process $\mathcal{A}_{\text{loc}}^m$ is carried by the graph of ζ if and only if there is an increasing sequence T_n of stopping times such that $\lim_n T_n = \infty$ a.s., and so that for all n, $A^{T_n} \in \mathcal{A}^m$ is carried by the graph of ζ.*

Obviously, $A \in \mathcal{A}^m$ is carried by the graph of ζ if and only if the corresponding P-measure ν_A has the property $\nu_A(H) = 0$ for all $H \in b\mathcal{P}$ with $H_\zeta = 0$ a.s. on $\{\zeta < \infty\}$. We say in these circumstances that ν is carried by the graph of ζ.

(4.4) Lemma. *$A \in \mathcal{A}_{\text{loc}}^m$ is carried by the graph of ζ if and only if its compensator $A^\sim$ is carried by $[\![\zeta]\!]$.*

Proof. Suppose A is carried by $[\![\zeta]\!]$. Let T_n be stopping times such that for each n, $B := A^{T_n} \in \mathcal{A}^m$ and B is carried by $[\![\zeta]\!]$. Since the dual predictable projection has the property $(1_{[\![0,T_n]\!]} A)^\sim = 1_{[\![0,T_n]\!]} A^\sim$, $B^\sim$ is indistinguishable from $A^\sim$ stopped at T_n. However, if $H \in b\mathcal{P}$ and $H_\zeta = 0$, $0 = \int_0^\infty H_t \, dB_t = \int_0^\infty H_t \, dB_t^\sim$, from which it follows that $A^\sim$ is carried by $[\![\zeta]\!]$. To go the other direction, simply reverse the steps in the argument.

Note that if ζ is predictable, $A \in \mathcal{A}^p$ is carried by ζ if and only if A has the form $F1_{[\![\zeta,\infty[\![}$ for some $F \in L^1(\mathcal{F}_{\zeta-})$. However, if ζ is not predictable, no such simple characterization is possible. For example, if ζ is the first jump time of a Poisson process starting at 0, it is not hard to see that $A \in \mathcal{A}^p$ is carried by $[\![\zeta]\!]$ if and only if $A_t(\omega)1_{[\![0,\zeta[\![}(t,\omega) = F(t)1_{[\![0,\zeta[\![}(t,\omega)$ for some deterministic function F, right continuous with left limits on $\mathbf{R}^+$, vanishing at 0, having locally bounded variation and satisfying $\int_0^\infty e^{-t}|dF(t)| < \infty$. Intuitively, the signed measure dA_t can be spread out over the *least* predictable set Γ containing the graph of ζ, if such a set Γ exists.

(4.5) Lemma. *Let $A \in \mathcal{A}_{\text{loc}}^m$, and let C denote $\epsilon_\zeta^\sim$, the compensator of (the finite part of) unit mass at ζ. Then A is carried by the graph of ζ if and only if $A^\sim \ll C$.*

Proof. We assume first that $A \in \mathcal{A}^m$. By (4.4), there is no loss of generality assuming $A \in \mathcal{A}^p$, and we shall so assume. Suppose that A is carried by the graph of ζ. Considering the P-measures ν_A, ν_C generated by A and C respectively, we may make a Lebesgue-Radon-Nikodym decomposition of A with respect to C on the σ-field $\mathcal{P}$, say $dA = dB + Z\,dC$ with $B \perp C$, $B \in \mathcal{A}^p$, and $Z \in \mathcal{P}$ with $\mathbf{P} \int_0^\infty |Z_t|\,dC_t < \infty$. Then $Z\,dC$ is carried by the graph of ζ, for if $H \in b\mathcal{P}$ vanishes at ζ, $\mathbf{P} \int_0^\infty H_t Z_t\,dC_t = \mathbf{P} \int_0^\infty H_t Z_t\,\epsilon_\zeta(dt)$ by the defining property of the compensator, and the last term clearly vanishes. It follows now that B is carried by the graph of ζ. Let $[\![0,\infty[\![= \Gamma \cup \Gamma^c$ be a decomposition of $[\![0,\infty[\![$ such that $\Gamma \in \mathcal{P}$, with C not charging $\Gamma \in \mathcal{P}$, and B not charging Γ^c. In particular, $\mathbf{P}\,1_\Gamma(\zeta)1_{\{\zeta<\infty\}} = \mathbf{P} \int_0^\infty 1_\Gamma(t)\epsilon_\zeta(dt) = 0$, and therefore 1_Γ vanishes at ζ. It follows that $\mathbf{P} \int_0^\infty H_t 1_\Gamma(t)\,dB_t = 0$ for all $H \in b\mathcal{P}$. As B lives on Γ, this proves that B is evanescent. Hence $A \ll C$. Conversely, if $A \ll C$, then for $H \in b\mathcal{P}$ with $H_\zeta = 0$ a.s. on $\{\zeta < \infty\}$, $0 = \mathbf{P}H_\zeta 1_{\{\zeta<\infty\}} = \mathbf{P} \int_0^\infty H_t\,dC_t$ implies $\mathbf{P} \int_0^\infty H_t\,dA_t = 0$, hence A is carried by the graph of ζ.

(4.6) Definition. *A process M defined on $[\![0,\zeta[\![$ is a semimartingale on $[\![0,\zeta[\![$ provided it is the restriction to $[\![0,\zeta[\![$ of some globally defined semimartingale $\bar{M}$. It is a local semimartingale on $[\![0,\zeta[\![$ provided there is an increasing sequence (T_n) of stopping times with $\lim_n T_n \geq \zeta$ and M^{T_n} a semimartingale on $[\![0,\zeta[\![$ for every n.*

Given a process M defined on $[\![0,\zeta[\![$, there are two particularly simple extensions of M. First, if $M_{\zeta-}$ exists a.s. on $\{\zeta < \infty\}$, we may set $M'_t = M_t$ for $t < \zeta$, $= \lim_{r\uparrow\uparrow\zeta} M_r$ for $t \geq \zeta$. This is called the (unique) extension of M that stops at $\zeta-$. Second, if we set $M''_t = M_t$ for $t < \zeta$, $= 0$ for $t \geq \zeta$, then M'' is called the (unique) extension of M vanishing on $[\![\zeta,\infty[\![$. It is easy to see that M is a semimartingale on $[\![0,\zeta[\![$ if and only if M'' is a (global) semimartingale, and in case $M_{\zeta-}$ exists a.s. on $\{\zeta < \infty\}$, an equivalent condition is that M' is a (global) semimartingale.

Suppose for the moment that M is a semimartingale on $[\![0,\zeta[\![$. Let us define $\mathcal{P}_\zeta$ to be the trace of $\mathcal{P}$ on $[\![0,\zeta[\![$. (This is what was called the *natural* σ-field in [GS84].) We may then construct stochastic integrals $H \cdot M$ for $H \in b\mathcal{P}_\zeta$ as follows. Let $\bar{H} \in b\mathcal{P}$ be an arbitrary bounded extension of H and let $\bar{M}$ denote an arbitrary extension extension of M as a semimartingale. By the localization theorem for semimartingale stochastic integrals [DM80, VIII.23], the values of $\bar{H} \cdot \bar{M}$ on $[\![0,\zeta[\![$ do not depend on the particular extensions $\bar{H}$ and $\bar{M}$, up to evanescence. Thus it is meaningful to let $H \cdot M$ denote the restriction of $\bar{H} \cdot \bar{M}$ to $[\![0,\zeta[\![$. It is important to note that we don't need to know whether there is a global martingale extension of M to construct $H \cdot M$.

We may now set down a criterion for a given process M on $[\![0,\zeta[\![$ to be an $\mathfrak{H}^1$ martingale on $[\![0,\zeta[\![$, using only conditions involving M strictly prior to ζ. Recall [DM80, VII.23] or [Pr90,V.2] that a (globally defined) semimartingale X is of class $\mathfrak{H}^1$ if and only if it has at least one decomposition $X = N + A$ with N a local martingale, A an optional process of locally finite variation, such that $\mathbf{P}[N,N]^{1/2} + \mathbf{P} \int_0^\infty |dA_t| < \infty$. In this case, X is in fact a special semimartingale, and the decomposition can be supposed to be the canonical decomposition with $A \in \mathcal{A}^p$. See [DM80, §98c].

A theorem of Yor [DM80, §104] characterizes the class of $\mathfrak{H}^1$ (global) semi-martingales X as those for which the map $H \to H \cdot X$ of $b\mathcal{P}$ into semimartingales has the property

$$(4.7) \qquad\qquad \sup_{t>0, |H| \le 1} \|(H \cdot X)_t\|_1 < \infty.$$

The next result is our preliminary characterization of martingales on $[\![0,\zeta[\![$ and their closing values. It is preliminary because the procedure it specifies for the closing value involves the canonical decomposition of a global semimartingale, which is not a construction intrinsic to $[\![0,\zeta[\![$. The final version of the theorem is given in the next section. See (5.7).

(4.8) Theorem. *Let M be a process defined on $[\![0,\zeta[\![$ and satisfying the con-*

ditions

(4.8-i) $M_{\zeta-}$ *exists a.s. on* $\{\zeta < \infty\}$;

(4.8-ii) M *is a semimartingale on* $[\![0, \zeta[\![$ *whose extension* M' *stopping at* $\zeta-$ *is of class* $\mathfrak{H}^1$. *(Recall:* $M'_t = M_{\zeta-}$ *for* $t \geq \zeta$.)

Let μ denote the functional $H \to \mu(H) := \mathbf{P}(H \cdot M')_{\zeta-} 1_{\{\zeta < \infty\}}$, $H \in b\mathcal{P}$. Then μ defines a signed $\mathbf{P}$-measure on $([\![0, \infty[\![, \mathcal{P})$, and M is a martingale on $[\![0, \zeta[\![$ if and only if μ is carried by the graph of ζ. Moreover, in this case, the closing value $\overline{M}_\zeta$ of M is given by

$$(4.9) \qquad \overline{M}_\zeta = M_{\zeta-} - H_\zeta 1_{\{\zeta < \infty\}},$$

where H is a predictable version of the relative density dA/dC, and $A \in \mathcal{A}^p$ is the unique element of $\mathcal{A}^p$ for which $N := M' - A$ is a martingale.

Proof. Under (4.8-ii), M' is necessarily a special semimartingale by VII.98 c) of [DM80], and if $M' = N + A$ is a decomposition of M' into the sum of an $\mathfrak{H}^1$ global martingale N and $A \in \mathcal{A}^p$, then for $H \in b\mathcal{P}$, $H \cdot N$ is a martingale satisfying $\|H \cdot N\|_{\mathfrak{H}^1} \leq \|H\|_\infty \|N\|_{\mathfrak{H}^1} < \infty$, hence $|\mu(H)| \leq \|H\|_\infty (\|N\|_{\mathfrak{H}^1} + \mathbf{P} \int_0^\infty |dA_t|) = \|H\|_\infty \|M'\|_{\mathfrak{H}^1}$. Since $N \in \mathfrak{H}^1$, $\Delta N_\zeta \in L^1(\mathbf{P})$. As stochastic integrals satisfy $\Delta(H \cdot N) = H\Delta N$, we have

$$(4.10) \quad \mathbf{P} H \cdot N_{\zeta-} 1_{\{\zeta < \infty\}} = \mathbf{P} H \cdot N_\zeta 1_{\{\zeta < \infty\}} - \mathbf{P}\Delta(H \cdot N)_\zeta 1_{\{\zeta < \infty\}}$$

$$= -\mathbf{P} H_\zeta \Delta N_\zeta 1_{\{\zeta < \infty\}}.$$

Since $A \in \mathcal{A}^p$, $\mu(H) = -\mathbf{P} H_\zeta \Delta N_\zeta 1_{\{\zeta < \infty\}} + \mathbf{P} \int_0^\infty H_t \, dA_t$, from which it is clear that μ is a signed $\mathbf{P}$-measure on $([\![0, \infty[\![, \mathcal{P})$.

Write $\mu = \mu_N + \mu_A$, where $\mu_N(H) := \mathbf{P} H \cdot N_{\zeta-} 1_{\{\zeta < \infty\}}$ and $\mu_A(H) := \mathbf{P} \int_0^\infty H_t \, dA_t$. By (4.10), μ_N is carried by the graph of ζ, in the sense of (4.3). Therefore, the condition that μ is carried by the graph of ζ is equivalent to the condition that μ_A is carried by the graph of ζ. Suppose that μ is indeed carried by the graph of ζ so that $A \ll C$ by (4.5). Let $H \in \mathcal{P}$ be a version of dA/dC.

Define then $\bar{M} := M' - H_\zeta 1_{[\zeta,\infty[}$. Because A has integrable total variation, we know that $\infty > \mathbf{P} \int_0^\infty |dA_t| = \mathbf{P} \int_0^\infty |H_t| \, dC_t = \mathbf{P} \int_0^\infty |H_t| \, \epsilon_\zeta(dt) = \mathbf{P}|H_\zeta|$. It suffices to establish that $\bar{M}$ is a uniformly integrable (globally defined) martingale. But, $\bar{M} = N + A - H_\zeta 1_{[\zeta,\infty[}$, and noting that the process $H_\zeta 1_{[\zeta,\infty[}$ is the distribution function of the random measure $H * \epsilon_\zeta$, whose compensator is $H * \epsilon_\zeta^\sim = H * C = A$, it is clear that $\bar{M}$ is a martingale as claimed. Its uniform integrability is a consequence of the fact that $H_\zeta \in L^1$ (proved above) and $\mathbf{P} \sup_{t<\zeta} |M_t| \in L^1$ by the $\mathfrak{H}^1$ hypothesis on M'. For the converse, suppose M is a martingale on $[0,\zeta[$. Let $\bar{M}$ denote the minimal extension of M. Because M' and $\bar{M}$ are identical on $[0,\zeta[$, for $H \in b\mathcal{P}$, the localization theorem [DM80,VIII.10] gives $\mu(H) = \mathbf{P}(H \cdot M')_{\zeta-} = \mathbf{P}(H \cdot \bar{M})_{\zeta-}$, and by the same reasoning employed in the first paragraph of the proof, the latter term is equal to $-\mathbf{P}H_\zeta \Delta \bar{M}_\zeta$, from which it is evident that μ is carried by the graph of ζ.

The following simple examples illustrate the use of the preceding proposition in computing the minimal extension of M.

In the first case, let ζ denote the time of first jump of a Poisson process with rate 1, with its natural augmented filtration. Let f be an absolutely continuous function on $\mathbf{R}^+$ satisfying $\int_0^\infty |f'(t)|e^{-t} \, dt < \infty$, and let $M_t = f(t)1_{\{t<\zeta\}}$. Note that $\sup_{t<\zeta} |M_t| \le \int_0^\zeta |f'(t)| \, dt$, and the expectation of the latter term is $\int_0^\infty |f'(t)|e^{-t} \, dt < \infty$. Hence M is dominated by an integrable random variable. Observe that M is a martingale with minimal extension $\bar{M}_t = f(t \wedge \zeta) - f'(\zeta)1_{\{t\ge\zeta\}}$. (This is well known in case $f(t) = t$. Let $\tilde{M}_t = t \wedge \zeta - 1_{\{t\ge\zeta\}}$. Then $\bar{M} = H \cdot \tilde{M}$, where $H_t := f'(t)$, proving the general case.) When we form the measure μ of (4.8), because M has bounded variation, $\mu(H)$ reduces to $\mu(H) = \mathbf{P} \int_{[0,\zeta[} H_t \, df(t)$, and since every predictable process H_t on $[0,\zeta]$ coincides with some deterministic function $h(t)$ for $t \le \zeta$, the latter integral may be written in the form $\mathbf{P} \int_0^\infty e^{-t}h(t)f'(t) \, dt$. On the other hand, for $H_t := h(t)1_{[0,\zeta]}$, $\mu_C(H) = \mathbf{P} H(\zeta) = \int_0^\infty e^{-t}h(t) \, dt$ from which it is clear that $(d\mu/d\mu_C)_t = f'(t)$, just as asserted by the theorem.

For a different type of example, let X denote uniform motion to the right on $\mathbf{R}^+$, starting at the origin and killed with probability α $(0 < \alpha < 1)$ as it passes through the state 1. Let $R := 1$, so that $[\![\zeta]\!] \subset [\![R]\!]$. Obviously, R is predictable. Observe that $\mathcal{F}_{1-}$ is the trivial σ-field. Thus, if T is a stopping time, T is constant on $\{T < 1\}$. It follows that if, as in section 2, we set $\lambda = {}^p 1_{[\![\zeta]\!]}$, then λ_t is constant for $t < 1$, and for such t, $\lambda_t = \mathbf{P}\{t = \zeta\} = 0$. On the other hand, $\lambda_1 = \mathbf{P}\{1 = \zeta\} = 1 - \alpha$, so that $\lambda = (1-\alpha)1_{[\![\zeta]\!]}$. By (2.2), $\zeta_p = \infty$ a.s., and in particular, ζ is not predictable. Note that in this example, $C := \epsilon_\zeta \tilde{} = \alpha \epsilon_R$, as C and ϵ_ζ clearly generate the same potential. In order for $A \in \mathcal{A}^p$ to satisfy $A \ll C$, it is necessary and sufficient that $dA = \beta \, d\epsilon_R$ for some scalar β. A simple argument based on (4.8) shows then that the martingales M on $[\![0, \zeta[\![$ have the form $M = \beta_1 1_{[\![0,\zeta[\![} + \beta_2 1_{[\![R,\infty[\![} 1_{\{R < \zeta\}}$, for arbitrary constants β_1, β_2.

The third example is based on a standard linear Brownian motion B killed at an independent exponential time ζ with parameter 1 to give us a new process X. We seek a characterization of those functions h on $\mathbf{R}$ for which $h \circ X$ is a martingale on $[\![0, \zeta[\![$. By results from [CJPS80], $h \circ X$ is a semimartingale over X if and only if h is locally a difference of convex functions, and so h has a second derivative ν (in the sense of Schwartz) that can be identified locally with a signed measure. It is clear that the functional C in (4.8) is the additive functional $dC_t = 1_{[\![0,\zeta[\![}(t) \, dt$, and by Tanaka's formula, $2A_t = \int \nu(dx) L_t^x$, where L_t^x are local times for X, chosen jointly continuous in (x, t). It follows then from (4.8) that $h \circ X$ is a martingale on $[\![0, \zeta[\![$ if and only if ν is (locally) absolutely continuous with respect to Lebesgue measure ℓ on $\mathbf{R}$, and the closing value is given by $h(X_{\zeta-}) - g(X_{\zeta-})$, where g is a version of $d\nu/d(2\ell)$. (It is easy to verify this directly without using (4.8).) As a particular case, we find that X_t^2 is a martingale on $[\![0, \zeta[\![$.

The third example above is a special case of a more general problem that will be dealt with in a future publication, in which the underlying processes involved are (right) Markovian, and on asks for a characterization of those functions h for which $h \circ X$ is a martingale on $[\![0, \zeta[\![$.

Note too that the last example above could have been discussed in the framework of the theory of *grossissement de filtration* [Je80]. Indeed, it is apparent from the statement of (4.8) that there is a close formal connection between that theory and ours.

5. Natural decomposition

Given a special semimartingale Z on $[\![0, \zeta[\![$, its canonical decomposition $Z = M + A$, with M a local martingale and A a predictable process of locally integrable variation, is not necessarily intrinsic to $[\![0, \zeta[\![$. The following decomposition of such a process Z has the virtue of being intrinsic. It is not generally the same as the decomposition $Z = M + A$ above. We need to make use of some ideas and results from [GS84]. See also [Sh88]. Recall that the class $\mathcal{P}_\zeta$ of *natural* processes (relative to the fixed time ζ) means the trace of $\mathcal{P}$ on $[\![0, \zeta[\![$. Define the *natural projection* $M \to {}^n M$ of $p\mathcal{M}$ onto $p\mathcal{P}_\zeta$ by the formula

$$(5.1) \qquad\qquad {}^n M := {}^p(M 1_{[\![0,\zeta[\![}) 1_{[\![0,\zeta[\![} / \rho.$$

Note that a natural process Z is not in general predictable unless ζ is a predictable time.

We shall say that a global semimartingale Z *lives on* $\Lambda := \{\rho > 0\}$ in case $1_\Lambda \cdot Z = Z - Z_0$ and $Z_\zeta \in \mathcal{F}_{\zeta-}$. It is easy to see that Z lives on Λ if and only if

(5.2-i) Z stops at ζ (or equivalently, $1_{[\![0,\zeta]\!]} \cdot Z = Z - Z_0$);

(5.2-ii) ΔZ vanishes off $[\![\zeta]\!] \cap \Lambda$. That is, by (2.4), $\Delta Z_{\zeta_p} = 0$, where ζ_p denotes the predictable part of ζ.

(5.2-iii) $\Delta Z_\zeta \in \mathcal{F}_{\zeta-}$.

Note that if a process Z is the minimal extension of a uniformly integrable martingale on $[\![0, \zeta[\![$, then Z is a semimartingale that lives on Λ.

Call a process A a *natural process of locally finite variation* in case

(5.3-i) A is right continuous with left limits and $A_0 = 0$;

(5.3-ii) A is of locally finite variation;

(5.3-iii) $A_t = A_{\zeta-}$ a.s. for all $t \geq \zeta$;

(5.3-iv) the restriction of A to $[\![0, \zeta[\![$ is in $\mathcal{P}_\zeta$.

Such a process A is obviously a semimartingale living on Λ.

The *dual natural projection* $A \to A^n$ of a locally integrable process A of locally integrable variation is defined provided A is carried by Λ and A possesses a dual predictable projection, and is given in this case by the formula

$$(5.4) \qquad dA_t^n := 1_{[\![0,\zeta[\![}(t)\, dA_t^p / \rho_t.$$

Clearly, A^n satisfies the conditions (5.3), and therefore A^n lives on Λ. It was proved in [GS84] that, if A is carried by Λ, then $A^n = (A^p)^n$ and $A^p = (A^n)^p$, and consequently $A^n - A^p$ is a local martingale. It is almost immediate from properties of dual predictable projections that A^p is carried by Λ if A is, and thus, so is $A^n - A^p$.

(5.5) Theorem. *Let Z be a semimartingale of class $\mathfrak{H}^1$ living on Λ. Then Z has a unique decomposition of the form $Z = Z_0 + N + B$, where N is the minimal extension of an $\mathfrak{H}^1$ martingale on $[\![0, \zeta[\![$, $N_0 = 0$, and B is a natural process of integrable variation. (This decomposition will be called the* **natural decomposition** *of Z.)*

Proof. Let $Z = Z_0 + M + A$ be the canonical decomposition of Z into the sum of a global $\mathfrak{H}^1$ martingale M and a predictable process A of integrable total variation. Since $1_\Lambda \cdot Z = Z - Z_0$, $Z - Z_0 = 1_\Lambda \cdot M + 1_\Lambda \cdot A$, and by uniqueness of the canonical decomposition, $M = 1_\Lambda \cdot M$ and $A = 1_\Lambda \cdot A$. As A is predictable, $A_\zeta \in \mathcal{F}_{\zeta-}$. Hence A lives on Λ. Let $B := A^n$, the dual natural projection of A described above. Then $A - B$ is a martingale. Let $N := M + A - A^n$, so that N is a martingale and $Z = Z_0 + N + B$. Obviously, N stops at ζ, and $\Delta Z_\zeta = \Delta N_\zeta + \Delta B_\zeta \in \mathcal{F}_{\zeta-}$ gives $\Delta N_\zeta \in \mathcal{F}_{\zeta-}$. As to uniqueness of the decomposition, it suffices, by stopping, to verify that if N is simultaneously a uniformly integrable martingale satisfying $\Delta N_\zeta \in \mathcal{F}_{\zeta-}$ and also a natural process of integrable variation, then N is a.s. constant. Since N is natural, its dual predictable projection N^p exists and $N^p - N$ is a martingale. It follows that $N^p = N + (N^p - N)$ is also a martingale, and because it is predictable and of locally bounded variation, it must be a.s. constant, hence evanescent. Finally, use the fact that $N = N^n = (N^p)^n$, as discussed prior to the statement of (5.5), to see that N is evanescent.

(5.6) Corollary. *Let Z be a semimartingale of class $\mathfrak{H}^1$ that stops at $\zeta-$, and let $Z = N + B$ be its natural decomposition, as in (5.5). Then N and B also stop at $\zeta-$.*

Proof. That B stops at $\zeta-$ is clear from its construction as a dual natural projection. The result follows at once.

(5.7) Corollary. *Given a process M defined on $[\![0,\zeta[\![$, let M' denote the extension of M that stops at $\zeta-$. Assume that M' is a semimartingale of class $\mathfrak{H}^1$. Then M' lives on Λ, and if $M' = N + B$ is the decomposition described in (5.5), then M is a martingale on $[\![0,\zeta[\![$ if and only if $B \ll D := 1_{[\![0,\zeta[\![}\epsilon_\zeta^\sim$. The closing value of M is given in this case by $\overline{M}_\zeta := M_{\zeta-} - H_\zeta$, where $H \in \mathcal{P}$ satisfies $dB = H 1_{[\![0,\zeta[\![}\, dD$.*

Proof. It is clear that M' lives on Λ. Let μ denote the signed **P**-measure on $([\![0,\infty[\![,\mathcal{P})$ defined as in (4.8) by $\mu(H) := \mathbf{P}(H\cdot M)_{\zeta-}$. Because $\Delta N_\zeta = 0$, we find $\mu(H) = \mathbf{P}\int_0^\infty H_t\, dB_t = \mathbf{P}\int_0^\infty H_t\, dB_t^p$ for $H \in b\mathcal{P}$. According to (4.8), μ lives on the graph of ζ if and only if $B^p \ll C := \epsilon_\zeta^\sim$. The corollary will therefore be proved once we establish the following result.

(5.8) Lemma. *Let B be an arbitrary natural process of integrable total variation. Then $B^p \ll C := \epsilon_\zeta^\sim$ with $dB_t^p = H_t\, dC_t$, $H \in \mathcal{P}$, if and only if $B \ll D := (1_{[\![0,\zeta[\![}\epsilon_\zeta^\sim)^n$, with $dB_t = H_t 1_{[\![0,\zeta[\![}(t)\, dD_t$*

Proof. Assume first that $B^p \ll C$ with $dB_t^p = H_t \, dC_t$, $H \in \mathcal{P}$. We also use the notation $B^p = H * C$. Using the definition (5.4) of dual natural projection together with the fact that C and D are carried by $\Lambda = \{\rho > 0\}$, we have

$$dB_t = d(B^p)_t^n = d(H * C)_t^n = 1_{[0,\zeta[}(t) H_t \, dC_t/\rho_t = H_t \, dD_t = 1_{[0,\zeta[}(t) H_t \, dD_t.$$

Therefore, $dB_t = 1_{[0,\zeta[}(t) H_t \, dD_t$, with $1_{[0,\zeta[} H \in p\mathcal{P}_\zeta$. Conversely, starting with $dB_t = 1_{[0,\zeta[}(t) H_t \, dD_t = H_t \, dD_t$, with an $H \in p\mathcal{P}$, essentially the same reasoning gives $dB_t^p = d(H * D)_t^p = H_t \, dD_t^p = H_t \, d(1_{[0,\zeta[}/\rho * C)_t^p = H_t^p(1_{[0,\zeta[}/\rho)_t \, dC_t = H_t \, dC_t$, the last equality because C is carried by Λ.

For any set $\Gamma \in \mathcal{P}$ with $\mathbf{P} \int_0^\infty 1_\Gamma(t) \, dC_t = 0$, $\mathbf{P} \int_0^\infty 1_\Gamma(t) \, dB_t^p = 0$, and therefore $\mathbf{P} \int_0^\infty 1_\Gamma(t) \, dB_t = 0$. But, by definition of D,

$$\mathbf{P} \int_0^\infty 1_\Gamma(t) \, dD_t = \mathbf{P} \int_0^\infty 1_\Gamma(t) 1_{[0,\zeta[}(t)/\rho(t) \, dC_t = 0.$$

(5.9) Lemma. *The semimartingale $\mathfrak{H}^1$ norm is equivalent to the norm $\|M\| :=$ $\|N\|_{\mathfrak{H}^1} + \mathbf{P} \int_0^\infty |dB_t|$, in which $M = N + B$ is the natural decomposition of M.*

Proof. As was proved in [DM80,VII.98, remarque c)], the $\mathfrak{H}^1$ norm, defined as the infimum of the quantities $j(W, A) := \|W\|_{\mathfrak{H}^1} + \mathbf{P} \int_0^\infty |dA_t|$, in which $M = W + A$ is some decomposition of M as the sum of a martingale and a process of locally bounded variation, is equivalent in the case of special semimartingales to the norm $j(W, A)$ in which $M = W + A$ is the canonical decomposition with A predictable. It follows that $\|M\| \geq const\|M\|_{\mathfrak{H}^1}$. On the other hand, $\mathbf{P} \int_0^\infty |dB_t| = \mathbf{P} \int_0^\infty |dA_t|$, since $A = B^p$ and $dB = 1_{[0,\zeta[}/\rho \, dA$ implies $|dB| = 1_{[0,\zeta[}/\rho |dA|$, whence $|dA| = |dB|^p$. Finally, $N = W + (A - B)$ gives $\|N\|_{\mathfrak{H}^1} \leq \|W\|_{\mathfrak{H}^1} + \mathbf{P} \int_0^\infty |dA_t| + \mathbf{P} \int_0^\infty |dB_t| \leq 2\|M\|_{\mathfrak{H}^1}$.

REFERENCES

[CJPS80] E. Çinlar, J. Jacod, P. Protter and M. J. Sharpe, *Semimartingales and Markov processes*, Z. Wahrscheinlichkeitstheorie verw. Gebiete **54** (1980), 161–219.

[DM75] C. Dellacherie and P.-A. Meyer, *Probabilités et Potentiel*, (2ième édition); Chapitres I–IV, Hermann, Paris, 1975.

[DM80] ———, *Probabilités et Potentiel*, (2ième édition); Chapitres V–VIII, Hermann, Paris, 1980.

[GS84]   R. K. Getoor and M. J. Sharpe, *Naturality, standardness and weak duality for Markov processes*, Z. Wahrscheinlichkeitstheorie verw. Gebiete **67** (1984), 1–62.

[Ja79]   J. Jacod, *Calcul Stochastique et Problèmes de Martingales*, Lecture Notes in Mathematics **714**, Springer, Berlin Heidelberg New York, 1979.

[Je80]   T. Jeulin, *Semi-Martingales et Grossissement d'une Filtration*, Lecture Notes in Mathematics **833**, Springer, Berlin Heidelberg New York, 1980.

[Ma77]   B. Maisonneuve, *Une mise au point sur les martingales locales définies sur un intervalle stochastique*, Séminaire de Probabilités XI (Univ. Strasbourg), Lecture Notes in Math. **581**, Springer, Berlin Heidelberg New York, 1977, pp. 435–445.

[MS81]   P.-A. Meyer and C. Stricker, *Sur les martingales au sens de L. Schwartz*, Mathematical Analysis and applications, Part B, Essays dedicated to L. Schwartz, edited by L. Nachbin, Academic Press, San Diego, 1981.

[Sh75]   M. J. Sharpe, *Homogeneous extension of random measures*, Séminaire de Probabilités IX (Univ. Strasbourg), Lecture Notes in Math. **465**, Springer, Berlin Heidelberg New York, 1975, pp. 496–514.

[Sh80]   _____, *Local times and singularities of continuous local martingales*, Séminaire de Probabilités XIV (Univ. Strasbourg), Lecture Notes in Math. **784**, Springer, Berlin Heidelberg New York, 1980, pp. 76–101.

[Sh88]   _____, *General Theory of Markov Processes*, Academic Press, San Diego, 1988.

[Ya82]   J.-A. Yan, *Martingales locales sur un ouvert droit optionnel*, Stochastics (1982), 161–180.

[Zh82]   W.-A. Zheng, *Semimartingales in predictable random open sets*, Séminaire de Probabilités XVI (Univ. Strasbourg), Lecture Notes in Math. **920**, Springer, Berlin Heidelberg New York, 1982, pp. 370–379.

Department of Mathematics 0112
University of California, San Diego
9500 Gilman Drive,
La Jolla, CA 92093-0112
USA

Internet: msharpe@euclid.ucsd.edu

Construction of Markov Processes from Hitting Distributions Without Quasi-Left-Continuity

by

C.T. SHIH

Let K be a compact metric space, Δ a fixed point in K, and $\mathcal{D}$ the family of all closed sets in K containing Δ. Given a family $\{H_D(x,\cdot) : D \in \mathcal{D}, x \in K\}$ of measures on K satisfying very general conditions, we consider the problem of constructing a right process $(X_t; P^x)$ on K with Δ as the adjoined death point such that, for all x and D, the hitting distribution $P^x[X(T_D) \in \cdot, T_D < \infty]$, where $T_D = \inf\{t \geq 0 : X_t \in D\}$, is the given $H_D(x,\cdot)$.

In [11] we proved that under the hypotheses of Markov property (or consistency), nearly Borel measurability, intrinsic right continuity, transience, and quasi-left-continuity, there does exist a right process (which is a Hunt process because of the quasi-left-continuity) with the prescribed hitting distributions. This result is restated in section 1 below for the purpose of comparison. As indicated it deals with the construction of transient processes. But using the theorem in [12], one can extend it to the construction of recurrent processes with little additional effort; see Theorem 3 at the end of section 1.

In the context of the theory of right processes, the only essential remaining unnecessary condition in this construction problem is that of q.l.c. (quasi-left-continuity). In this article we show that a reasonable result can be obtained without the q.l.c. Briefly, with the q.l.c. replaced by a necessary condition requiring the existence of path left limits and a weak predictability condition, with nearly Borel measurability replaced by Borel measurability (to avoid measurability complications in the compactification referred to below), with the transience condition slightly strengthened, and with certain (hidden) holding points added to K, one can construct a right process on the expanded state space (contained in an appropriate compactification of K) that has the prescribed hitting distributions for the original sets D.

We believe that through this result one gains insight into the role of the q.l.c. in this problem. Furthermore, an open problem is posed at the end of the article: is it possible, based on this result, to define a different time scale so that one can avoid adding the hidden holding points, and obtain a right process on the original state space with the prescribed hitting distributions? An example on the positive side will be given.

In proving this result we found a gap in [11] in the proof of the convergence of the time scale when holding points exist (the proofs of [11] in this case were somewhat sketchy). Here a new proof is presented which also works for the present situation.

The papers on this problem are, chronologically, Meyer [8], Knight and Orey [7], Dawson [3], Boboc, Constantinescu and Cornea [2], Hansen [5], and [6], Shih [9] and [10], Bliedtner and Hansen [1], Taylor [13], Graveraux and Jacod [4] and Shih [11]. See [9], [10], and [11] for comments on many of them. Most of these papers treat strong Feller processes, including [1], [2], [5], [6], [8] and [13]. which deal with constructing Markov processes corresponding to axiomatic potential theories of harmonic functions. [9] and [10] deal with Feller processes in general. [11] does away with the Feller conditions on the prescribed hitting distributions, using completely new proofs. The paper [7] by Orey and Knight was one of two articles that introduced us to this problem, and it is appropriate for us to devote this work to the memory of Professor Orey.

Finally, it is our pleasure to thank Pat Fitzsimmons for some very stimulating conversations that inspired this study.

1. Statement of Results

It is convenient and useful to state the hypotheses and theorems of [11] and of this article side by side.

As stated earlier K is a compact metric space, Δ a fixed point in K and $\mathcal{D} = \{D : D$ is a closed set in K $\Delta \in D\}$. Let $d(x,y)$ be the metric on $K, \mathcal{B}$ the σ-algebra of Borel sets of K, and $\mathcal{B}^*$ that of universally measurable sets of K. $\mathcal{C}(K)$ denotes the space of real continuous functions on K. $f \in \mathcal{B}$ (resp. $f \in b\mathcal{B}$), e.g., means that f is a real (resp. bounded real) $\mathcal{B}$-measurable function on K.

Let $\{H_D(x,\cdot) : D \in \mathcal{D}, x \in K\}$ be a family of measures on $(K, \mathcal{B})$ (and thus on $(K, \mathcal{B}^*)$). We introduce first the following hypotheses used in [11].

H1) $H_D(x,\cdot)$ is a probability measure concentrated on D for all D and x, and is the point mass ϵ_x if $x \in D$; $H_D(\cdot, B) \in \mathcal{B}^*$ for all D and $B \in \mathcal{B}^*$.

H2) (Markov property, or consistency) If $D \subset D'$,

$$H_D(x,\cdot) = \int H_{D'}(x, dy) H_D(y,\cdot)$$

for all x, i.e. $H_D f = H_D f = H_{D'} H_D f$ for all $f \in b\mathcal{B}$, where $H_D f(x) = \int H_D(x, dy) f(y)$.

To state H3), we need to define the nearly Borel sets (w.r.t. $\{H_D(x,\cdot)\}$). A set B in K is nearly Borel if for every probability measure μ on $\mathcal{B}$ there exist B_1, B_2 in $\mathcal{B}$ with $B_1 \subset B \subset B_2$ such that for all compact $C \subset B_2 - B_1, \int \mu(dx) H_{C \cup \Delta}(x, C) = 0$ (note that singletons $\{\Delta\}$ and $\{x\}$ are usually written as Δ and x for convenience). The family $\mathcal{B}^n$ of nearly Borel sets turns out to be a σ-algebra; obviously $\mathcal{B} \subset \mathcal{B}^n \subset \mathcal{B}^*$.

H3) (Nearly Borel measurability) $H_D(\cdot, B) \in \mathcal{B}^n$ for all D and $B \in \mathcal{B}$.

H4) (Quasi-left-continuity) For any x and sequence $D_n \downarrow D$ (D_n decreasing to D), $H_{D_n}(x, \cdot)$ converges weakly to $H_D(x, \cdot)$, i.e. $H_{D_n}f(x) \to H_D f(x)$ for all $f \in \mathcal{C}(K)$.

While H4) is not necessarily satisfied in a right process, the following hypothesis H4A) is necessary in a right process whose paths have left limits. H4A) is assumed in Theorem 2 of [11] in place of H4).

H4A) For any x and $D_n \downarrow D$ the following are satisfied:

H4A.1) $H_{D_n}(x, \cdot)$ converges weakly;

H4A.2) if compact sets $F_m \uparrow K - D$, then for $\epsilon > 0$ there is $\delta > 0$ such that for all m for which the weak limit

$$\nu_m(dy, dz) = w\text{-}\lim_m w\text{-}\lim_n H_{D_n}(x, dy)H_{F_m \cup D}(y, dz)$$

exists, $\nu_m\{(y, z) : 0 < d(y, z) < \delta\} < \epsilon$.

H5) (Intrinsic right continuity) For any x and increasing sequence D_n, and for any $F \in \mathcal{D}$ and $f \in b\mathcal{B}^*$, if $H_{D_n}(x, dy)$ converges weakly to $\mu(dy)$, then

$$H_{D_n}(x, dy)1_{\{H_F f(y) \in da\}} \to \mu(dy)1_{\{H_F f(y) \in da\}}$$

vaguely as subprobability measures on the locally compact $(K - F) \times (-\infty, \infty)$, (vague convergence means convergence of integrals of all real continuous functions vanishing at infinity).

This hypothesis was written in [11] in the following form. The two versions are equivalent under other conditions in any of the theorems stated in this section, which imply that for the (W_n) below, $W_\infty = \lim_n W_n$ exists a.s. and $\lim_n H_F f(W_n)$ exists a.s. on $\{W_\infty \notin F\}$.

Alternative version of H5): For any x and increasing sequence D_n, for any $F \in \mathcal{D}$ and $f \in b\mathcal{B}^*$, and with (W_n) denoting the nonhomogeneous reversed Markov chain (under a single probability measure P) satisfying $P(W_n \in \cdot) = H_{D_n}(x, \cdot)$ and $P(W_n \in \cdot \mid W_{n+1} = y) = H_{D_n}(y, \cdot)$, if $W_\infty = \lim_n W_n$ exists a.s. then $H_F f(W_n) \to H_F f(W_\infty)$ a.s. on $\{W_\infty \notin F\}$.

H6) (Transience) For any x and D with $x \notin D$, there exists a compact neighborhood C of x such that $\int H_D(x, dy)H_{C \cup \Delta}(y, C) < 1$.

We can now restate

Theorem 1 of [11]. *Let* $\{H_D(x, \cdot) : D \in \mathcal{D}, x \in K\}$ *be a family of measures on* $(K, \mathcal{B})$ *satisfying H1), H2), H3), H4), H5) and H6). Then there exists a right process* $(X_t; P^x)$ *on* K, *with* Δ *as the death point, such that starting at any* x *the hitting distribution of any* $D \in \mathcal{D}$ *is* $H_D(x, \cdot)$.

For the interested reader, we refer to [11] for a number of remarks on the various hypotheses. The q.l.c. in the form of H4) actually implies the usual q.l.c. of (X_t); therefore the process in the theorem is a Hunt process. Note that either by the transience hypothesis or by the fact $\{\Delta\} \in \mathcal{D}$ and $H_\Delta(x, \Delta) = 1$, which implies the lifetime $T_\Delta = \inf\{t \geq 0 : X_t = \Delta\}$ is finite a.s. P^x for all x, the process is transient. At the end of this section we will state a result in constructing recurrent processes.

Next we make changes of some of the hypotheses and introduce the following ones.

H3A) (Borel Measurability) $H_D(\cdot, B) \in \mathcal{B}$ for all D and $B \in \mathcal{B}$.

H4B) For any x and $D_n \downarrow D$, the following are satisfied:

H4B.1) = H4A.1);

H4B.2) = H4A.2);

H4B.3) with (W_n) denoting the nonhomegeneous Markov chain satisfying $P(W_n \in \cdot) = H_{D_n}(x, \cdot)$ and $P(W_n \in \cdot \mid W_{n-1} = y) = H_{D_n}(y, \cdot)$, if compact $F_m \uparrow K - D$, then a.s. on

$$\{W_\infty = \lim_n W_n \text{ exists}; \ \rho(\cdot) \equiv \rho(\omega, \cdot) = w\text{-}\lim_m w\text{-}\lim_n H_{F_m \cup D}(W_n, \cdot) \text{ exists}\}$$

we have $\rho\{W_\infty\} = 0$ or 1, and $\rho\{W_\infty\} = 0$ implies $\rho(\cdot) \neq H(W_\infty, \cdot)$, where $H(y, \cdot)$ is defined below.

For any $y \in K - \Delta, H(y, \cdot)$ denotes weak limit of $H_F(y, \cdot)$ as $F \uparrow K - y$, or of $H_{(K-B(y,1/k))\cup\Delta}(y, \cdot)$ as $k \to \infty$, where $B(y, \delta) = \{z : d(y, z) < \delta\}$; this weak limit exists as an easy consequence of H2), and by H5) we have $H(y, \{y\}) = 0$

or 1. The seemingly strange H4B.3) is added to avoid technical complications in the compactification below. See a remark after Theorem 1 for its meaning. The limit W_∞ and the weak limit ρ actually always exist a.s.

H6A) (Transience) For any x and neighborhood U of x, if D_n increases with $\inf_n H_{D_n}(x, K - U) > 0$, there exists a compact neighborhood C of x such that

$$\sup_n \int H_{D_n}(x, dy) H_{C \cup \Delta}(y, C) < 1 .$$

Before Theorem 1 can be stated, we need to introduce an appropriate compactification of K.

Compactification of K. Fix a countable family $\{C_i, i \geq 1\}$ such that $\Delta \in C^\circ$ (interior of C_i) for i odd and $\{K - C_i, i \text{ odd}\}$ contains a fundamental sequence of neighborhoods of x for each $x \neq \Delta$, and such that $C_i - \Delta$ is compact for i even and $\{(K - C_i) \cup \Delta, i \text{ even}\}$ is a fundamental sequence of neighborhoods of Δ. For each i choose g_i in the unit ball of $C(K)$ with $g_i = 0$ on C_i for i odd but with $g_i = 0$ on $C_i - \Delta, g_i(\Delta) = 1$ for i even, and with g_i strictly positive on $K - C_i$. Let $\{f_j, j \geq 1\}$ be dense in the unit ball of $C(K)$. Define a new metric $\hat{d}$ on K by

$$\hat{d}(x, y) = d(x, y) + \sum_{i,j} 2^{-i-j} \mid g_i(x) H_{C_i} f_j(x) - g_i(y) H_{C_i} f_j(y) \mid .$$

It is standard knowledge that the completion $(\hat{K}, \hat{d})$ of $(K, \hat{d})$ is compact, and $\mathcal{B} = \hat{\mathcal{B}} \cap K$ where $\hat{\mathcal{B}}$ is the Borel σ-algebra of $\hat{K}$, in particular $K \in \hat{\mathcal{B}}$. (We needed H3A), which implies that the functions $H_{C_i} f_j$ are in $\mathcal{B}$.) The mapping $\hat{\pi} : \hat{K} \to K$ denotes the continuous extension of the mapping $x \to x$ from $(K, \hat{d})$ to (K, d). (We note that, quite apart from $\hat{\pi}$, the symbol π will denote a certain ordinal in the sequel.) Each function $g_i H_{C_i} f_j$ is uniformly continuous on $(K, \hat{d})$ and so has a continuous extension on $\hat{K}$. Denote

$$G_i = \hat{\pi}^{-1}(K - C_i), i \text{ odd}; \quad G_i = \hat{\pi}^{-1}((K - C_i) \cup \Delta), i \text{ even} .$$

Since g_i has a continuous extension $g_i \circ \hat{\pi}$ on $\hat{K}$, which is strictly positive on G_i, $H_{C_i} f_j$ has a continuous extension on G_i, still denoted $H_{C_i} f_j$. Since $\{f_j\}$ is dense in the unit ball of $C(K)$, we obtain, for $x \in G_i - K$, a probability measure $H_{C_i}(x, \cdot)$ on $(K, \mathcal{B})$ (not $(\hat{K}, \hat{\mathcal{B}})$) such that for all j

(1.1) $x_n \to x$ in $(\hat{K}, \hat{d})$ iff $d(\hat{\pi}(x_n), \hat{\pi}(x)) \to 0$ and, for all i with $x \in G_i$,

$\qquad H_{C_i}(x_n, \cdot) \to H_{C_i}(x, \cdot)$ weakly (as measures on K).

Define (the set of "regular" points of $\hat{K}$ to be)

$$\hat{K}_r = \{x \in \hat{K} : \text{ for all } C_i \subset C_\ell, \text{ with } x \in G_\ell,$$

$$H_{C_i}(x, \cdot) = \int H_{C_\ell}(x, dy) H_{C_i}(y, \cdot)\}.$$

Of course $K \subset \hat{K}_r$.

We have defined for $x \in K - \Delta$ the probability measure $H(x, \cdot)$ to be the weak limit of $H_D(x, \cdot)$ as $D \uparrow K - x$, and observed that, as a consequence of H5 (see [11], Corollary 2.4), $H(x, \{x\}) = 0$ or 1. We also denote $H(\Delta, \cdot) = \epsilon_\Delta(\cdot)$. Let

(1.2) $$H = \{x \in K : H(x, \{x\}) = 0\}.$$

H is (to be) the set of holding points in K, and $K - H - \Delta$ the set of instantaneous points. We now define the hidden holding points. For $x \in \hat{K}_r - K$

$$H(x, \cdot) = \underset{C_i \uparrow (K - \hat{\pi}(x)) \cup \Delta}{w - \lim} H_{C_i}(x, \cdot)$$

exists (because of the equations defining $\hat{K}_r$) as a probability measure on K. Define

$$\hat{K}_0 = K \cup \{x \in \hat{K}_r - K : H(x, \{x\}) = 0\},$$

$$\hat{\mathcal{B}}_0 = \hat{\mathcal{B}} \cap \hat{K}_0.$$

The points in $\hat{K}_0 - K$ are called hidden holding points (although some of them may turn out to be branching points for the process in Theorem 1 or Theorem 2 in this article). Note that for $x \in \hat{K}_0 - K$, as $C_i \uparrow (K - \hat{\pi}(x)) \cup \Delta$, the measure

$H_{C_i}(x, \cdot)$ converges strongly to $H(x, \cdot)$, i.e. $H_{C_i}f(x) \to Hf(x) = \int H(x, dy)f(y)$ for all $f \in b\mathcal{B}$. From this it follows that the definition

$$(1.3) \qquad H_D(x, \cdot) = \int H(x, dy)H_D(y, \cdot), \quad x \in \hat{K}_0 - K, D \in \mathcal{D}$$

agrees with the previously defined measure when D is some C_i and $x \in G_i$. From (1.3) and H2) we have

$$(1.4) \qquad H_D f(x) = H_{D'} H_D f(x) \text{ for all } x \in \hat{K}_0, f \in b\mathcal{B}, D \subset D' \text{ in } \mathcal{D}.$$

Theorem 1. Let $\{H_D(x, \cdot) : D \in \mathcal{D}, x \in K\}$ be a family of measures on $(K, \mathcal{B})$ satisfying H1), H2), H3A), H4B), H5) and H6A). Then there exists a right process $(X_t; P^x)$ on $(\hat{K}_0, \hat{d})$, with Δ as the death point, such that starting at any $x \in \hat{K}_0$ the hitting distribution of any $D \in \mathcal{D}$ is $H_D(x, \cdot)$. A.s. $(P^x$ for all $x \in \hat{K}_0)$ the path (is right $\hat{d}$-continuous and) has left $\hat{d}$-limits. Each $x \in \hat{K}_0 - K$ is a holding or branching point, from which a jump is made into K (with distribution $H(x, \cdot)$). (X_t) satisfies the additional property that a.s. the path is left $\hat{d}$-continuous at all t for which $X_t \in \hat{K}_0 - K$.

Remarks. (i) Which points in $\hat{K}_0$ are branching points depends on the function $e(x)$ in Theorem 2 defined under the conditions of Theorem 1; these points of course can be deleted from $\hat{K}_0$. (ii) As said above, the reason to replace H3) by the stronger H3A) is to avoid measurability complications that would arise in the compactification above. (iii) Concerning H4B), H4B.1) = H4A.1) reflects the requirement that the paths of (X_t) have left limits (in (K, d)); the weak limit is of course (to be) $P^x[Y \in \cdot]$ where $Y = \lim_n X(T_{D_n})$. Note Y is either $X(T-)$ or $X(T)$ where $T = \lim_n T_{D_n}$. The weak limits ν_m in H4B.2) = H4A.2) actually always exists and so does $\nu = w\text{-}\lim_m \nu_m$; see [11], section 2. $\nu_m(dy, dz)$ is $P^x[Y \in dy, X(S_m) \in dz]$ where $S_m = \inf\{t \geq T : X_t \in F_m \cup D\}$, and $\nu(dy, dz)$ is $P^x[Y \in dy, X(S) \in dz]$ where $S = \lim_m S_m$. Now if $d(Y, X(S_m)) \to 0$, then $X(S) = Y \in D$ and so $S_m = S$ for all m. It is thus clear that H4B.2) must be valid. The d-limit W_∞ and the weak limit $\rho(\cdot)$ in H4B.3) always exist a.s., from arguments in [11], section 2. Of course $\rho(\cdot) = \rho(\omega, \cdot)$ is

$$P^x(X(S) \in \cdot \mid \vee_n \mathcal{F}(T_{D_n}))$$

where $\mathcal{F}(t)$ denotes the usual right continuous complete filtration generated by (X_t). The first part of H4B.3), that $\rho\{W_\infty\} = 0$ or 1, means that the time T_D is predictable on the accessible part of T_D. It is assumed to avoid technical complications in the above compactification. The second part of H4B.3), that $\rho\{W_\infty\} = 0$ implies $\rho(\cdot) \neq H(W_\infty, \cdot)$, is assumed because if $\rho\{W_\infty\} = 0$ then

$$X(S) \neq X(T-) = \lim_n X(T_{D_n}),$$

and so, provided $T < S$ (which is possible), the path should be placed, for $t \in [T, S)$, at a hidden holding point y with $\hat{\pi}(y) = W_\infty$ (but $y \neq W_\infty$) and $H(y, \cdot) = \rho(\cdot)$; but if $H(y, \cdot) = H(W_\infty, \cdot)$ then according to our compactification one must have $y = W_\infty$, a contradiction. Thus the second part is also assumed to avoid technical complications in the compactification. (iv) H6A) is (still) a reasonable transience assumption. It is equivalent to H6) under the q.l.c. H4); see [11], section 2.

Theorem 1 (resp. Theorem 1 of [11]) is proved via Theorem 2 (resp. Theorem 2 of [11]) below, after a function $e(x)$ is defined on $\hat{K}_0$ (resp. on K) and shown to satisfy H7A) (resp. H7)) to be stated next. $e(x)$ is meant to be the expected lifetime $P^x[T_\Delta]$ ($P^x[\Phi]$ denotes the expectation w.r.t. P^x for a real random variable Φ). Under H7) define $e_D(x) = e(x) - H_D e(x)$, which is of course to be $P^x[T_D]$. As a trivial consequence of H7.1), $e_D(x) > 0$ for $x \notin D$; thus $e_D \geq 0$, since $e_D(x) = 0$ for $x \in D$. From this and the equality $e_D(x) = e_{D'}(x) + H_{D'} e_D(x)$ for $D \subset D', e_D(x)$ is decreasing in D. Define also

$$h(x) = \begin{cases} \inf \{e_D(x) : D \in \mathcal{D}, x \notin D\}, & x \in K - \Delta \\ 0, & x = \Delta. \end{cases}$$

Then for $x \neq \Delta$, $h(x)$ is the decreasing limit of $e_D(x)$ as $D \uparrow K - x$. $h(x)$ will be the expected holding time at x if $x \in H$. (H was defined in (1.2) and is (to be) the set of holding points.)

Let e be a real-valued function on K; we state hypothesis

H7) e is a nonnegative $\mathcal{B}^*$-measurable function with $e(\Delta) = 0$ and satisfies the following:

H7.1) for any $x \neq \Delta$, neighborhood U of x and increasing sequence D_n, if $\inf_n H_{D_n}(x, K - U) > 0$ then $\inf_n e_{D_n}(x) > 0$;

H7.2) $h(x) = 0$ if $x \in K - H$;

H7.3) if $D_n \downarrow D$ and compact sets $F_m \uparrow K - D$, then for any x

$$\varlimsup_m \varlimsup_n \int H_{D_n}(x, dy) e_{F_m \cup D}(y) = 0 \; ;$$

H7.4) for any x and $e > 0$ there is $a > 0$ such that, for all compact $C \subset \{y : e(y) > a\}$, $H_{C \cup \Delta} e(x) < \epsilon$.

Theorem 2 of [11]. *Let* $\{H_D(x, \cdot) : D \in \mathcal{D}, x \in K\}$ *be a family of measures on* $(K, \mathcal{B})$ *satisfying H1) , H2) , H3) , H4A) and H5), and e be a real-valued function on K satisfying H7). Then there exists a unique right process $(X_t; P^x)$ on K, with Δ as the death point, such that starting at any x its expected lifetime is $e(x)$ and its hitting distribution of any $D \in \mathcal{D}$ is $H_D(x, \cdot)$.*

Note that H4) and H6) are not assumed in the theorem. The process constructed does not have to be quasi-left-continuous. See [11] for remarks on H7.1) through H7.4); they are all necessary for a right process whose pathes have left limits and whose expected lifetime $e(x)$ starting at any x is finite.

For Theorem 2 of this article, we consider a real-valued function e on $\hat{K}_0$. Under H7A) below we (again) define $e_D(x) = e(x) - H_D e(x)$, for $x \in \hat{K}_0$ and $D \in \mathcal{D}$ (recall the definition of $H_D(x, \cdot)$ in (1.3)). From H7A.1) $e_D(x) > 0$

for $x \in K - D$, and so $e_D \geq 0$ on K. Consequently if $D \subset D', e_D(x) = e_{D'}(x) + H_{D'} e_D(x) \geq e_{D'}(x)$ for all $x \in \hat{K}_0$. Define

$$h(x) = \begin{cases} \inf \{ e_D(x) : D \in \mathcal{D},\ \hat{\pi}(x) \in K - D \}, & \hat{\pi}(x) \neq \Delta \\ \inf \{ e_D(x) : D \in \mathcal{D},\ D - \Delta \text{ is } d\text{-compact} \}, & \hat{\pi}(x) = \Delta . \end{cases}$$

Note this definition agrees with the previous definition of h (under H7)) when $x \in K$. Note also $h(x)$ is the decreasing limit of $e_D(x)$ as $D \uparrow K - \hat{\pi}(x)$ if $\hat{\pi}(x) \neq \Delta$, and as $D - \Delta \uparrow K - \Delta$ through d-compacts if $\hat{\pi}(x) = \Delta$. Let us state hypothesis

H7A) e is a non-negative $\hat{\mathcal{B}}_0$-measurable function on $\hat{K}_0$ with $e(\Delta) = 0$ and satisfies the following:

H7A.1) for any $x \in K - \Delta$, d-neighborhood U of x and increasing sequence D_n in $\mathcal{D}$, if $\inf_n H_{D_n}(x, K - U) > 0$, then $\inf_n e_{D_n}(x) > 0$;

H7A.2) $h(x) = 0$ for $x \in K - H$;

H7A.3) for any $x \in \hat{K}_0$ and $\epsilon > 0$ there is $a > 0$ such that, for all d-compact $C \subset \{ y \in K : e(y) > a \}$, $H_{C \cup \Delta} e(x) < \epsilon$.

H7A.4) $h(x) \geq 0$ for $x \in \hat{K}_0 - K$;

H7A.5) for any random sequence (Y_n) in K such that $Y_\infty = \hat{d}\text{-}\lim_n Y_n$ exists a.s. in $\hat{K}_0$ and $\lim_n e(Y_n)$ exists a.s., and such that for all D and $\Lambda \subset \{ \hat{\pi}(Y_\infty) \notin D \}$, and also for all D with $D - \Delta$ d-compact and $\Lambda \subset \{ \hat{\pi}(Y_\infty) \notin D - \Delta \}$

(1.5) $\qquad P[H_D(Y_n, \cdot); \Lambda] \to P[H_D(Y_\infty, \cdot); \Lambda]$ strongly ,

we have $e_D(Y_n) \to e_D(Y_\infty)$ a.s. on $\{ \hat{\pi}(Y_\infty) \notin D; \lim_n e_D(Y_n) \text{ exists} \}$ for all D, and a.s. on $\{ \hat{\pi}(Y_\infty) \notin D - \Delta; \lim_n e_D(Y_n) \text{ exists} \}$ for all D with $D - \Delta$ d-compact.

Theorem 2. *Let* $\{ H_D(x, \cdot) : D \in \mathcal{D},\ x \in K \}$ *be a family of measures on* $(K, \mathcal{B})$ *satisfying H1), H2), H3A), H4B) and H5), and* e *be a real-valued function on* $\hat{K}_0$ *satisfying H7A). Then there exists a right process* $(X_t; P^x)$ *on* $(\hat{K}_0, \hat{d})$*, with* Δ *as the death point, such that starting at any* $x \in \hat{K}_0$ *its expected lifetime is* $e(x)$ *and its hitting distribution of any* $D \in \mathcal{D}$ *is* $H_D(x, \cdot)$*. A.s. the path (is right*

$\hat{d}$-continuous and) has left $\hat{d}$-limits. Each $x \in \hat{K}_0 - K$ is a holding point (when $h(x) > 0$) or a branching point (when $h(x) = 0$), from which a jump is made into K (with distribution $H(x, \cdot)$). The process satisfies the additional property that a.s. the path is left $\hat{d}$-continuous at all t for which $X_t \in \hat{K}_0 - K$, and it is unique if this additional property is required.

Remarks. (i) Note that H6A) is not assumed. (ii) H7A.1) and H7A.2) involve only the values of e and K; they and H7A.3) are necessary for the same reasons as are H7.1), H7.2) and H7.4) (note that in H7A.1), with $H_{D_n}(x, \cdot)$ regarded as a measure on $\hat{K}_0$, we have $H_{D_n}(x, K - U) = H_{D_n}(x, \hat{K}_0 - \hat{\pi}^{-1}(U))$). (iii) H7A.3), H7A.4) and H7A.5) involve the values of e on $\hat{K}_0 - K$. H7A.4) is of course necessary. H7A.5) makes it possible to use the same $\hat{K}_0$ as the state space for (X_t) regardless of what e is; otherwise a refined compactificaton is needed, involving the function e in the definition of $\hat{d}$. It also permits the "additional property" in the theorem to be satisfied. The presence of (Y_n) in H7A.5) may look a little strange at first. It is almost the same as, but stronger than, the more appealing statement that for any $x_n \in K$, $x \in \hat{K}_0$ such that $\hat{d}(x_n, x) \to 0$ and $\lim_n e(x_n)$ exists and such that

$$H_D(x_n, \cdot) \to H_D(x, \cdot) \text{ strongly}$$

for all D with $\hat{\pi}(x) \notin D$ if $\hat{\pi}(x) = \Delta$ or for all D with $D - \Delta$ d-compact if $\hat{\pi}(x) = \Delta$, we have $e_D(x_n) \to e_D(x)$ for all such D for which $e_D(x_n)$ converges. (iv) H7.3) is in general not satisfied; if it is, there is no need to use the points in $\hat{K}_0 - K$.

Next we state a result on constructing recurrent Markov processes from hitting distributions. This is an extenstion of Theorem 1 of [11]. Let $K, \Delta, \mathcal{B}, \mathcal{B}^*, \mathcal{D}$ be as above. Consider a fix (countable) open covering $\{E_m\}$ of $K - \Delta$ (a special but important case being $E_m \uparrow K - \Delta$) such that each E_m has a compact closure in $K - \Delta$. Denote $\mathcal{D}' = \{D \in \mathcal{D} : D^c \subset E_m \text{ for some } m\}$. Let $\{H_D(x, \cdot) : D \in \mathcal{D}', x \in K\}$ be a family of measures on $(K, \mathcal{B})$. We consider the same hypotheses H1) (see however the first sentence following Theorem 3 for $H_D(x, \cdot)$ being always a probability measure), H2), H3) (using an obvious definition of $\mathcal{B}^n$), H4), H5), of course with D, D', D_n, F in $\mathcal{D}'$, and a new hypothesis

H6B) (Local transience) For any x and D with $x \notin D$ and $D^c \subset E_m$, there exists a compact neighborhood C of x such that

$$\int H_D(x, dy) H_{C \cup (K - E_m)}(y, C) < 1 .$$

Theorem 3. *Let $\{H_D(x, \cdot) : D \in \mathcal{D}', x \in K\}$ be a family of measures on $(K, \mathcal{B})$ satisfying H1), H2), H3), H4), H5) and H6B). Then there exists a right process on K with Δ as the death point, such that starting at any x its hitting distribution of any $D \in \mathcal{D}'$ is $H_D(x, \cdot)$.*

Here the hitting distribution $P^x[X(T_D) \in \cdot]$ has a slightly different meaning from before: We still define $T_D = \inf \{t \geq 0 : X_t \in D\}$, but use the convention $X(T_D) = \Delta$ when $T_D = \infty$; thus $P^x[X(T_D) \in \cdot]$ is always a probability measure. The theorem is easily proved as follows. Using Theorem 1 of [11], or its proof, one obtains for each m a right process $(X_t^m; P_x^m)$ on K (actually a Hunt process) such that starting at any $x \in E_m$ its distribution of any D in $\mathcal{D}'_m = \{D \in \mathcal{D}' : D^c \subset E_m\}$ is $H_D(x, \cdot)$, and each $x \notin E_m$ is an absorbing point. That is, with $K - E_m$ considered as a single point Δ_m one obtains (X_t^m) up to the exit time from E_m, then easily attaches the correct (conditional) exit distribution at this exit time. The exit time of (X_t^m) from E_m is finite. Any two processes (X_t^m) and (X_t^n) when *stopped* (not *killed*) at the exit time from $E_m \cap E_n$ have the same hitting distributions and so are time-changes of each other. By the theorem of [12] one obtains a (global) right process $(X_t; P^x)$ on K, with Δ as the death point, which, when stopped at the exit time from any E_m, is a time-change of (X_t^m); so for each m its hitting distribution of any $D \in \mathcal{D}'_m$ agrees with that of (X_t^m), i.e. $H_D(x, \cdot)$, when starting at x. Thus (X_t, P^x) is as desired. (The process constructed is a standard process.)

Theorem 1 of this article can be extended in a similar manner.

(2.1) and (2.2) imply that (2.3) will follow if for $\hat{\pi}(x) \notin F_k(r) - \Delta$

$$H_{F_k(r)}(x, F_k(r) - \Delta) - \int H_D(x, dy) H_{F_k(r)}(y, F_k(r) - \Delta) \geq 0 .$$

But this last inequality follows from the fact that, for any F,

$$H_F(x, F - \Delta) - \int H_D(x, dy) H_F(y, F - \Delta)$$
$$= \int H(x, dz) \left[H_F(z, F - \Delta) - \int H_D(z, dy) H_F(y, F - \Delta) \right]$$

which is nonnegative because the above function of z is nonnegative on K, as a relatively easy consequence of H1) and H2) (see [11], display (2.1)). It remains to prove H7A.5). Thus assume (Y_n) is a random sequence in K such that $Y_\infty = \hat{d}\text{-}\lim_n Y_n$ exists a.s. and $\lim_n e(Y_n)$ exists a.s., and such that (1.5) holds. Since e is bounded, we claim that to prove H7A.5) it suffices to show $e(Y_n) \to e(Y_\infty)$ a.s. For assuming this and assuming, for some D, $\lim_n e_D(Y_n) > e_D(Y_\infty)$ (say) a.s. on $\Lambda = \{\hat{\pi}(Y_\infty) \notin D, \lim_n e_D(Y_n) \text{ exists }\}$ and $P(\Lambda) > 0$ (or for some D with $D - \Delta$ d-compact this is true with D in the definition of Λ replaced by $D - \Delta$), we must have

$$\lim_n P[H_D e(Y_n); \; \Lambda] = \lim_n P[e(Y_n) - e_D(Y_n); \; \Lambda]$$
$$< P[e(Y_\infty) - e_D(Y_\infty); \; \Lambda] = P[H_D e(Y_\infty); \; \Lambda] .$$

But this contradicts (1.5). To prove $e(Y_n) \to e(Y_\infty)$ a.s., observe that for each k there are at most countably many values of r for which $P[\hat{\pi}(Y_\infty) \notin \partial(F_k(r) - \Delta)] > 0$. If $\hat{\pi}(Y_\infty) \in (F_k(r) - \Delta)^\circ$, then since $d(Y_n, \hat{\pi}(Y_\infty)) \to 0$

$$H_{F_k(r)}(Y_n, F_k(r) - \Delta) \to 1 = 1_{F_k(r) - \Delta}(\hat{\pi}(Y_\infty)) .$$

Now assume $\Lambda_1 = \{\lim_n e(Y_n) > e(Y_\infty)\}$ (say) has positive probability. Then it is clear from (2.1) and (2.2) that for some k, r_1 we must have

$$\overline{\lim_n} P\left[\int_{r_1}^1 H_{F_k(r)}(Y_n, F_k(r) - \Delta) \, dr; \Lambda \right] > P\left[\int_{r_1}^1 H_{F_k(r)}(Y_\infty, F_k(r) - \Delta) \, dr; \Lambda \right]$$

where $\Lambda = \Lambda_1 \cap \{\hat{\pi}(Y_\infty) \notin F_k(r_1) - \Delta\}$. But by (1.4), and (1.5) with $D = F_k(r_1)$

$$P\left[\int_{r_1}^1 H_{F_k(r)}(Y_n, F_k(r) - \Delta)\, dr;\, \Lambda\right]$$

$$= P\left[\int H_D(Y_n, dy) \int_{r_1}^1 H_{F_k(r)}(y, F_k(r) - \Delta)\, dr;\, \Lambda\right]$$

$$\to P\left[\int H_D(Y_\infty, dy) \int_{r_1}^1 H_{F_k(r)}(y, F_k(r) - \Delta)\, dr;\, \Lambda\right]$$

$$= P\left[\int_{r_1}^1 H_{F_k(r)}(Y_\infty, F_k(r) - \Delta);\, \Lambda\right]$$

which contradicts $P(\Lambda_1) > 0$. ∎

3. The Trajectory Process Z_∞

In sections 3 through 6 we prove Theorem 2. The general approach is the same as in the proof of Theorem 2 of [11], and a number of theorems and proofs in [11] will be used. In this section we essentially restate what are proved in [11], sections 3 and 4. They construct a stochastic process Z_∞ describing the trajectories of the sought-for process (X_t), that is, (X_t) without its time scale. Z_∞ is defined as the projective limit process of a sequence of discrete time Markov chains Z_n on $(K, \mathcal{B})$ (or $(K, \mathcal{B}^*)$), whose time parameter ranges over ordinals up to a certain fixed countable ordinal, such that Z_{n+1} is a refinement of Z_n, i.e. Z_n is imbedded in Z_{n+1}, for all n. Sections 4 and 5 will deal with the time scale, and in section 6 we define (X_t) and prove that it has the desired properties. New proofs are needed in those sections.

To define the Z_n fix a sequence of finite open coverings $\mathcal{U}_1 \subset \cdots \subset \mathcal{U}_n \subset \cdots$ of K such that each $\mathcal{U}_n$ is closed under (finite) union and intersection and for all x

$$U_{nx} = U(n, x) = \bigcap \{U \in \mathcal{U}_n : x \in U\}$$

has d-diameter $< 1/n$. Let

$$\mathcal{D}_n = \{(K - U) \cup \Delta : U \in \mathcal{U}_n\}, \quad \mathcal{D}_\infty = \bigcup_n \mathcal{D}_n,$$

$$D_{nx} = D(n, x) = (K - U_{nx}) \cup \Delta.$$

Then $\mathcal{D}_n$ is closed under union and intersection; $\mathcal{D}_n \uparrow \mathcal{D}_\infty$; $D(n,x)$ is the largest set in $\mathcal{D}_n$ not containing x; and if $y \notin D(n,x)$ then $d(x,y) < 1/n$ and $D(n,x) \subset D(k,y)$ for $k \geq n$.

Let π denote the (countable) ordinal ω^ω; here ω is the first infinite ordinal and $\omega^\omega = \lim_n \omega^n$. $\mathcal{L}$ will denote the set of limit ordinals $\alpha \leq \pi$. For $0 < \alpha \leq \pi$, $\alpha \notin \mathcal{L}$, $\alpha-$ denotes its predecessor $\alpha - 1$; if $\alpha \in \mathcal{L}$, $\alpha-$ is not an ordinal, but will have a meaning below when used.

For each n define a Markov chain $Z_n = (Z_{n\alpha} = Z(n,\alpha), \alpha < \pi; Q_x^n, x \in K)$ on $(K, \mathcal{B}^*)$ with one-step transition probability $Q_n(x,B) = H_{D(n,x)}(x,B)$ and satisfying, for $\alpha \in \mathcal{L}$,

$$Q_x^n(Z_{n\alpha} \in \cdot \mid Z_{n\gamma}, \gamma < \alpha) = w\text{-}\lim_{\gamma \uparrow \alpha} H_{D(n,\alpha-)}(Z_{n\gamma}, \cdot)$$

where $D(n,\alpha-) = \lim_{\beta \uparrow \alpha} \bigcap_{\beta < \gamma < \alpha} D(n, Z_{n\gamma})$; this weak limit (again as a probability measure on K) exists a.s. (Q_x^n for all x) by martingale arguments. We also denote

$$D(n,\alpha-) = D(n, Z(n,\alpha-)) \quad 0 < \alpha \notin \mathcal{L}.$$

It turns out that the path $\alpha \to Z_{n\alpha}$ has left d-limits (at all $\alpha \in \mathcal{L}$) a.s., and $Z_{n\alpha} \equiv \Delta$ for $\alpha \geq \omega^{|\mathcal{D}_n|}$ a.s. where $|\mathcal{D}_n|$ is the cardinality of $\mathcal{D}_n$. We sometimes need the convention $Z_{n\pi} \equiv \Delta$. With $\tau_D = \tau_D^n = \inf\{\alpha : Z_{n\alpha} \in D\}$ ($\inf \emptyset = \pi$ in this context), we have the following fact about hitting distributions of Z_n, which is of basic importance:

$$(3.1) \qquad Q_x^n[Z(n,\tau_D) \in \cdot] = H_D(x,\cdot), \quad x \in K, D \in \mathcal{D}_n.$$

Since for $m < n$ we have $\mathcal{D}_m \subset \mathcal{D}_n$, the above implies that Z_m can be imbedded in Z_n, i.e. there exists an increasing family of stopping times $\sigma(n,m,\beta)$, $\beta < \pi$, in Z_n such that $(Z_{n,\sigma(n,m,\beta)}, \beta < \pi; Q_x^n, x \in K)$ is equivalent to Z_m. Indeed, these stopping times (valid also when $n = m$) are defined by: $\sigma(n,m,0) = 0$,

$$\sigma(n,m,1) = \tau_{D(m,y)}^n \quad \text{if } Z(m,0) = y$$

$$\sigma(n,m,\beta+1) = \sigma(n,m,\beta) + \sigma(n,m,1) \circ \theta_{n,\sigma(n,m,\beta)}$$

where $\theta_{n,\alpha} = \theta(n,\alpha)$ are the obvious shift operators for Z_n, and for $\beta \in \mathcal{L}$

$$\sigma(n,m,\beta) = \lim_{\gamma \uparrow \beta} \left(\sigma(n,m,\gamma) + \tau^n_{D(m,\beta-)} \circ \theta_{n,\sigma(n,m,\gamma)} \right),$$

here $D(m,\beta-) = \lim_{\gamma' \uparrow \beta} \bigcap_{\gamma' < \gamma < \beta} D(m, Z_{n,\sigma(n,m,\gamma)})$. Below we will also denote for $\beta \in \mathcal{L}$

$$\sigma(n,m,\beta-) = \lim_{\gamma \uparrow \beta} \sigma(n,m,\gamma).$$

In general $\sigma(n,m,\beta-) < \sigma(n,m,\beta)$.

Let $Z_\infty = (Z_{n\alpha} = Z(n,\alpha),\ \alpha < \pi,\ n \leq 1;\ P^x, x \in K)$ be the projective limit process of the Z_n: for each n, $Z_n = (Z_{n\alpha}, \alpha < \pi; P^x, x \in K)$ is equivalent to the previous $(Z_{n\alpha}; Q^n_x)$, and for $m \geq n$, with $\sigma(n,m,\beta)$ defined as above for the new Z_n, we have for every sample point (not just a.s.)

(3.2) $$Z(n,\sigma(n,m,\beta)) = Z(m,\beta) \text{ for all } \beta.$$

The sample space of Z_∞ is denoted Ω (which can be taken as a certain subspace of the product space $\Pi_{n,\alpha} K_{n\alpha}$ where each $K_{n\alpha}$ is a copy of K), and a typical sample point denoted ω (the first infinite ordinal ω is now forgotten).

We denote by $T = T(\omega)$ the set of "(trajectory) times" (n,α), $n \geq 1$, $\alpha < \pi$ and $T' = T'(\omega)$ the of "times" $(n,\alpha-)$, $n \leq 1$, $0 < \alpha \leq \pi$, and by T_n, T'_n their respective subsets of $(n,\alpha), (n,\alpha-)$ with n fixed. Of course $T \subset T', T_n \subset T'_n$. Regarded as dependent on ω, $T'(\omega)$ has relations "$<$" and "$=$" defined as follows: for $m \geq n$.

(i) $(n,\alpha) < (m,\beta)$ (resp. $=,>$) if and only if $\alpha < \sigma(n,m,\beta)$ (resp. $=,>$).

(ii) $(n,\alpha) < (m,\beta-)$ (resp. $>$) where $\beta \in \mathcal{L}$, if and only if
$\alpha < \sigma(n,m,\beta-)$ (resp. $\geq$).

(iii) $(n,\alpha-) < (m,\beta)$ (resp. $>$) where $\alpha \in \mathcal{L}$, if and only if
$\alpha \geq \sigma(n,m,\beta)$ (resp $>$).

(iv) $(n,\alpha-) < (m,\beta-)$ (resp. $=,>$) where $\alpha \in \mathcal{L}, \beta \in \mathcal{L}$, if and only if $\alpha < \sigma(n,m,\beta-)$ (resp. $=,>$).

Observe that for $\beta \in \mathcal{L}$, there may be many (n, α) with $(m, \beta-) < (n, \alpha) < (m, \beta)$. Note that with the "=" in $T'(\omega)$, we have $T_n(\omega) \uparrow T(\omega), T_n'(\omega) \uparrow T'(\omega)$.

The trajectory of ω is the function $(n, \alpha) \rightarrow Z_{n\alpha}(\omega)$ on $T(\omega)$. From (3.2) and the existence of left limits in Z_n, it is clear that almost every path (and we will assume every path) has no d-oscillation.

Next we define σ-algebras generated by the $Z_{n\alpha}$ (regarded as taking values in $(K, \mathcal{B}^*)$). Let

$$\mathcal{H}(n, \alpha) = \sigma(Z_{n\gamma}, \gamma \geq \alpha), \; \mathcal{H}(n\alpha-) = \sigma(Z_{n\gamma}, \gamma < \alpha), \; \alpha \leq \pi$$

$$\mathcal{G} = \sigma(Z_{n\alpha}, n \geq 1, \alpha < \pi), \; \mathcal{F} = \bigcap_\mu \mathcal{G}^\mu$$

where $\mathcal{G}^\mu$ is the P^μ-completion of $\mathcal{G}, \mu$ any probability measure on $K, P^\mu = \int \mu(dx) P^x$. For a Z_n-stopping time τ (i.e. one w.r.t. $\{\mathcal{H}(n, \alpha), \alpha < \pi\}$), $H(n, \tau)$ and $H(n, \tau-)$ are defined in the usual way. Finally let

$$\mathcal{G}(m, \beta) = \bigvee_{n \geq m} \mathcal{H}(n, \sigma(n, m, \beta)), \; \mathcal{G}(m, \beta-) = \bigvee_{n \geq m} \mathcal{H}(n, \sigma(n, m, \beta-)-),$$

$$\mathcal{G}((m, \beta)-) = \bigvee_{n \geq m} \mathcal{H}(n, \sigma(n, m, \beta)-) \, ,$$

and $\mathcal{F}(m, \beta), \mathcal{F}(m, \beta-), \mathcal{F}((m, \beta)-)$ be their rejective completions in $\mathcal{F}$ w.r.t. all the measures P^μ (in the usual way). The Markov property of Z_∞ can be stated as follows: For $\emptyset \in b\mathcal{F}$, $x \in K$, stopping time τ w.r.t. $\{\mathcal{F}(m, \beta), \beta < \pi\}$

$$P^x(\Phi \circ \theta_{m\tau} \mid \mathcal{F}(m, \tau)) = P^{Z(m, \tau)}(\Phi) \; \text{on} \; \{\tau < \pi\}$$

where $\theta_{m\beta}$ are the obvious shift operators for Z_∞.

We begin to deal with the time scale. Define

$$e(n, \alpha) = e_{D(n, \alpha)}(Z(n, \alpha)) = e_{D(n, Z(n, \alpha))}(Z(n, \alpha)), \; \alpha < \pi$$

$$e(n, \alpha-) = \lim_{\gamma \uparrow \alpha} e_{D(n, \alpha-)}(Z(n, \gamma)), \; \alpha \in \mathcal{L} \, .$$

The above limit exists a.s. by martingale arguments. Note $e(n,0) = e_{D(n,y)}(y)$ if $Z(n,0) = y$ and $e(n,\alpha) = e(n,0) \circ \theta_{n\alpha}$. (We remark that $e(n,\alpha-), \alpha \leq \pi$, was written as $e(n,\alpha)$ in [11].) Next define

$$R(n,\alpha) = \sum_{0 < \gamma \leq \alpha} e(n,\gamma-) = \sum_{\gamma < \alpha} e(n,\gamma) + \sum_{\gamma \leq \alpha, \gamma \in \mathcal{L}} e(n,\gamma-), \ \beta \leq \pi$$

$$R(n,\alpha-) = \sum_{0 < \gamma < \alpha} e(n,\gamma-) = \lim_{\gamma \uparrow \alpha} R(n,\gamma), \ \beta \in \mathcal{L}$$

and for $m \leq n$

(3.3) $\qquad R(n,m,\beta) = R(n,\sigma(n,m,\beta)), \ \beta \leq \pi$

$$R(n,m,\beta-) = R(n,\sigma(n,m,\beta-)-) = \lim_{\gamma \uparrow \beta} R(n,m,\gamma), \ \beta \in \mathcal{L} \,.$$

We have the following facts: for all x

(3.4) $\qquad\qquad\qquad P^x(R(n,\pi)) = e(x)$

$$P^x(R(n,\tau_D^n)) = e_D(x), \ D \in \mathcal{D}_n$$

$$P^x(R(n,m,\beta)) = P^x(R(m,\beta))$$

and

(3.5) $\qquad \{R(n,\pi), n \geq 1\}$ is uniformly integrable w.r.t. P^x.

4. Convergence of Approximating Times

We will define times $S(m,\beta)$ for the sought-for process (X_t) which have the following meaning when (X_t) has no holding points (even hidden ones): $S(m,0) = 0$; $S(m,1) = T_{D(m,y)}$ if $X_0 = y$; and $S(m,\beta)$ is the βth iterate of $S(m,1)$ (in the same manner that $\sigma(n,m,\beta)$ is the βth iterate of $\sigma(n,m,1)$). If holding points (including hidden ones) exist the $S(m,\beta)$ have the same meaning, except that holding times appear in their average (i.e. expected) values—we will omit the last step of constructing the desired (X_t) by diffusing, so to speak, these averaged holding times (see the beginning of section 6) .

In [11] under the conditions of Theorem 2 and assuming $H = \emptyset$ (condition $NH)$), (recall H is the set of holding points in K), it was proved that for all x

$$\{R(n, m, \beta), n \geq m\} \text{ converges in } P^x - \text{measure uniformly in } \beta,$$

i.e. for $\varepsilon > 0$ there is n such that $P^x(|R(n, m, \beta) - R(n', m, \beta)| > \varepsilon) < \varepsilon$ for all $n' > n$ and all β. It follows that there exist positive integer functions $n_k = n_k(x) \to \infty$, $\mathcal{B}^*$-measurable in x, such that a.s.

$$S(m, \beta) = \lim_k R(n_k, m, \beta) \text{ exists for all } m, \beta.$$

In the case $H \neq \emptyset$ (condition (H) in [11]), the definition of $S(m, \beta)$ and their convergence were more involved, and the proof was somewhat sketchy. In examining the present situation we have found a gap in that proof. What we do below is to produce a new definition and convergence proof for $S(m, \beta)$, which work for the present situation as well as the case (H) in [11].

We first locate where holding times occur. The symbol $(\bar{n}, \bar{\alpha})$ will (always) denote an increasing sequence $\{(n_j, \alpha_j)\}$ in $\mathcal{T} = \mathcal{T}(\omega)$ where n_j is strictly increasing. Its limit, also denoted $(\bar{n}, \bar{\alpha})$, is regarded to always exist. Two sequences $(\bar{n}, \bar{\alpha}) \leq (\bar{n}', \bar{\alpha}')$ if and only if for all j there exists ℓ such that $(n_j, \alpha_j) \leq (n'_\ell, \alpha'_\ell)$; if also $(\bar{n}', \bar{\alpha}') \leq (\bar{n}, \bar{\alpha})$, we say they are equal. Note $(\bar{n}, \bar{\alpha}) = (k, \gamma)$ (a constant (k, γ) in $\mathcal{T}$ is also regarded as a sequence) if and only if $(n_j, \alpha_j) = (k, \gamma)$ for all large j.

Let

$$U[m, x] = \cup \{U(m, y) : d(x, y) \leq 1/m\}, \ D[m, x] = (K - U[m, x]) \cup \Delta.$$

Note that $x \in U(m, x) \subset B(x, 1/m) \subset U[m, x] \subset B(x, 2/m)$, and $D[m, x] \in \mathcal{D}_m$, $D[m, x] \uparrow K - x$ for $x \neq \Delta$, $D[m, \Delta] \uparrow K$. Define for $m \leq n$

(4.1) $e(n, \alpha, m) = e_{D[m, Z(n, \alpha)]}(Z(n, \alpha))$

(note $e(n, \alpha, m)$ is comparable to $h(n, \alpha, m)$ in [11] but not the same), and define for a sequence $(\bar{n}, \bar{\alpha})$

$$(4.2) \qquad e(\bar{n}, \bar{\alpha}, m) = \varliminf_j e(n_j, \alpha_j, m).$$

Theorem 4.1. *A.s.* *(*P^μ *for all probability measure μ on K), $e(\bar{n}, \bar{\alpha}, m)$ is well-defined by (4.2), i.e. $(\bar{n}, \bar{\alpha}) = (\bar{n}', \bar{\alpha}')$ implies $e(\bar{n}, \bar{\alpha}, m) = e(\bar{n}', \bar{\alpha}', m)$; in particular the $\liminf$s are limits.*

Proof. Suppose not. Then there exists μ with $\int \mu(dx)e(x) < \infty$ such that with positive P^μ-outer-measure the statement is false. It is easy to see that with $D = D[m, y]$ for some y (note there are only finitely many such D, and if $d(y_j, y) \to 0$ then $D[m, y_j] = D[m, y]$ for large j), there exists a probability measure ν with $\int e \, d\nu < \infty$ (which may be taken as $P^\mu[Z(m, \beta) \in \cdot]$ for some β), reals $a < b$ and integers $m = n_1 < \cdots < n_j < \cdots$ such that

$$(4.3) \qquad \varliminf_j e(n_j, \tau_j, m) < a < b < \varlimsup_j e(n_j, \tau_j, m)$$

with positive P^ν-measure, where $\tau_1 = 0$ and for $j \geq 2, \tau_j$ is the Z_{n_j}-stopping time defined successively by

$$\tau_j = \inf\left\{\alpha : (n_j, \alpha) > (n_{j-1}, \tau_{j-1}) \text{ and } e_D(Z(n_j, \tau_j)) > b; \text{ or } \alpha = \tau_D^{n_j}\right\}$$

for j even, and for j odd τ_j is defined in the same way except with "$> b$" replaced by "$< a$". But (4.3) is impossible because $\{e_D(Z(n_j, \tau_j))\}$ is a positive supermartingale (with $P^x[e_D(Z(n_1, \tau_1))] \leq \int e \, d\nu < \infty$). (Incidentally, this argument is behind the existence of the limits defining $e(n, \alpha-)$ for $\alpha \in \mathcal{L}$.) ∎

We now define (the (average) "holding time" at $(\bar{n}, \bar{\alpha})$ as)

$$(4.4) \qquad h(\bar{n}, \bar{\alpha}) = \lim_m e(\bar{n}, \bar{\alpha}, m).$$

Since $(\bar{n}, \bar{\alpha}, m)$ is decreasing in m, the above theorem implies that a.s. $h(\bar{n}, \bar{\alpha})$ is defined for all $(\bar{n}, \bar{\alpha})$.

We will say that a constant (sequence) (m, β) and a nonconstant sequence $(\bar{n}, \bar{a})$ (in $T = T(\omega)$) are *inseparable* if (m, β) is the least element in T such that $(n_j, a_j) < (m, \beta)$ for all j, and if $d\text{-}\lim_j Z(n_j, a_j) = Z(m, \beta)$ (i.e. $Z(\cdot, \cdot)$ is left d-continuous at (m, β)).

Theorem 4.2. (i) *For any m, β the set $\Gamma(m, \beta) = \{$ there exists $(\bar{n}, \bar{a})$ insepa-rable from (m, β), i.e. $Z(\cdot, \cdot)$ is left d-continuous at $(m, \beta)\}$ is in the σ-algebra $\mathcal{F}((m, \beta)-)$. (ii) A.s. on $\Gamma(m, \beta), h(m, \beta) = h(\bar{n}, \bar{a})$ where $(\bar{n}, \bar{a})$ is inseparable from (m, β).*

Proof. (i) It suffices to show the following: for any probability measure μ on K and $D \in \mathcal{D}_m$, the set $\Gamma = \{$there exists $(\bar{n}, \bar{a})$ inseparable from $(m, \tau_D^m)\}$ is in $\mathcal{G}((m, \tau_D^m)-)^\mu$ (the superscript μ indicates P^μ-completion), and a.s. P^μ on $\Gamma, h(m, \tau_D^m) = h(\bar{n}, \bar{a})$ where $(\bar{n}, \bar{a})$ is inseparable from (m, τ_D^m). It is easy to see that there exist countably many (increasing) sequences $(\bar{n}^i, \bar{\tau}^i)$ of stopping times (meaning each τ_j^i is a $Z(n_j^i)$-stopping time) such that $(\bar{n}^i, \bar{\tau}^i) \leq (m, \tau_D^m)$ and $d\text{-}\lim_j Z(n_j^i, \tau_j^i) \in D$ for all i, and a.s. P^μ on $\Gamma, (m, \tau_D^m)$ is inseparable from some $(\bar{n}^i, \bar{\tau}^i)$ (that is nonconstant) . (One considers the hitting times (in T) of a sequence $D_n \downarrow D$ with $D_n \in \mathcal{D}_\infty$ and $D \subset D_n^\circ$, their limit, and iterates of this limit, etc.) Fix one such $(\bar{n}^i, \bar{\tau}^i)$ and denote it by $(\bar{n}, \bar{\tau})$. For any $(\bar{n}, \bar{a})$, denote by $\langle \bar{n}, \bar{a} \rangle$ the least (k, γ) in T with $(\bar{n}, \bar{a}) \leq (k, \gamma)$, if such (k, γ) exists. We claim

(4.5) $\langle \bar{n}, \bar{\tau} \rangle$ exists a.s. P^μ .

For if not, then on a set of positive P^μ-measure, $d\text{-}\lim Z(k, \gamma) = d\text{-}\lim_j Z(n_j, \tau_j)$ as (k, γ) decreases to $(\bar{n}, \bar{\tau})$ (in an obvious sense), with $(k, \gamma) < (m, \tau_D^m)$. This implies that, with

$$F_N = K - \cup \{U(N, y) : y \in D\} ,$$

the weak limits

$$\nu_N^*(dy, dz) = w\text{-}\lim_j P^\mu[Z(n_j, \tau_j) \in dy] H_{F_N \cup D}(y, dz)$$

(which exist and indeed have an obvious meaning in Z_∞) satisfy

$$\lim_{\delta \downarrow 0} \sup_N \nu_N^* \{(y, z) : 0 < d(y, z) < \delta\} > 0 \,.$$

From this and arguments in the proof of [11], Theorem 3.4, one can find $x \in K, D_n \downarrow D$ such that with the above F_N, (H4B.2) is contradicted, (we omit the details). Thus (4.5) holds. Now denote $W_j^* = Z(n_j, \tau_j), W_\infty^* = d\text{-}\lim_j W_j^*$, and with the above F_N,

$$\rho^*(\cdot) = \rho^*(\omega, \cdot) = w\text{-}\lim_N \; w\text{-}\lim_j H_{F_N \cup D}(W_j^*, \cdot) \,.$$

$\rho^*(\cdot)$ exists a.s. P^μ (by martingale arguments and (H2); see (again) [11], section (2)); actually it is obviously the conditional distribution of $Z\langle \bar{n}, \bar{\tau}\rangle$ given $\bigvee_j \mathcal{G}(n_j, \tau_j)$, which will be denoted $\mathcal{G}(\bar{n}, \bar{\tau})$. Note that W_j^*, W_∞^* and ρ^* resemble the W_j, W_∞ and ρ in (H4B.3), which we now use to show that

$$(4.6) \qquad \rho^*\{W_\infty^*\} = 0 \text{ or } 1 \text{ a.s. } P^\mu \,.$$

For suppose the contrary; then (again) by the arguments in the proof of [11], Theorem 3.4, one can find $x \in K, D_n \downarrow D$, such that, with the above F_N, the (W_j) defined in (H4B.3) satisfies the condition that $W_\infty = d\text{-}\lim_j W_j$ exists a.s. and the weak limit ρ exists a.s., but such that $0 < \rho\{W_\infty\} < 1$ on a set of positive probability. Thus (H4B.3) is contradicted. Now (4.6) implies $Z\langle \bar{n}, \bar{\tau}\rangle = W_\infty^*$ (equivalently $(\bar{n}, \bar{\tau})$ is inseparable from (m, τ_D^m)) with conditional probability 0 and 1 resp. Let $\Gamma_i = \{W_j^* \notin D \text{ for all } j; \rho^*\{W_\infty^*\} = 1\}$ (noting $(\bar{n}, \bar{\tau}) = (\bar{n}^i, \bar{\tau}^i)$). Then $\Gamma_i \in \mathcal{G}((m, \tau_D^m)-)$. Since the symmetric difference between Γ and $\bigcup_i \Gamma_i$ has P^μ-measure $0, \Gamma \in \mathcal{G}((m, \tau_D^m)-)^\mu$. μ being arbitrary, we have $\Gamma \in \mathcal{F}((m, \tau_D^m)-)$.

(ii) For the above $(\bar{n}, \bar{\tau})$ write $\Gamma_0 = \{W_j^* \notin D \text{ for all } j; \rho^*\{W_\infty^*\} = 1\}$. It suffices to show $h(\bar{n}, \bar{\tau}) = h(m, \tau_D^m)$ a.s. P^μ on Γ_0. This will follow from the definitions of $h(\bar{n}, \bar{\alpha})$ and $h(m, \tau_D^m) = h(Z(m, \tau_D^m))$ if for all $C \in \mathcal{D}_\infty$ we have a.s. P^μ on $\Gamma_0 \bigcap \{W_\infty^* \notin C\}$, and for all $C \in \mathcal{D}_\infty$ with $C - \Delta$ d-compact we have a.s. P^μ on $\Gamma_0 \bigcap \{W_\infty^* \notin C - \Delta\}$,

$$e_C(W_j^*) \to e_C(W_\infty^*) \,.$$

To prove this convergence for all C in $\mathcal{D}$ (not necessarily in $\mathcal{D}_\infty$) we use H7A.5). Let $P(\cdot) = P^\mu(\cdot \mid \Gamma_0), Y_j = W_j^*$ and $Y_\infty = d\text{-}\lim_j Y_j = W_\infty^*$. A supermartingale argument (as used in the proof of Theorem 4.1) shows that $e_C(Y_j)$ converges a.s. on $\{Y_\infty \notin C\}$, and a.s. on $\{Y_\infty \notin C - \Delta\}$ if $C - \Delta$ is d-compact. In particular with $C = \Delta$ we have $e(Y_j)$ converges a.s. Thus by H7A.5) we need to show that for all $C \in \mathcal{D}$ and $\Lambda \subset \Gamma_0 \cap \{W_\infty^* \notin C\}$, and for all $C \in \mathcal{D}$ with $C - \Delta$ d-compact and $\Lambda \subset \Gamma_0 \bigcap W_\infty^* \notin C - \Delta\}$,

$$P^\mu[H_C f(W_j^*); \Lambda] \to P^\mu[H_C f(W_\infty^*); \Lambda]$$

for all $f \in bB$. But this follows from the fact $H_C f(W_j^*) \to H_C f(W_\infty^*)$ a.s. P^μ on $\Gamma_0 \bigcap \{W_\infty^* \notin C\}$ (resp. on $\Gamma_0 \bigcap \{W_\infty^* \notin C - \Delta\}$ if $C - \Delta$ is d-compact), which can be established as follows. By replacing C by a slightly larger $C_1 \in \mathcal{D}_\infty$ and f by $H_C f$ we may assume $C \in \mathcal{D}_\infty$. Let $C \in \mathcal{D}_{n_0}$ and $\mu_1 = P^\mu[Z(n_{j_0}, \tau_{j_0}) \in \cdot]$ for some large j_0. Use the upcrossing lemma on the martingale $\{H_C f(Z(n, \alpha \wedge \tau_C^n)), \alpha < \pi\}$ w.r.t. P^{μ_1} where $n \geq n_0$, and let $n \to \infty$ to obtain the desired a.s. P^μ convergence of $H_C f(W_j^*)$. That its limit is $H_C f(W_\infty^*)$ follows from the fact $\Gamma_0 \in \mathcal{G}(\bar{n}, \bar{\tau})$. ∎

As the above proof shows, the holding times $h(m, \beta)$ and $h(\bar{n}, \bar{\alpha})$ for two inseparable (m, β) and $(\bar{n}, \bar{\alpha})$ should count as one, and we will identify $(\bar{n}, \bar{\alpha})$ with (m, β) when counting different holding times along a trajectory.

Theorem 4.3. *A.s. there are at most countably many $(\bar{n}, \bar{\alpha})$ such that $h(\bar{n}, \bar{\alpha}) > 0$. Furthermore, for a fixed μ, there exist countably many (increasing) sequences $(\bar{n}^i, \bar{\tau}^i) = \{(n_j^i, \tau_j^i)\}$ of stopping times (meaning τ_j^i is a $Z(n_j^i)$-stopping time for all i, j) such that a.s. P^μ*

$$\{(\bar{n}, \bar{\alpha}) : h(\bar{n}, \bar{\alpha}) > 0\} \subset \{(\bar{n}^i \bar{\tau}^i), i \geq 1\}$$

(and such that a.s. P^μ if $h(\bar{n}^i, \bar{\tau}^i) > 0$ then $(\bar{n}^i, \bar{\tau}^i) \neq (\bar{n}^\ell, \bar{\tau}^\ell)$ for $\ell \neq i$).

Proof. We may assume $\int \mu(dx)e(x) < \infty$ in proving both statements. Fix $m_j \uparrow \infty$, e.g. $m_j = j$. For a fixed ω and a sequence $(\bar{n}, \bar{\alpha})$ with $h(\bar{n}, \bar{\alpha}) > 0$, there exists a subsequence of $(\bar{n}, \bar{\alpha})$, still denoted $(\bar{n}, \bar{\alpha})$ such that $e(n_j, \alpha_j, m_j) \rightarrow h(\bar{n}, \bar{\alpha})$ (obviously any faster subsequence of $(\bar{n}, \bar{\alpha})$ also satisfies this). Therefore, for $\varepsilon > 0$, it is not difficult to define countably many sequences $(\bar{n}^i, \bar{\tau}^i)$ of stopping times such that a.s. P^μ

$$\lim_j e(n^i_j, \tau^i_j, m_j) = h(\bar{n}^i, \bar{\tau}^i)$$

and $h(\bar{n}, \bar{\alpha}) > \varepsilon$ implies $(\bar{n}, \bar{\alpha}) = (\bar{n}^i, \bar{\tau}^i)$ for some i, and such that, a.s. P^μ, if $h(\bar{n}^i, \bar{\tau}^i) > 0$ then $(\bar{n}^i, \bar{\tau}^i) \neq (\bar{n}^\ell, \bar{\tau}^\ell)$ for $\ell \neq i$. Note the inseparable $(\bar{n}, \bar{\alpha})$'s have been identified. The $(\bar{n}^i, \bar{\tau}^i)$ may be required to satisfy the following. Let

$$\sigma^i_j = \tau^i_j + \tau^{n^i_j}_{D[m_j, y]} \circ \theta(n^i_j, \tau^i_j)$$

where $y = Z(n^i_j, \tau^i_j)$, and let $I_{ij} = \{(k, \gamma) \in \mathcal{T} : (n^i_j, \tau^i_j) \leq (n, \gamma) < (n^i_j, \sigma^i_j)\}$. Then for any ℓ we have a.s. P^μ, $I_{ij}, 1 \leq i \leq \ell$, with $h(\bar{n}^i, \bar{\tau}^i) > 0$, are disjoint for all large j. The countability assertion then follows from the fact that, as can be easily seen,

$$(4.7) \qquad \int \mu(dx)e(x) \geq \lim_j P^\mu \left[\sum_{i \leq \ell} e(n^i_j, \tau^i_j, m_j) \right]$$

for all ℓ, and consequently $\int e d\mu$ dominates $P^\mu \left[\sum h(\bar{n}, \bar{\alpha})\right]$ where the sum is over all $(\bar{n}, \bar{\alpha})$ with $h(\bar{n}, \bar{\alpha}) > \varepsilon$. Now let $\varepsilon \downarrow 0$ to obtain the assertion of the theorem. ∎

Define (the total (average) holding times up to, but not including, the trajectory time (m, β) as)

$$H(m, \beta) = \sum_{(\bar{n}, \bar{\alpha}) < (m, \beta)} h(\bar{n}, \bar{\alpha}), \quad m \geq 1, \ 0 \leq \beta \leq \pi .$$

Note that if $(\bar{n}, \bar{\alpha})$ is inseparable from (m, β), then $h(\bar{n}, \bar{\alpha})$ is not included in the above sum.

Theorem 4.4. *The functions $H(m, \beta)$ satisfy the following properties:* (i) *A.s.* $(m, \beta) \rightarrow H(m, \beta)$ *is well-defined and increasing on* $\mathcal{T}$, *with* $H(m, 0) = 0$.

(ii) $H(m,\beta) \in \mathcal{F}((m,\beta)-)$. (iii) $x \to P^x[H(m,\beta)]$ is in $\mathcal{B}^*$. (iv) (Additivity)
A.s. $H(m,\beta+\beta') = H(m,\beta) + H(m,\beta') \circ \theta_{m\beta}$. (v) $P^x[H(m,1)] \le e_{D(m,x)}(x)$,
$P^x[H(m,\beta)] \le P^x[R(m,\beta)] \le e(x)$, and $P^x[H(m,\tau_D^m)] \le e_D(x)$ for $D \in \mathcal{D}_m$.

Proof. All except (v) are quite obvious or routine, using the observations in the preceding proof. To prove (v), we show that for $D \in \mathcal{D}_m, e_D(x) \ge P^x[H(m,\tau_D^m)]$, (the second inequality then follows from this and the Markov property of Z_∞). As in the proof of Theorem 4.3 there exist $(\bar{n}^i,\bar{\tau}^i) \le (m,\tau_D^m)$ such that a.s. P^x, if $(\bar{n},\bar{\alpha}) < (m,\tau_D^m)$ and $h(\bar{n},\bar{\alpha}) > 0$ then $(\bar{n},\alpha) =$ some $(\bar{n}^i,\bar{\tau}^i)$, such that a.s. P^x, if $h(\bar{n}^i,\bar{\tau}^i) > 0$ and $(\bar{n}^i,\bar{\tau}^i) < (m,\tau_D^m)$ then $(\bar{n}^\ell,\bar{\tau}^\ell) \ne (\bar{n}^i,\bar{\tau}^i)$ for $\ell \ne i$, and such that a.s. P^x

$$\lim_j e(n_j^i, \tau_j^i, m_j) = h(\bar{n}^i, \bar{\tau}^i)$$

where $m_j = j$. As seen in the proof of Theorem 4.3 it suffices to show for all ℓ

$$e_D(x) \ge \lim_j P^x \left[\sum_{i \le \ell} e(n_j^i, \tau_j^i, m_j) \cdot 1_{\{(\bar{n}^i,\bar{\tau}^i)<(m,\tau_D^m)\}} \right]$$

which is similar to (4.7). By a similar argument to the one justifying (4.7) we have

$$e_D(x) \ge \lim_j P^x \left[\sum_{i \le \ell} e'(n_j^i, \tau_j^i, m_j) \right]$$

where (for this proof only) $e'(n,\alpha,m') = e_{D[m',y]\cup D}(y)$, with $y = Z(n,\alpha)$. So it suffices to show that, with $(\bar{n},\bar{\tau})$ denoting a fixed $(\bar{n}^i,\bar{\tau}^i)$, a.s. P^x on $\{(\bar{n},\bar{\tau}) < (m,\tau_D^m)\}$

$$\lim_j e(n_j, \tau_j, m_j) = \lim_j e'(n_j, \tau_j, m_j).$$

If $W_\infty^* = d\text{-}\lim_j W_j^* \notin D$, where $W_j^* = Z(n_j,\tau_j)$, then obviously $e(n_j,\tau_j,m_j) = e'(n_j,\tau_j,m_j)$ for large j, so that the above equality holds. So we may assume $W_\infty^* \in D$. Now $\{(\bar{n},\bar{\tau}) < (m,\tau_D^m)\}$ equals a.s. P^x the set $\{W_j^* \notin D$ for all j; $\rho^*\{W_\infty^*\} = 0\}$ (see the proof of Theorem 4.2). If is clear that a.s. P^x we have the following: the weak limit of $H_{D[m_j,W_j^*]\cup D}(W_j^*, \cdot)$ exists and is just $\rho^*(\cdot)$. Therefore, a.s. P^x on $\{W_j^* \notin D$ for all j; $\rho^*\{W_\infty^*\} = 0\}$

$$\lim_j e(n_j, \tau_j, m_j) = \lim_j e_{D[m_j, W_j^*]}(W_j^*)$$

$$= \lim_j [e_{D[m_j, W_j^*] \cup D}(W_j^*) + \int H_{D[m_j, W_j^*] \cup D}(W_j^*, dy) e_{D[m_j, W_j^*]}(y)]$$

$$= \lim_j [e'(n_j, \tau_j, m_j) + \int_{D - D[m_j, W_j^*]} H_{D[m_j, W_j^*] \cup D}(W_j^*, dy) e_{D[m_j, W_j^*]}(y)]$$

$$= \lim_j e'(n_j, \tau_j, m_j)$$

because the last integral converges to 0 a.s. P^x, which follows from an easy argument using H7A.3). ∎

We now define $\bar{e}_D(x)$ for $D \in \mathcal{D}_\infty$, and $\bar{e}(x)$ by

$$\bar{e}_D(x) = e_D(x) - P^x[H(m, \tau_D^m)], \ D \in \mathcal{D}_m; \ \bar{e}(x) = \bar{e}_\Delta(x).$$

By the additivity of $H(m, \beta)$, we have $\bar{e}(x) = \bar{e}_D(x) + H_D \bar{e}(x)$.

Note $\bar{e}_D \geq 0$ by Theorem 4.3.(v). Define $\bar{e}(n, \alpha-)$, similar to $e(n, \alpha-)$ in (3.4), by

(4.8)
$$\bar{e}(n, \alpha) = \bar{e}_{D(n,\alpha)}(Z(n, \alpha)) = \bar{e}_{D(n, Z(n,\alpha))}(Z(n, \alpha)), \alpha < \pi$$

$$\bar{e}(n, \alpha-) = \lim_{\gamma \uparrow \alpha} \bar{e}_{D(n, \alpha-)}(Z(n, \gamma)), \alpha \in \mathcal{L}.$$

The above limit exists a.s. for the same reason as does that of $e(n, \alpha-)$; actually this limit equals (a.s. P^μ)

$$e(n, \alpha-) - \lim_{\gamma \uparrow \alpha} P^\mu \left(\sum h(\bar{k}, \bar{\beta}) \mid \mathcal{F}(n, \gamma) \right)$$

where the sum is over $(n, \gamma) \leq (\bar{k}, \bar{\beta}) < (n, \alpha)$, and the last limit exists by supermartingale convergence. Similar to (3.5) and (3.6), define

(4.9) $\bar{R}(n, \alpha) = \sum_{0 < \gamma \leq \alpha} \bar{e}(n, \gamma-) = \sum_{\gamma < \alpha} \bar{e}(n, \gamma) + \sum_{\gamma \leq \alpha, \gamma \in \mathcal{L}} \bar{e}(n, \gamma-), \ \alpha \leq \pi$

$\bar{R}(n, \alpha-) = \sum_{0 < \gamma < \alpha} \bar{e}(n, \gamma-) = \lim_{\gamma \uparrow \alpha} \bar{R}(n, \gamma), \ \alpha \in \mathcal{L}$

and then, for $m \leq n$

(4.10) $\bar{R}(n,m,\beta) = \bar{R}(n,\sigma(n,m,\beta)),\ \beta < \pi$

$\bar{R}(n,m,\beta-) = \bar{R}(n,\sigma(n,m,\beta-)-) = \lim_{\gamma\uparrow\beta} \bar{R}(n,m,\gamma),\ \beta \in \mathcal{L}$

and finally similar to $e(n,m,\beta)$ in [11], for $m \leq n,\ 0 < \beta < \pi$

(4.11) $\bar{e}(n,m,\beta) = \bar{R}(n,m,\beta) - \bar{R}(n,m,\beta-) = \sum_{(m,\beta-)\leq(n,\alpha-)<(m,\beta)} e(n,\alpha-)\ .$

The following facts follow from the corresponding facts in (3.7) (or from their proofs), and/or from definitions: for all x

(4.12) $\bar{e}(x) = P^x[\bar{R}(n,\pi)]$

$\bar{e}_D(x) = P^x[\bar{R}(n,\tau_D^n)],\ D \in \mathcal{D}_n$

$P^x[\bar{R}(n,m,\beta)] = P^x[\bar{R}(m,\beta)] = P^x[R(m,\beta)] - P^x[H(m,\beta)]$

and the next fact is immediate from (3.8), since $0 \leq \bar{e}(n,\alpha-) \leq e(n,\alpha-)$: for all x

(4.13) $\{\bar{R}(n,\pi),n \geq 1\}$ is uniformly integrable w.r.t. P^x .

To prove the next theorem, we need to define $\bar{e}(n,\alpha,m)$ similar to $e(n,\alpha,m)$ in (4.1)

$$\bar{e}(n,\alpha,m) = \bar{e}_{D[m,Z(n,\alpha)]}(Z(n,\alpha))\ .$$

Of course $\bar{e}(n,\alpha,m) \leq e(n,\alpha,m)$.

Theorem 4.4. *For each x and m, $\{\bar{R}(n,m,\beta),n \geq m\}$ converges in P^x-measure uniformly in β.*

Proof. Based on the following lemma, the proof is the same as that of [11], Theorem 5.2, which is based on Lemma 5.1 there. ∎

Lemma 4.5. *For each $x,\ \varepsilon \geq 0$*

$$\lim_{m}\ \sup_{n\geq m}\ P^x[\sup_{\beta} e(n,m,\beta) > \varepsilon] = 0\ .$$

Proof. Suppose not. Then as shown in [11], sublemmas 5.1.1 and 5.1.2, using now the facts stated in (4.10) and (4.11), there exist an (increasing) sequence $(\bar{n}, \bar{\tau})$ of stopping times, positive integers $m_j \to \infty$ (even with $n_{j-1} < m_j < n_j$, but this is unnecessary), and $c > 0$ such that

$$P^x[\lim_j \bar{e}(n_j, \tau_j, m_j) > c] > c$$

and such that $D[n_j, Z(n_j, \tau_j)]$ is increasing in j a.s. P^x. The latter fact implies that, with

$$\sigma_j = \tau_j + \tau_{D[n_j, Z(n_j, \tau_j)]}^{n_j} \circ \theta(n_j, \tau_j),$$

(n_j, σ_j) is decreasing and $(\bar{n}, \bar{\tau}) < (n_j, \sigma_j)$ for all j. Now a.s. P^x

$$h(\bar{n}, \bar{\tau}) \geq \lim_j e(n_j, \tau_j, m_j)$$

$$= \lim_j \left(\bar{e}(n_j, \tau_j, m_j) + P^x \left[\sum_{(n_j, \tau_j) \leq (\bar{k}, \bar{\gamma}) < (n_j, \sigma_j)} h(\bar{k}, \bar{\gamma}) \mid \mathcal{F}(n_j, \tau_j) \right] \right)$$

$$\geq \lim_j \left(\bar{e}(n_j, \tau_j, m_j) + P^x(h(\bar{n}, \bar{\tau}) \mid \mathcal{F}(n_j, \tau_j)) \right).$$

where the equality follows from the definition of $\bar{e}_D(y)$ and the Markov property of Z_∞. But the above conditional expectation converges to $h(\bar{n}, \bar{\tau})$ since it is in $\bigvee_j \mathcal{F}(n_j, \tau_j)$. It follows that $\lim_j e(n_j, \tau_j, m_j) = 0$ a.s. P^x and we have a contradiction. ∎

Because of Theorem 4.4, there exist, for each $x \in K$, integers $n_k = n_k(x)$ such that $n_1 = 1$ and for $k \geq 2$

$$n_k =$$

$$\inf \{n > n_{k-1} : \sup_{n' > n} \sup_{m \leq n} \sup_\beta P^x[|\bar{R}(n, m, \beta) - \bar{R}(n', m, \beta)| > 2^{-k}] < 2^{-k}\}.$$

Obviously $n_k(x)$ is in $\mathcal{B}^*$. We now define

$$\bar{R}_k(m, \beta) = \bar{R}(n_k(Z_{m0}), m, \beta)$$

$$\bar{S}(m, \beta) = \lim_k \bar{R}_k(m, \beta)$$

$$S(m, \beta) = \bar{S}(m, \beta) + H(m, \beta).$$

as $S_{(n,\alpha)} \downarrow t$. Suppose not. Then for some x and Λ with positive P^x-outer-measure we have, for $\omega \in \Lambda$, there exists $t = t(\omega)$ in $S(\omega)$, $t = \inf(S(\omega) \cap (t, \infty))$ such that $X_t(\omega) \neq \hat{d}$-$\lim Z_{n\alpha}(\omega)$ as $S_{n\alpha}(\omega) \downarrow t$. From this one can obtain the following. There exist $F = C_i$, $f = f_k$ (where the C_i, f_k are as in the definition of $\hat{d}$), reals $a < b$, a compact set $V \subset \{H_F f < a\}$ and increasing $D_j \in \mathcal{D}$ with $D_j - \Delta \subset \{H_F f > b\}$ (or with "$< a$" and "$> b$" interchanged), and a random time T in (X_t) and decreasing random times (n_j, σ_j) in $\mathcal{T}$ with n_j strictly increasing and independent of ω, (T and (n_j, σ_j) can be made stopping times), such that on a set $\Lambda_0 \subset \{X_T \notin F\}$ with $P^x(\Lambda_0) > 0$ we have

$$T \notin S; \ S(n_j, \sigma_j) \downarrow T; \ X_T = d\text{-}\lim_j Z(n_j, \sigma_j);$$
$$X_T \in V; \ Z(n_j, \sigma_j) \in D_j \text{ for all } j.$$

Let $T_1 = \sup(S \cap [0, T))$ and $S(m_j, \beta_j) \uparrow T_1$ (it will be seen later that $T_1 = T$ a.s. P^x on Λ_0). Then a.s. P^x on $\Lambda_0, d\text{-}\lim_j Z(m_j, \beta_j) = X_T$; for otherwise $T \in S$. Now consider the $\{V \cup \Delta; D_j, j \geq 1\}$-refinement $\tilde{Z}_\infty$ of Z_∞. $\tilde{Z}_\infty$ is defined from the sequence $\tilde{\mathcal{D}}_n$ where $\tilde{\mathcal{D}}_n$ is the smallest family containing $\mathcal{D}_n$ and the sets $V \cup \Delta, D_1, \cdots, D_n$ and closed w.r.t. union and intersection. Let Λ_1 be the set of ω in Λ_0 such that there exist in $\tilde{Z}_\infty$ infinitely many visits to $\cup_j D_j$ immediately to the left of $\lim_j(n_j, \sigma_j)$ (the meaning of this is clear). By Lemma 6.1, a.s. P^x on Λ_1 there cannot be infinitely many visits to $V \cup \Delta$. Now using the analysis in the proof of Theorem 4.2 with $V \cup \Delta$ being the set D there one obtains the conclusion that a.s. P^x on Λ_1 there exists $(k, \gamma) \in \tilde{\mathcal{T}}$ with $(m_j, \tilde{\beta}_j) < (k, \gamma) < (n_j, \sigma_j)$ for all j, and

$$\tilde{Z}(k, \gamma) = d\text{-}\lim_j Z(m_j, \beta_j) = d\text{-}\lim Z(n_j, \sigma_j) = X_T.$$

Furthermore this $\tilde{Z}(k, \gamma)$ is not in $H \cup \Delta$; for otherwise we must have $T \in S$ (incidentally $T_1 = T$ a.s. P^x on Λ_1). But now applying the first part of this proof to $\tilde{Z}_\infty$ we have $\hat{d}\text{-}\lim Z(n_j, \sigma_j) = \tilde{Z}(k, \gamma) = X_T$, and this implies $P^x(\Lambda_1) = 0$. On $\Lambda_0 - \Lambda_1$, which has positive P^x-measure, by changing x we may assume $\tilde{\sigma}_j = \tilde{\tau}_{D_j}^{n_j}$.

(6.3) $D - \Delta \subset K - H$ and $\sup \{e_{D[n,y]}(y) : y \in D\} \to 0$.

Consider the $\{D\}$-refinement $\tilde{Z}_\infty$ of Z_∞. We claim that a.s. P^x on Λ there are
no visits to D in $\tilde{Z}_\infty$ between $(n_j, \tilde{\tau}_j)$ for j large. This is because under (6.2)
each such visit takes an amount of "time" $\geq b$, and under (6.3) a sequence of
such visits leads to $\tilde{h}(\overline{n}, \tilde{\tau}) = h(\overline{n}, \overline{\tau}) = 0$. Now by shifting x we may assume
$(\overline{n}, \tilde{\tau}) \leq (1, \tilde{\tau}_D^1)$ in $\tilde{T}$, and of course we may assume $\Lambda \in \tilde{\mathcal{G}}(\overline{n}, \tilde{\tau})$ (the obvious σ-
algebra in $\tilde{Z}_\infty$). As in the proof of Theorem 4.2, let F_m be compact, $F_m \uparrow K - D$
(as in that proof), and

$$\rho^*(\cdot) = \rho^*(\omega, \cdot) = w\text{-}\lim_m \ w\text{-}\lim_j \ H_{F_m \cup D}(W_j^*, \cdot) \ ;$$

then $\rho^*\{W_\infty^*\} = 0$ or 1. Since a.s. P^x on Λ we have assumed $\hat{W}_\infty^* \in \hat{K}_r - \hat{K}_0$ so
that $H(\hat{W}_\infty^*, \{W_\infty^*\}) > 0$, it is clear that $\rho^*\{W_\infty^*\} > 0$ and therefore $\rho^*\{W_\infty^*\} =$
1 a.s. P^x on Λ. Using a martingale argument and the fact that $\rho^*(\cdot)$ is the
conditional distribution of $\tilde{Z}\langle \overline{n}, \tilde{\tau} \rangle$ given $\tilde{\mathcal{G}}(\overline{n}, \tilde{\tau})$ (see the proof of Theorem 4.2),
one can easily prove the following: for any C_i, f_k (in the definition of $\hat{d}$)

$$H_{C_i} f_k(W_j^*) \to H_{C_i} f_k(W_\infty^*)$$

a.s. P^x on $\Lambda \cap \{W_j^* \notin C_i\}$. From this and (1.1) if follows that $\hat{W}_\infty^* = W_\infty^* \in$
$K \subset \hat{K}_0$ a.s. P^x on Λ, and we have a contradiction.   ∎

Theorem 6.3. *A.s. the path X_t is right $\hat{d}$-continuous.*

Proof. For $y \in K - H - \Delta$ and $F \in \mathcal{D}$ with $y \notin F$, applying hypothesis H5) to $x =$
$y, D_n = D(n, y)$ and $f \in b\mathcal{B}$ we have $H_F f(Z(n, 1)) \to H_F f(y)$ a.s. P^y. It follows
from this and the definition of $\hat{d}$ that X_t restricted to $t \in \mathcal{S}$ is right $\hat{d}$-continuous at
$t = 0$, a.s. P^y for all y (this being trivial if $y \in H \cup \Delta$). From the Markov property
of Z_∞, a.s. the path X_t is right $\hat{d}$-continuous when restricted to $\mathcal{S}$. So we need
to show that a.s. for all $t \notin \mathcal{S}(\omega)$ with $t = \inf (\mathcal{S}(\omega) \cap (t, \infty)), X_t = \hat{d}\text{-}\lim Z_{n\alpha}$

$d\text{-}\lim_j W_j^* = \hat{\pi}(\hat{W}_\infty^*)$. We show $\hat{W}_\infty^* \in \hat{K}_r$ a.s. P^x. Suppose not. Then for some $C_i \subset C_l$ and f_k (where again C_i, f_k are as in the definition of $\hat{d}$)

$$(6.1) \qquad H_{C_i} f_k(\hat{W}_\infty^*) < \int H_{C_l}(\hat{W}_\infty^*, dy) H_{C_i} f_k(y)$$

(or with "<" replaced by ">") on a set $\Lambda \subset \{\hat{W}_\infty^* \notin C_l\}$ with $P^x(\Lambda) > 0$. Consider the measures

$$\mu_j(\cdot) = P^x[H_{C_l}(W_j^*, \cdot); \Lambda], \mu(\cdot) = P^x[H_{C_l}(\hat{W}_\infty^*, \cdot); \Lambda].$$

Since $H_{C_l}(W_j^*, \cdot) \to H_{C_l}(\hat{W}_\infty^*, \cdot)$ weakly a.s. P^x on Λ (see (1.1)), we have $\mu_j \to \mu$ weakly. Now for all $f \in b\,B$, $\int f \, d\mu_j = P^x[H_{C_l} f(W_j^*); \Lambda]$ converges (see the proof of Lemma 6.1); it follows that $\mu_j \to \mu$ strongly. (This is probably standard knowledge. It may be proved as follows: it suffices to show $\mu_j(C) \to \mu(C)$ for any compact C; if not, then we may assume $\sup_j \mu_j(C) < \mu(C)$ for some C, and one can easily obtain a subsequence μ_{j_N} and open sets $U_N \downarrow C$ such that with $A = \bigcup_{N \text{odd}} (U_N - U_{N+1})$ one has

$$\mu_{j_N}(A) > 2\varepsilon \text{ for } N \text{ odd}; \ \mu_{j_N}(A) < \varepsilon \text{ for } N \text{ even}$$

for some $\varepsilon > 0$, which is a contradiction.) It then follows that

$$P^x[H_{C_i} f_k(\hat{W}_\infty^*); \Lambda] = \lim_j P^x[H_{C_i} f_k(W_j^*); \Lambda]$$

$$= \lim_j P^x[H_{C_l} H_{C_i} f_k(W_j^*); \Lambda] = \lim_j \int H_{C_i} f(y) \mu_j(dy)$$

$$= \int H_{C_i} f_k(y) \mu(dy) = P^x[H_{C_l} H_{C_i} f_k(\hat{W}_\infty^*); \Lambda].$$

But this contradicts (6.1).

(B) Next we show that a.s. P^x (for any x) the $\hat{d}$-limits in question are in fact in $\hat{K}_0$. Suppose not. Then there exists $(\bar{n}, \bar{\tau})$ in $\mathcal{T}$ with $h(\bar{n}, \bar{\tau}) > 0$ when $(\bar{n}, \bar{\tau})$ is nonconstant such that $\hat{W}_\infty^* = \hat{d}\text{-}\lim_j W_j^* \in \hat{K}_r - \hat{K}_0$ on a set Λ with $P^x(\Lambda) > 0$, where $W_j^* = Z(n_j, \tau_j)$. We may assume $W_\infty^* = \hat{\pi}(\hat{W}_\infty^*) \in D$ where $D \in \mathcal{D}$ and either

$$(6.2) \qquad D - \Delta \subset H \text{ and } b = \inf\{h(y) : y \in D - \Delta\} > 0$$

provided this $\hat{d}$-limit exists, and $X_t(\omega) = \Delta$ otherwise;

(iv) if t does not satisfy (i), (ii) or (iii) , and $s = \sup(S(\omega) \cap [0, t))$, let $X_t(\omega) = X_s(\omega)$.

The following lemma is needed in examining the path behavior of (X_t).

Lemma 6.1. *For $D \in \mathcal{D}$, $f \in b\mathcal{B}^*, \delta > 0$, the following is valid a.s.: $(n, \alpha) \rightarrow H_D f(Z_{n\alpha})$, where $(n, \alpha) \in T$ with $Z_{n\alpha} \notin B(D, \delta)$, has no oscillation. (Here $B(D, \delta) = \{y : \text{dist}(y, D) < \delta\}$, with $\text{dist} = d$-distance.) The same statement holds if $D - \Delta$ is d-compact and $B(D, \delta)$ is replaced by $B(D - \Delta, \delta)$.*

Proof. By replacing D by a suitable $D_1 \in \mathcal{D}_\infty$ with $D \subset D_1$ and f by $f_1 = H_D f$ we may assume $D \in \mathcal{D}_\infty$. If $D \in \mathcal{D}_m$ and $n \leq m$, then for any probability measure ν on K, $\{H_D f(Z(n, \alpha \wedge \tau_D^n)), \alpha < \pi\}$ is a martingale w.r.t. P^r, and since $T_n \uparrow T$, the upcrossing lemma implies that a.s.-P^r, $H_D f(Z(n, \alpha \wedge \tau_D^n))$, $(n, \alpha) \in T$, has no oscillation. Now for a fixed μ, a.s. P^μ there can be at most finitely many visits to D sandwiched by visits to $K - B(D, \delta)$. From these one clearly has the first statement. The second statement follows from the same reasoning, noting that when Δ is reached, a trajectory stays at Δ forever. ∎

Theorem 6.2. *In (iii) of the definition of (X_t), a.s. the $\hat{d}$-limits exist for all such t and are points in $\hat{K}_0$. In particular, (X_t) takes values in $\hat{K}_0$.*

Proof. By Lemma 6.1, with D running through $\{C_i\}$ and f through $\{f_k\}$ in the definition of $\hat{d}$, we see that a.s. all the $\hat{d}$-limits in (iii) exist (in $\hat{K}$). Since the proof is long, it is divided into parts (A) and (B).

(A) We first show that a.s. P^x (for any x) these $\hat{d}$-limits are in $\hat{K}_r$. Because these times t at which a $\hat{d}$-limit is in question correspond to a (nonconstant) trajectory limit time $(\bar{n}, \bar{\alpha}) > 0$, a.s. P^x all these times t are among countably many sequences $(\bar{n}^i, \bar{\tau}^i)$ of stopping times, in the sense that $t = \lim_j S(n_j^i, \tau_j^i)$ for some i. Now the $\hat{d}$-limits in question are among $\hat{d}\text{-}\lim_j Z(n_j^i, \tau_j^i)$. Fix one such sequence $(\bar{n}, \bar{\tau})$ and denote $W_j^* = Z(n_j, \tau_j)$, $\hat{W}_\infty^* = \hat{d}\text{-}\lim_j W_j^*$ and $W_\infty^* =$

Let $n \geq 2m$ (so that $1/m \geq 2/n$). It is easy to see that if $\tilde{\tau}_F^n < \tilde{\tau}_D^n$, i.e. if $\tilde{Z}(n, \tilde{\tau}_F^n) \in F - D$,

$$(n, \tilde{\tau}_F^n + \tilde{\tau}_{D[m, \tilde{Z}(n, \tilde{\tau}_F^n)]}^n \circ \tilde{\theta}(n, \tilde{\tau}_F^n)) = (n, \tilde{\alpha})$$

for some $(n, \alpha) \in \mathcal{T}$. Since $S(1, \tau_D^1) = 0$ implies $S(1, \tau_D^1) \circ \theta(n, \alpha) = 0$, we have

$$P^x[S(1, \tau_D^1) = 0; \ \tilde{Z}(n, \tilde{\tau}_F^n) \in F_0]$$

$$\leq P^x[\tilde{Z}(n, \tilde{\tau}_F^n) \in F_1; \ S(1, \tau_D^1) \circ \tilde{\theta}(n, \tilde{\tau}_{D[m, \tilde{Z}(n, \tilde{\tau}_F^n)]}^n) \circ \tilde{\theta}(n, \tilde{\tau}_F^n) = 0] .$$

But from the definition of $u(y)$, the latter probability

$$= \int_{F_1} P^x[\tilde{Z}(n, \tilde{\tau}_F^n) \in dy] H_{D[m, y]} u(y) < a/2$$

in view of Lemma 6.2.(vii) .   ∎

6. The Process (X_t)

We now define the process (X_t) and show that it satisfies the desired properties. However, as said earlier, when holding points or hidden holding points exist, the holding times appear in their average values. We omit the last step of constructing the correct right process, which is more or less routine; see however a sketch of it in [10], section 5.

6.1. Definition of (X_t) and its path behavior

Definition. Define $X_t(\omega), \omega \in \Omega, t \geq 0$, as follows:

(i) if $t = S_{n\alpha}(\omega)$, let $X_t(\omega) = Z_{n\alpha}(\omega)$;

(ii) if $t \notin \mathcal{S}(\omega)$ but $t = \inf(\mathcal{S}(\omega) \cap (t, \infty))$, let

$$X_t(\omega) = d\text{-}\lim_{S_{n\alpha}(\omega) \downarrow t} Z_{n\alpha}(\omega)$$

provided this d-limit exists, and $X_t(\omega) = \Delta$ otherwise;

(iii) if t does not satisfy (i) or (ii), but $t = \sup(\mathcal{S}(\omega) \cap [0, t))$, let

$$X_t(\omega) = \hat{d}\text{-}\lim_{S_{n\alpha}(\omega) \uparrow t} Z_{n\alpha}(\omega)$$

work here we need the following lemma, from which it follows that the functions $\bar{e}_D(x)$ defined for $\tilde{Z}_\infty$ (using $\tilde{h}(\bar{n}, \bar{\alpha})$) are the same for Z_∞ if $D \in \mathcal{D}_\infty$. ∎

Lemma 5.4. $\tilde{Z}_\infty$ (defined above) has the same (positive) holding times as Z_∞, in the sense that a.s., for $(\bar{n}, \bar{\alpha})$ in $\mathcal{T}, \tilde{h}(\bar{n}, \tilde{\alpha}) = h(\bar{n}, \bar{\alpha})$, and for $(\bar{n}, \bar{\alpha})$ in $\tilde{\mathcal{T}}$, there exists $(\bar{n}', \bar{\alpha}')$ in $\mathcal{T}$ with $(\bar{n}', \tilde{\alpha}') = (\bar{n}, \bar{\alpha})$ (which may be inseparable in $\tilde{Z}_\infty$) such that $h(\bar{n}', \bar{\alpha}') = \tilde{h}(\bar{n}, \bar{\alpha})$.

Proof. If $(\bar{n}, \bar{\alpha})$ is in $\mathcal{T}$, then by Theorem 4.1 applied to $\tilde{Z}_\infty$ and the definitions of $h(\bar{n}, \bar{\alpha})$ and $\tilde{h}(\bar{n}, \tilde{\alpha})$, we have $\tilde{h}(\bar{n}, \bar{\alpha}) = h(\bar{n}, \bar{\alpha})$. If $(\bar{n}, \bar{\alpha})$ is in $\tilde{\mathcal{T}}$ and nonconstant, then there exists $(\bar{n}', \bar{\alpha}')$ in $\mathcal{T}$ with $(\bar{n}', \tilde{\alpha}') = (\bar{n}, \bar{\alpha})$ in $\tilde{\mathcal{T}}$, and so $h(\bar{n}', \bar{\alpha}') = \tilde{h}(\bar{n}', \tilde{\alpha}') = \tilde{h}(\bar{n}, \bar{\alpha})$. If $(\bar{n}, \bar{\alpha})$ is a constant (m, β) in $\tilde{\mathcal{T}}$ and $(m, \beta) \neq (m_1, \tilde{\beta}_1)$ for any (m_1, β_1) in $\mathcal{T}$ then $\tilde{Z}(\cdot, \cdot)$ is left d-continuous at (m, β), and so (m, β) is inseparable from a nonconstant $(\bar{k}, \bar{\gamma})$ in $\tilde{\mathcal{T}}$; consequently, with $(\bar{n}', \bar{\alpha}')$ in $\mathcal{T}$ satisfying $(\bar{n}', \tilde{\alpha}') = (\bar{k}, \bar{\gamma})$, we have $h(\bar{n}', \bar{\alpha}') = \tilde{h}(\bar{k}, \bar{\gamma}) = h(m, \beta) = h(\bar{n}, \bar{\alpha})$ by Theorem 4.2. ∎

Lemma 5.5. $H_{D_j}(x, D) < a/2$ for all j.

Proof. Again for a fixed $D_j = F$ consider the $\{F\}$-refinement $\tilde{Z}_\infty$ of Z_∞, Since $F \in \tilde{\mathcal{D}}_n$ for all n, $P^x[\tilde{Z}(n, \tau_F^n) \in \cdot] = H_F(x, \cdot)$ where $\tilde{\tau}_F^n = \inf\{\alpha : \tilde{Z}(n, \alpha) \in F\}$. Thus we need to show

$$P^x[\tilde{Z}(n, \tilde{\tau}_F^n) \in D] > a/2 .$$

Since $P^x[S(1, \tau_D) = 0] = a$ (see (6.1)), this will follow if

$$P^x[S(1, \tau_D^1) = 0, \ \tilde{Z}(n, \tilde{\tau}_F^n) \in F_0 \cup F_1 \cup F_2] < a/2$$

(recall F_0, F_1, F_2 are resp. C_j, C_j', C_j''). From Lemma 5.4, a.s. P^x on $\{S(1, \tau_D^1) = 0\}$ there are no positive holding times $\tilde{h}(\bar{n}, \bar{\alpha})$ with $(\bar{n}, \bar{\alpha}) < (1, \tilde{\tau}_D^1)$; consequently $\tilde{Z}(n, \tilde{\tau}_F^n) \notin F_1 \cup F_2$ because $h(y) > 0$ for $y \in F_1 \cup F_2$. Thus $P^x[S(1, \tau_D^1) = 0$; $\tilde{Z}(n, \tilde{\tau}_F^n) \in F_1 \cup F_2] = 0$. Let $m = m_j$ where the m_j are as in Lemma 6.2.

(iii) $P^x(\Lambda_1 \cap \{Z(n_j, \lambda_j) \notin C_j \cup C_j'\}) \to 0$

(iv) $P^x(\Lambda_2 \cap \{Z(n_j, \tau_j) \notin C_j' \cup C_j''\}) \to 0$

(v) $P^x(\Lambda_3 \cap \{Z(n_j, \tau_j) \notin C_j \cup C_j'\}) \to 0$

(vi) $\inf\{h(y) : y \in C_j' \cup C_j''\} > 0$ (recall $h(y) = \inf\{e_D(y) : y \notin D\}$ is the expected holding time at y)

(vii) $\sup\{H_{D[m_j, y]}u(y) : y \in C_j \cup C_j'\} < a/2$.

(Remarks: The use of n_j, rather than n, in (iii) is only for convenience. In (iv), C_j' can be omitted: for $Z(n_j, \tau_j)$ can not be in H, a.s. P^x on Λ_2. Although the sequence $(\bar{n}, \bar{\tau})$ mentioned before the statement of the lemma corresponds to a $\delta_1 > 0$, one can obtain with a little more care a sequence $(\bar{n}, \bar{\tau})$ as asserted in the lemma (i.e. corresponding to $\delta_1 = 0$.)

Using the C_j, C_j', C_j'' in the lemma we define

$$D_j = D \cup C_j \cup C_j' \cup C_j''.$$

For the rest of the proof of Theorem 5.1, we fix a D_j and denote $F = D_j, F_0 = C_j, F_1 = C_j'$ and $F_2 = C_j''$. Let $\tilde{Z}_\infty = (\tilde{Z}_{n\alpha} = \tilde{Z}(n, \alpha), n \geq 1, \alpha < \pi; P^x, x \in K)$ be the $\{F\}$-refinement of Z_∞ (thus the same P^x-notation is used) defined at the end of [11], section 4. It is constructed as Z_∞ except with each $\mathcal{D}_n$ replaced by $\tilde{\mathcal{D}}_n$ which is the smallest family containing $\mathcal{D}_n$ and F and closed w.r.t. union and intersection. Z_∞ being (obviously) imbedded in $\tilde{Z}_\infty$, all quantities in Z_∞ are defined in $\tilde{Z}_\infty$. Let $\tilde{T}$ denote the set of trajectory times in $\tilde{Z}_\infty$. Each $(n, \alpha) \in T$ is some $(n, \beta) \in \tilde{T}$ which we write as $(n, \tilde{\alpha})$; a sequence $(\bar{n}, \bar{\alpha})$ (with values) in T will be written as $(\bar{n}, \tilde{\bar{\alpha}})$, a slight abuse of notation. Note that if $(n_1, \alpha_1) < (n_2, \alpha_2)$ in $\tilde{T}$, there exist (k, γ) in T such that $(n_1, \alpha_1) \leq (k, \tilde{\gamma}) \leq (n_2, \alpha_2)$. $\tilde{h}(\bar{n}, \bar{\alpha})$ denotes the holding time at $(\bar{n}, \bar{\alpha})$ in $\tilde{T}$ for $\tilde{Z}_\infty$.

Lemma 5.3. $e_{D_j}(x) \downarrow 0$.

Proof. This is proved as in [11], based on (iii), (iv) and (v) of Lemma 5.2; see the first ten lines following display (6.5) in [11]. However, in order for that proof to

Define $\lambda_n = \inf\{\alpha : S(n,\alpha) > 0 \text{ or } \alpha = \tau_D^n\}$; λ_n is a stopping time w.r.t. $\{\mathcal{F}(n,\alpha), \alpha < \pi\}$. Let $\zeta_n = S(n,\lambda_n)$ and ζ be the decreasing limit of ζ_n. Denote

$$\Lambda_1 = \{S(1,D) > 0;\ \zeta = 0\}$$

$$\Lambda_2 = \{\zeta > 0;\ (n,\lambda_n-) \text{ is constant and } \lambda_n \notin \mathcal{L}, \text{ for all large } n\}$$

$$\Lambda_3 = \{\zeta > 0\} - \Lambda_2 .$$

A.s. P^x on Λ_1, since $\zeta_n - \zeta_{n+1} > 0$ for infinitely many n, the Markov property of Z_∞ implies that $u(Z(k,\gamma)) \to 0$ as (k,γ) decreases to $\lim_n (n,\lambda_n)$ (the meaning of this is clear and no elaboration is needed). Therefore by another application of the Markov property we have the following: for all $\delta > 0$ and for all sufficiently large m and n, with $Z(n,\lambda_n)$ written as y to simplify notation

$$(5.2) \qquad H_{D[m,y]}u(y) = P^y[S(1,\tau_D^1) \circ \theta(m,\tau_{D[m,y]}^m) = 0] < \delta$$

on Λ_1 excluding a subset of P^x-measure $< \delta$. Next we examine the trajectory behavior on $\Lambda_2 \cup \Lambda_3$ to the left of the "time" $\lim_n (n,\lambda_n)$. As in the proof of Theorem 4.2, for $\delta_1 > 0$ there exists an (increasing) sequence $(\bar{n},\bar{\tau})$ of stopping times such that $(m,\lambda_m-) \le (\bar{n},\bar{\tau}) < (m,\lambda_m)$ for all m and $h(\bar{n},\bar{\tau}) = \zeta$ on $\Lambda_2 \cup \Lambda_3$ excluding a subset Γ of P^x-measure $< \delta_1$. A.s. P^x on $\Lambda_2 - \Gamma, (n_j,\tau_j)$ is ultimately constant, and so $Z(n_j,\tau_j) \in H$ for j large. A.s. P^x on $\Lambda_3 - \Gamma, (n_j,\tau_j)$ is not ultimately constant; it is clear that $u(Z(k,\gamma)) \to 0$ as $(k,\gamma) \uparrow (\bar{n},\bar{\tau})$, and therefore the claim (5.2) holds with Λ_1 replaced by $\Lambda_3 - \Gamma$, with $Z(n,\lambda_n)$ replaced by $Z(n_j,\tau_j)$, with $n = n_j$ large and m large but dependent on n ($m = m_j$ may have to be much larger than n_j). Summing over these arguments we have

Lemma 5.2. *There exist increasing sequences of compact sets C_j, C_j' and C_j'', integers $m_j \to \infty$, and an (increasing) sequence $(\bar{n},\bar{\tau})$ of stopping times such that*

  (i) $C_j \subset K - (D \cup H),\ C_j' \cup C_j'' \subset H - D,\ C_j' \cap C_j'' = \emptyset$

  (ii) *a.s. P^x on $\Lambda_2 \cup \Lambda_3, (m,\lambda_m-) \le (\bar{n},\bar{\tau}) < (m,\lambda_m)$*

    for all m, and $h(\bar{n},\bar{\tau}) = \zeta$

Theorem 4.6. *The functions $S(m, \beta)$ satisfy the following properties:* (i) *A.s.* $(m, \beta) \to S(m, \beta)$ *is well-defined on T and increasing, with $S(m, 0) = 0$.* (ii) $S(m, \beta) \in \mathcal{F}((m, \beta)-)$. (iii) $\bar{R}_k(m, \beta) \to \bar{S}(m, \beta)$ *a.s. and in expectation w.r.t. any P^x.* (iv) *(Additivity) A.s.* $S(m, \beta+\beta') = S(m, \beta)+S(m, \beta') \circ \theta_{m\beta}$. (v) *For all* $x, P^x[\bar{S}(m, \beta)] = P^x[\bar{R}(m, \beta)]$ *and so* $P^x[S(m, \beta)] = P^x[R(m, \beta)]$, *in particular* $P^x[S(m, \pi-)] = e(x)$; *also for* $D \in \mathcal{D}_m, P^x[S(m, \tau_D^n)] = e_D(x)$. (vi) *Define*

$$S = \{S(m, \beta) : m \geq 1, \ \beta < \pi\}, \bar{S} = \{\bar{S}(m, \beta) : m \geq 1, \beta < \pi\},$$

$$S_\Delta = \sup S, \bar{S}_\Delta = \sup \bar{S} ;$$

then $P^x(\bar{S}_\Delta) = \bar{e}(x), P^x(S_\Delta) = e(x)$, *and $\bar{S}$ is dense in $[0, \bar{S}_\Delta(\omega)]$, (and so the gaps in S occur exactly at "times" $(\bar{n}, \bar{\alpha})$ when $h(\bar{n}, \bar{\alpha}) > 0$).*

Proof. Most of these are immediate from Theorem 4.4, the definition of n_k, Theorem 4.3, and the facts (4.10) and (4.11). For the additivity of $\bar{S}(m, \beta)$, see the proof of [11], Theorem 5.3. The denseness of $\bar{S}$ is a consequence of Lemma 4.5. ∎

5. Proof of $S(m, 1) > 0$

We will establish the important property that $S(m, 1) > 0$ a.s. P^x for all $x \neq \Delta$ (of course $x \in K$) and $m \geq 1$. Thus by the additivity of $S(m, \beta)$ we have a.s. $S(m, \beta)$ is strictly increasing on T (until Δ is reached). The proof below is also valid for the situation in [11] and is an improvement over the proof in [11] in the case (H) when $H \neq \emptyset$.

Theorem 5.1. $P^x[S(m, 1) > 0] = 1$ *for all $x \in K - \Delta$, $m \geq 1$.*

To prove the theorem, assume $m = 1$ for convenience. Fix $x \neq \Delta$ and denote $D = D(1, x)$. Let $u(y) = P^y[S(1, \tau_D^1) = 0]$. We need to show $u(x) = 0$. Assume

$$(5.1) \qquad\qquad\qquad u(x) = a > 0 .$$

We will find an increasing sequence D_j in $\mathcal{D}$ such that $H_{D_j}(x, D) > a/2$ but $e_{D_j}(x) \downarrow 0$. Thus H7A.1) will be contradicted, with $U = K - D$.

Now apply hypothesis H5) (alternative version) to the present x, D_j, F, f. Note that the (W_j) in H5) is $W_j = \tilde{Z}(n_j, \tilde{\tau}_{D_j}^{n_j}) = \tilde{Z}(n_j, \tilde{\sigma}_j) = Z(n_j, \sigma_j)$. Since

$$W_\infty = d\text{-}\lim_j W_j = X_T$$

and

$$H_F f(W_\infty) = H_F f(X_T) < a < b \leq \lim_j H_F f(Z(n_j, \sigma_j)) = \lim_j H_F f(W_j)$$

a.s. P^x on $\Lambda_0 - \Lambda_1$, we have a contradiction to H5). ■

We remark that this proof is similar to the proof of [11], Theorem 7.4, which establishes the strong Markov property of (X_T) in the case when $H = \emptyset$.

6.2. Hitting Distributions of (X_t)

For $D \in \mathcal{D}$, define $T_D = \inf \{t \geq 0 : X_t \in D\}$. Note that $T_D \leq S_\Delta < \infty$ a.s.

Theorem 6.4. *For all x and $D \in \mathcal{D}$, $P^x[X(T_D) \in \cdot, T_D < \infty] = H_D(x, \cdot)$.*

Proof. Consider the $\{D\}$-refinement $\tilde{Z}_\infty$ of Z_∞. Since $P^x[\tilde{Z}(1, \hat{\tau}_D^1) \in \cdot] = H_D(x, \cdot)$, it suffices to show $X(T_D) = \tilde{Z}(1, \hat{\tau}_D^1)$ a.s. P^x. Define T as follows: if $(1, \tilde{\tau}_D^1) = (m, \tilde{\beta})$ for some $(m, \beta) \in \mathcal{T}$ let $T = S(m, \beta)$; otherwise, with $(\tilde{n}, \tilde{\alpha})$ denoting the largest (nonconstant) sequence in $\mathcal{T}$ such that $(\tilde{n}, \tilde{\alpha}) < (1, \tilde{\tau}_D^1)$ (note $(\tilde{n}, \tilde{\alpha})$ is inseparable from $(1, \tilde{\tau}_D^1)$ in $\tilde{\mathcal{T}}$), let $T = \lim_j S(n_j, \alpha_j)$. We show that $X(T) = \tilde{Z}(1, \tilde{\tau}_D^1)$ and $T = T_D$ a.s. P^x. If $(1, \tilde{\tau}_D^1) = (m, \tilde{\beta})$ for some $(m, \beta) \in \mathcal{T}$, then $X_T = Z(m, \beta) = \tilde{Z}(1, \tilde{\tau}_D^1)$, and in particular $T_D \leq T$. If not, there are two cases to consider: (i) $h(\tilde{n}, \tilde{\alpha}) > 0$; (ii) $h(\tilde{n}, \tilde{\alpha}) = 0$. In case (i), $X(T) = \hat{d}\text{-}\lim_j Z(n_j, \alpha_j)$ a.s. P^x by definition; by the analysis in the proof of Theorem 4.2 applied to $\tilde{Z}_\infty$ (see also the end of the proof of Theorem 6.2)

$$\tilde{Z}(1, \tilde{\tau}_D^1) = \hat{d}\text{-}\lim_j \tilde{Z}(n_j, \tilde{\alpha}_j) = \hat{d}\text{-}\lim_j Z(n_j, \alpha_j)$$

a.s. P^x, and so $X(T) = \tilde{Z}(1, \tilde{\tau}_D^1)$ and $T_D \leq T$ a.s. P^x. In case (ii), $T = \inf(S \cap (T, \infty))$ and $X(T) = d\text{-}\lim_j Z(n_j', \alpha_j')$ where $S(n_j', \alpha_j') \downarrow T$. Here we must have $\tilde{Z}(1, \tilde{\tau}_D^1) \in K - H - \Delta$ and consequently $\tilde{Z}(\cdot, \cdot)$ is d-continuous at

$(1, \tilde{\tau}_D^1)$. It follows that $X(T) = \tilde{Z}(1, \tilde{\tau}_D^1)$, and again $T_D \leq T$. To complete the proof, it suffices to show $P^x[T_D < T] = 0$. Let $(\bar{n}, \bar{\alpha})$ be the largest sequence with $\lim\limits_j S(n_j, \alpha_j) \leq T_D$. Then $d\text{-}\lim\limits_j Z(n_j, \alpha_j) = X(T_D)$; for otherwise we must have $T_D = S(m, \beta)$ for some $(m, \beta) \in T$, which implies $T \leq T_D$ a.s. P^x. Now $(\bar{n}, \bar{\alpha}) < (1, \tilde{\tau}_D^1)$ in $\tilde{T}$. Again by the proof of Theorem 4.2, we cannot have $\tilde{Z}\langle \bar{n}, \bar{\alpha} \rangle = d\text{-}\lim\limits_j \tilde{Z}(n_j, \tilde{\alpha}_j)$ (recall $\langle \bar{n}, \bar{\alpha} \rangle$ is the least (k, γ) in $\hat{T}$ with $(\bar{n}, \bar{\alpha}) < (k, \gamma)$, which exists); for it implies $T_D = T$. But then $\tilde{Z}(\cdot, \cdot)$ has a jump at $(\bar{n}, \bar{\alpha})$, and it follows that $T_D = S(m, \beta)$ for some (m, β) and so $T_D = T$. Thus $P^x(T_D < T) = 0$.   ∎

6.3. Strong Markov property of (X_t)

We first define (X_t) starting at each $x \in \hat{K}_0 - K$ by

$$P^x[f(X_t)] = f(x), \qquad\qquad\qquad 0 \leq t < h(x)$$
$$= \int H(x, dy) P^y[f(X_{t-h(x)})], \quad t \geq h(x)$$

where $f \in b\hat{\mathcal{B}}_0$. Define for $a > 0$, $x \in \hat{K}_0$

$$U^a f(x) = \int_0^\infty e^{-at} P^x[f(X_t)]\, dt\,.$$

Note for $x \in \hat{K}_0 - K$

$$U^a f(x) = a^{-1}(1 - e^{-ah(x)}) f(x) + e^{-ah(x)} \int H(x, dy) U^a f(y)\,.$$

Define $(\mathcal{F}_t)$ to be the filtration generated by (X_t) that is right continuous and completed in $\mathcal{F}$ (which is as defined in section 3) w.r.t. all the measures P^μ (in the usual manner). To show the strong Markov property for the present (X_t), which would imply the strong Markov property for the legitimate right process corresponding to (X_t) (see the beginning of this section), it suffices to prove that, for any stopping time T w.r.t. $(\mathcal{F}_t)$, with $T \leq T_\Delta$ and satisfying the condition $T(\omega) \notin I$ for any open time interval I in which $X_t(\omega)$ is constant,

(6.4) $$P^x\left[\int_0^\infty e^{-at} f(X(T+t))\, dt \right] = P^x[U^a f(X(T))]$$

for all $x \in \hat{K}_0$, $a > 0$, $f \in C(\hat{K})$. We may assume (i) $X(T) \in (K - H) \cup \Delta$ a.s. P^x, or (ii) $X(T) \in H \cup (\hat{K}_0 - K) \cup \Delta$ a.s.P^x. The proof of case (i) (the difficult case) is essentially that of [11], Theorem 7.4. To prove (6.4) in case (ii), we may assume, in view of the analysis in the proof of Theorem 4.2 and the definition of (X_t), that there exists an (increasing) sequence $(\bar{n}, \bar{\tau})$ of stopping times in Z_∞ such that

$$T_j \equiv S(n_j, \tau_j) \uparrow T; \quad T_j = T \quad \text{if} \quad X(T) = \Delta \,;$$
$$X(T) = \hat{d}\text{-}\lim_j X(T_j) = \hat{d}\text{-}\lim_j Z(n_j, \tau_j) \,;$$

and $h(\bar{n}, \bar{\tau}) > 0$ except if $X(T) = \Delta$. By the Markov property of Z_∞, (6.4) holds if T is replaced by T_j. Therefore (6.4) will follow if

$$(6.5) \qquad P^x[U^a f(X(T_j))] \to P^x[U^a f(X(T))] \,.$$

To show (6.5) we first prove

$$(6.6) \qquad h(\bar{n}, \bar{\tau}) = h(X(T)) \quad \text{a.s.} \quad P^x \quad \text{on} \quad \{h(\bar{n}, \bar{\tau}) > 0\} \,.$$

This we do by applying H7A.5) to (Y_j) where $Y_j = X(T_j)$ and $Y_\infty = X(T)$ (and of course $P = P^x$). Now for all $D \in \mathcal{D}$, $e_D(Y_j)$ converges a.s. P^x on $\{\hat{\pi}(Y_\infty) \notin D\}$ and on $\{\pi(Y_\infty) \notin D - \Delta\}$ if $D - \Delta$ is d-compact by a supermartingale argument (see the proof of Theorem 4.1); in particular $e(Y_j)$ converges a.s. P^x. To verify condition (1.5): for the sets C_i in the definition of $\hat{d}$, we have

$$H_{C_i}(Y_j, \cdot) \to H_{C_i}(Y_\infty, \cdot) \quad \text{weakly}$$

a.s. P^x on $\{\hat{\pi}(Y_\infty) \notin C_i\}$ and on $\{\hat{\pi}(Y_\infty) \notin C_i - \Delta\}$ if $C_i - \Delta$ is d-compact: Thus the convergence in (1.5) holds if $D = C_i$ and "strongly" is replaced by "weakly". But by a martingale argument, for $f \in b\hat{B}_0$ and with Λ as in (1.5)

$$P^x[H_D f(Y_j); \Lambda] \quad \text{converges} \,.$$

Thus by a fact stated in the proof of Theorem 6.2 the strong convergence (1.5) holds with $D = C_i$ for all i, and it follows that (1.5) holds for all $D \in \mathcal{D}$. Now H7A.5) implies

$$e_D(X(T_j)) \to e_D(X(T))$$

a.s. P^x on $\{\hat{\pi}(Y_\infty) \notin D\}$ for any D and on $\{\hat{\pi}(Y_\infty) \notin D - \Delta\}$ if $D - \Delta$ is d-compact. This implies (6.6) from the definition of $h(\bar{n}, \bar{\tau})$. Next, from the path behavior of (X_t), the Markov property of Z_∞, the strong convergence (1.5) applied to sets $D = C_i$, the definition of $h(\bar{n}, \bar{\tau})$ and the continuity of f, it is easy to obtain

$$P^x[U^a f(X(T_j))] \to P^x \left[a^{-1}(1 - e^{-ah(\bar{n}, \bar{\tau})}) \int H(X(T), dy) U^a f(y) \right] .$$

But by (6.6) the right hand side is just $P^x[U^a f(X(T))]$. So (6.5) follows and the desired strong Markov property is proved.

To complete the proof of Theorem 2, we need to establish $P^x[T_\Delta] = e(x)$ for all $x \in \hat{K}_0$ and prove the uniqueness assertion. Since it is obvious that $T_\Delta = S_\Delta$ a.s., the former follows from Theorem 4.6 if $x \in K$. If $x \in \hat{K}_0 - K$, let $D \uparrow K - \hat{\pi}(x)$; then $e_D(x) \downarrow h(x)$, and the strong convergence of $H_D(x, \cdot)$ to $H(x, \cdot)$ implies

$$\int H_D(x, dy)e(y) \to \int H(x, dy)e(dy)$$

because of H7A.3). Thus

$$e(x) = e_D(x) + H_D e(x) \to h(x) + \int H(x, dy)e(y) .$$

But the right hand side is $P^x[T_\Delta]$ by the definition of P^x (for $x \in \hat{K}_0 - K$) at the beginning of subsection 6.3.

Finally, we prove the uniqueness assertion in Theorem 2. But this follows from the fact that, if (X_t) satisfies the "additional property" in Theorem 2, then, as can be easily seen, for all $x \in \hat{K}_0$, $f \in C(\hat{K})$, $a > 0$, $U^a f(x)$ is completely determined by the family $\{H_D(x, \cdot) : D \in \mathcal{D}, x \in \hat{K}_a\}$ and the function e; see a computation in [9], section 5.

7. An Example and an Open Problem

The following example illustrates how hidden holding points may arise.

Let K be the interval $[0, 1]$, with $\Delta = 1$. Let C denote the Cantor set, and let the points x in C that are at the left ends of the components of $K - C$, i.e. those with $(x, x + \delta)$ containing no points in C for some $\delta > 0$, be listed as $x_1, x_2, \cdots$. Let $C_0 = C - \{x_n, n \geq 1\}$, and denote the components of $K - C_0$ by $I_n = [x_n, y_n)$ (so $y_n \in C_0$).

Consider a right process in K with trajectories described as follows. Starting at $x \in C_0$ a particle moves to the right continuously inside C_0 (i.e. without skipping points in C_0), until it reaches Δ, and starting at $x \in I_n$ a particle moves to right (continuously inside I_n) until it reaches y_n, and thereon it moves according to what is described above. That is, the hitting distributions $H_D(x, \cdot)$ satisfy:

$$H_D(x, \cdot) = \text{point mass at } \min\,([x, 1] \cap D \cap C_0), \qquad x \in C_0\,,$$

$$= \text{point mass at } \min\,([x, 1] \cap D \cap (I_n \cup C_0))\,, \quad x \in I_n, \; n \geq 1\,.$$

It is easy to check that $\{H_D(x, \cdot) \colon D \in \mathcal{D}, \, x \in K\}$ satisfy all hypotheses of Theorem 1. It does not satisfy H4); H4) fails at point x_m, i.e. when $x < x_m$, $D = [x_m, 1]$, $D_n = [x_m - 1/n, 1]$. It is easy to see that the compactification $\hat{K}$ (only) adds a point $\hat{x}_n$ at each x_n; $\hat{x}_n$ is attached in $\hat{d}$-metric at the right end of $[0, x_n)$, (while $(I_n, \hat{d})$ is homeomorphic to (I_n, d)). Here $\hat{K}_0 = \hat{K}$. Define $e(x)$ on $\hat{K}_0$ as follows. Let $b_n > 0$ with $\Sigma b_n < \infty$, and let

$$e(x) \;=\; \sum_{x_n > x} b_n \qquad\qquad x \in C_0$$

$$= \sum_{x_n > x} b_n + y_m - x \quad x \in I_m, \; m \geq 1$$

$$= \lim_{y \uparrow x_n} e(y) \qquad\qquad x = \hat{x}_n, \; n \geq 1\,.$$

e satisfies H7A) for the family $\{H_D(x, \cdot)\}$. The process (X_t) constructed from $\{H_D(x, \cdot)\}$ and e can be described as follows. A particle starting at $x \in C_0$ moves to the right continuously but inside $C_0 \cup \{\hat{x}_n, n \geq 1\}$, spending an expected holding time b_n at $\hat{x}_n$ for each $x_n > x$, and spending zero total time in C_0 before reaching Δ. A particle starting at $x \in I_n$ moves to the right with speed 1 until it reaches y_n; thereon it moves as described above. (Finally, each $\hat{x}_n$ is a (hidden) holding point with $h(\hat{x}_n) = b_n$, and from $\hat{x}_n$ a jump is made to y_n with probablity 1.)

If e is a function defined as in section 2, the behavior of (X_t) constructed is similar to the above, where $b_n = h(\hat{x}_n)$ is exactly

$$h(\hat{x}_n) = \sum_k 2^{-k} \int_0^1 \varphi_{kn}(r)\, dr$$

where

$$\varphi_{kn}(r) \;= 1 \quad \text{if } x_n \in (F_k(r) - \Delta)^\circ$$
$$\text{but } (F_k(r) - \Delta) \cap C_0 \cap (x_n, 1] = \emptyset$$
$$= 0 \quad \text{otherwise}.$$

We now pose the following (general) question: Is it possible, for a family $\{H_D(x, \cdot)\}$ satisfying the hypotheses of Theorem 1, to define a function e satisfying H7A) but with no "excess time" (i.e. hidden holding times), so the resulting process moves entirely in the original state space K? Perhaps one could use the process constructed in Theorem 2 and try to define a different time scale which permits the deletion of all hidden holding times.

For the above example, it is possible to introduce a time scale under which the hidden holding points $\hat{x}_n$ need not be added. Regard C_0 as the interval $[0,1]$ in the natural way, i.e. regard $x = \sum 2\delta_n/3^n$, $\delta_n = 0$ or 1, in C_0 as $\eta(x) = \sum \delta_n/2^n$. Define

$$e'(x) \;= 1 - \eta(x) \qquad\qquad x \in C_0$$
$$= y_m - x + e'(y_m) \quad x \in I_m, \; m \geq 1.$$

Extend e' to $\hat{K}_0$ by defining

$$e'(\hat{x}_n) = \lim_{y \uparrow x_n} e'(y).$$

Then e' satisfies H7A), and $h(\hat{x}_n) = 0$ for all n. Thus the process constructed from $\{H_D(x, \cdot)\}$ and e' stays entirely inside K.

REFERENCES

[1] J. BLIEDTNER and W. HANSEN. Markov processes and harmonic spaces. Z. Wahrscheinlichkeitstheor. verw. Geb. **42** (1978), 309–325.

[2] N. BOBOC, C. CONSTANTINESCU and A. CORNEA. Semigroup of transitions on harmonic spaces. Rev. Roumaine Math. Pures Appl. **12** (1967), 763–805.

[3] D. A. DAWSON. The construction of a class of diffusions. Illinois J. Math. **8** (1964), 657–684.

[4] J. B. GRAVEREAUX and J. JACOD. Sur la construction des classes de processus de Markov invariantes par changement de temps. Z. Wahrscheinlichkeitstheor. verw. Geb. **52** (1980), 75–107.

[5] W. HANSEN. Konstruktion von Halbgruppen und Markoffschen Prozessen. Inventiones Math. **3** (1967), 179–214.

[6] W. HANSEN. Charakterisierung von Familien exzessiver funktionen. Inventiones Math. **5** (1968), 335–348.

[7] F. KNIGHT and S. OREY. Construction of a Markov process from hitting probabilities. J. Math. Mech. **13** (1968), 857–873.

[8] P. A. MEYER. Brelot's axiomatic theory of the Dirichlet problem and Hunt's theory. Ann. Inst. Fourier, Grenoble **13** (1963), 357–372.

[9] C. T. SHIH. Construction of Markov processes from hitting distributions. Z. Wahrscheinlichkeitstheor. verw. Geb. **18** (1971), 47–72.

[10] C. T. SHIH. Construction of Markov processes from hitting distributions II. Ann. Math. Stat. **42** (1971), 97–114.

[11] C. T. SHIH. Construction of right processes from hitting distributions. Seminar on Stochastic Processes 1983, Birkhäuser, Boston (1984), 189–256.

[12] C. T. SHIH. On piecing together locally defined Markov processes. Seminar on Stochastic Processes 1990, Birkhäuser, Boston (1991), 321–333.

[13] J. C. TAYLOR. The harmonic space associated with a "reasonable" standard process. Math. Ann. **233** (1978), 84–96.

C. T. SHIH
Department of Mathematics
University of Michigan
Ann Arbor, Michigan 48109

A CHARACTERIZATION OF BROWNIAN MOTION ON SIERPINSKI SPACES

by

ZORAN VONDRAČEK

1 Introduction

Brownian motion on the Sierpinski gasket enjoys several properties of a linear Brownian motion. For example, each point is regular for itself, there exists a jointly continuous version of the local time which satisfies the occupation density formula. Moreover, the process is uniquely determined (up to a linear time-change) by the properties of rotation and translation invariance. All these facts are shown in [1].

In this note we give another characterization of Brownian motion on the Sierpinski gasket which is analogous to the following well-known fact of a linear Brownian motion: A linear diffusion in the natural scale is a time-change of Brownian motion. For the case of an open interval (a, b), the property of running in the natural scale can be written as $Q^x(Y_{\tilde{\zeta}-} = a) = P^x(X_{\zeta-} = a)$ and $Q^x(Y_{\tilde{\zeta}-} = b) = P^x(X_{\zeta-} = b)$ for all $x \in (a, b)$, where (Y_t, Q^x) denotes the diffusion and (X_t, P^x) Brownian motion, with $\tilde{\zeta}$ and ζ the corresponding lifetimes.

Let K denote the Sierpinski gasket. The natural boundary of K is formed by three vertices of the smallest triangle containing K. Let a_1, a_2, a_3 denote these vertices, (X_t, P^x) Brownian motion on $K^\circ = K \setminus \{a_1, a_2, a_3\}$ killed upon hitting the natural boundary, and (Y_t, Q^x) any diffusion on K°. We will show that if

$$P^x(X_{\zeta-} = a_j) = Q^x(Y_{\tilde{\zeta}-} = a_j) \qquad (1.1)$$

for $j = 1, 2, 3$ and for all $x \in K^\circ$, then Y is a time-change of X.

In the next section we set up notation and recall some known formulae. The precise statement of the result is given in Section 3. The last section contains proofs. We work on Sierpinski spaces, which are multidimensional analogs of the Sierpinski gasket.

2 Sierpinski spaces

For $N \geq 2$ let $a_1, a_2, \ldots, a_N$ be points in $\mathbf{R}^{N-1}$ satisfying $d(a_i, a_j) = 1$ for $i \neq j$, where d denotes the Euclidean distance. Let $F_i(x) = (x - a_i)/2 + a_i$ for $i = 1, 2, \ldots, N$. For each $w = w_1 w_2 \ldots w_m \in W_m = \{1, 2, \ldots, N\}^m$, let $F_w = F_{w_1} \circ F_{w_2} \circ \ldots \circ F_{w_m}$. For $m \geq 0$, $V_m = \cup_{w \in W_m} F_w(\{a_1, a_2, \ldots, a_N\})$, $V_m^\circ = V_m \setminus V_0$, where $V_0 = \{a_1, a_2, \ldots, a_N\}$. Let $V_* = \cup_{m \geq 0} V_m$. Then the Sierpinski space K^N is defined as the closure of V_* in $\mathbf{R}^{N-1}$. The topology on K^N is the relative topology inherited from $\mathbf{R}^{N-1}$. In the sequel we drop the superscript N on K^N. Let $K^\circ = K \setminus V_0$.

Further, for $w \in W_m$, let $K_w = F_w(K)$ and $B_w = F_w(V_0)$. Then K_w is a scaled copy of K. For $w, \tilde{w} \in W_m$, $K_w \cap K_{\tilde{w}} = B_w \cap B_{\tilde{w}}$ is either empty or precisely one point. This property is referred to as finite ramification of K. Let $K_w^\circ = K_w \setminus B_w$. For $m \geq 0$, let $\mathcal{W}_m$ be a family of subsets of K° defined as follows: For $w \in W_m$, $K_w^\circ \in \mathcal{W}_m$. If $w, \tilde{w} \in W_m$ such that $B_w \cap B_{\tilde{w}} \neq \emptyset$, then $K_w^\circ \cup K_{\tilde{w}}^\circ \cup (B_w \cap B_{\tilde{w}}) \in \mathcal{W}_m$ and these are the only sets belonging to $\mathcal{W}_m$. Let $\mathcal{W} = \cup_{m \geq 0} \mathcal{W}_m$. Then it is easy to see that $\mathcal{W}$ is a basis for the topology on K°.

For $p \in V_m^\circ$ and $j \geq m$, let $V_{j,p} = \{q \in V_* : d(p, q) = 2^{-j}\}$. Points in $V_{j,p}$ are j-neighbors of p. It is easily seen that there are $2(N-1)$ points in $V_{j,p}$. Also, there are exactly two elements $w, \tilde{w} \in W_j$ such that $B_w \cup B_{\tilde{w}} = V_{j,p} \cup \{p\}$. Let us denote $B_w = \{u_1^{(j)}, u_2^{(j)}, \ldots, u_{N-1}^{(j)}, p\}$ and $B_{\tilde{w}} = \{v_1^{(j)}, v_2^{(j)}, \ldots, v_{N-1}^{(j)}, p\}$. Then $V_{j,p} = \{u_1^{(j)}, \ldots, u_{N-1}^{(j)}, v_1^{(j)}, \ldots, v_{N-1}^{(j)}\}$.

A continuous function $h : K \to \mathbf{R}$ is said to be harmonic if it satisfies the following mean value property: For every $p \in V_m^\circ$ and $j \geq m$

$$h(p) = \frac{1}{2(N-1)} \sum_{q \in V_{j,p}} h(q). \tag{2.1}$$

It is proved in [3] that given a function f on V_0, there is a unique harmonic function h on K such that $h_{|V_0} = f$. By h_j, $j = 1, 2, \ldots, N$, we shall denote harmonic function satisfying $h_j(a_i) = \delta_{ij}$. The following lemma is proved in [3]. By $1/2(uv)$ we denote the midpoint of $u, v \in \mathbf{R}^{N-1}$.

Lemma 1 *For $w \in W_m$, let $B_w = \{p_1, p_2, \ldots, p_N\}$, $K_w = F_w(K)$ and $C = (K_w \setminus V_m) \cap V_{m+1}$. Then for $f : B_w \cup C \to \mathbf{R}$ the following conditions are equivalent:*

(1) *For all $p \in C$,* $\quad f(p) = \dfrac{1}{2(N-1)} \displaystyle\sum_{q \in V_{m+1,p}} f(q)$,

(2) *For all $p \in C$, if $p = 1/2(p_i p_j)$, then*

$$f(p) = \frac{1}{N+2}\left(f(p_i) + f(p_j) + \sum_{k=1}^{N} f(p_k)\right).$$

From this lemma, it is easy to compute functions h_j's at points in V_*. We will need explicit values on V_1. Let $p = 1/2(a_{N-1}a_N)$, $V_{1,p} = \{u_1, \ldots, u_{N-1}, v_1, \ldots, v_{N-1}\}$ and assume that $u_j = 1/2(a_j a_{N-1})$, $v_j = 1/2(a_j a_N)$, $j = 1, 2, \ldots, N-2$. Then

$$h_j(p) = \begin{cases} \frac{1}{N+2}, & j = 1, 2, \ldots, N-2 \\ \frac{2}{N+2}, & j = N-1, N. \end{cases} \tag{2.2}$$

Suppose now that h is harmonic and $h(a_N) = 0$, $h(a_{jN}) = 0$, $j = 1, 2, \ldots, N-1$, where $a_{jN} = 1/2(a_j a_N)$. By Lemma 1,

$$0 = h(a_{jN}) \frac{1}{N+2}[h(a_j) + h(a_N) + \sum_{k=1}^{N} h(a_k)] = \frac{1}{N+2}[h(a_j) + \sum_{k=1}^{N} h(a_k)],$$

for $j = 1, 2, \ldots, N-1$. If $A = (\alpha_{ij})$ is the matrix with entries $\alpha_{ii} = 2/(N+2)$, $\alpha_{ij} = 1/(N+2)$, $i, j = 1, 2, \ldots, N-1$, then the system above can be written as $A(h(a_1), \ldots, h(a_{N-1}))^t = 0$. Since A is regular, it follows that $h(a_1) = \ldots = h(a_{N-1}) = 0$. The next lemma mimics the well-known fact that classical harmonic functions are analytic.

Lemma 2 *Suppose that h is harmonic on K and that for some $w \in W_m$, $h_{|B_w} \equiv 0$. Then $h \equiv 0$ in K.*

Proof: If $w = w_1 \ldots w_m$, let $w' = w_1 \ldots w_{m-1}$. The observation preceding the lemma shows that $h_{|B_{w'}} \equiv 0$. By repeating the argument, one gets $h_{|V_0} \equiv 0$. Now the maximum principle (which is known to hold) implies that $h \equiv 0$ on K. ∎

3 Statement of the result

Brownian motion on the Sierpinski gasket extended over the upper half-plane was constructed in [1]. A similar construction is possible for Sierpinski spaces in higher dimensions. Since we are interested in the behaviour of the process only until it hits V_0, we find convenient to kill it at that moment. Thus, let (X_t, P^x) be a

Brownian motion on K° killed upon hitting V_0. The process goes to the cemetary Δ when it dies. Let ζ denote its lifetime. Then $X_{\zeta-} = \lim_{t \to \zeta} X_t$ exists in V_0.

Suppose that F is a subset of K. We would like to define the hitting time T_F of X to F, and the value of X at this moment. Since X is the process on K° only, this does not make sense if $F \cap V_0 \neq \emptyset$. We will overcome this difficulty by pretending that Brownian motion had been stopped after hitting V_0. So, let $T_F = T_{F \cap K^\circ} \wedge \zeta$, where $T_{F \cap K^\circ}$ is the genuine hitting time to $F \cap K^\circ$, and let $X(T_F) = X(T_{F \cap K^\circ})$ if $T_{F \cap K^\circ} < \zeta$, $X(T_F) = X(\zeta-)$ if $\zeta < T_{F \cap K^\circ}$.

With this convention at hand, let for $j > 0$

$$T^j(X) = \inf\{t > 0 : X_t \in V_j \setminus \{X_0\}\}. \tag{3.1}$$

Then for $p \in V_m^\circ$, $j \geq m$, we have $X(T^j(X)) \in V_{j,p}$, P^p a.s., and

$$P^p(X(T^j(X)) = u) = \frac{1}{2(N-1)}, \tag{3.2}$$

for every $u \in V_{j,p}$ (see [1]). For $j = 1, 2, \ldots, N$ define

$$g_j(x) = P^x(X_{\zeta-} = a_j), \quad x \in K^\circ. \tag{3.3}$$

Then the strong Markov property easily implies that g_j's are harmonic functions on K°. Since $\lim_{x \to a_i} g_j(x) = \delta_{ij}$, each g_j can be extended to a harmonic function on K having boundary values $g_j(a_i) = \delta_{ij}$. Hence, $g_j = h_j$ where h_j's are functions defined in Section 2.

Let (Y_t, Q^x) be a transient, strong Markov process on K° with continuous paths up to its lifetime $\tilde{\zeta}$. We assume that $Y_{\tilde{\zeta}-} = \lim_{t \to \tilde{\zeta}} Y_t$ exists in V_0. The same convention for hitting times to $F \subset K$ is valid for Y. Now we may state our result.

Theorem 1 *Let (X_t, P^x) be a Brownian motion on K° killed while exiting K°, let (Y_t, Q^x) be a transient, strong Markov process on K° with continuous paths up to its lifetime $\tilde{\zeta}$. If*

$$P^x(X_{\zeta-} = a_j) = Q^x(Y_{\tilde{\zeta}-} = a_j), \tag{3.4}$$

for all $j = 1, 2, \ldots, N$, and for all $x \in K^\circ$, then there exists a continuous additive functional $A = (A_t)$ of X, which is strictly increasing and finite on $[0, \zeta)$, such that if $\tau = (\tau_t)$ is the right continuous inverse of X, then (X_{τ_t}, P^x) and (Y_t, Q^x) have same joint distributions.

From the potential-theoretical point of view this can be interpreted as follows: Knowledge of harmonic measures at every $x \in K^\circ$, completely determines the potential theory of Brownian motion.

4 Proof

In this section we give a proof of Theorem 1. We start by showing that X and Y have equal hitting distributions to the boundaries of scaled copies of K starting from inside.

Lemma 3 *For $m \geq 1$, let $B_w = F_w(V_0) = \{u_1, u_2, \ldots, u_N\}$, $K_w^\circ = K_w \setminus B_w$. If $T(X)$ and $T(Y)$ denote the hitting times to B_w for X and Y, respectively, then*

$$P^x(X_{T(X)} = u_j) = Q^x(Y_{T(Y)} = u_j), \qquad (4.1)$$

for all $j = 1, 2, \ldots, N$, and for all $x \in K_w^\circ$.

Proof: By conditioning on the first hitting to B_w, one gets

$$h_j(x) = \sum_{i=1}^{N} P^x(X_{T(X)} = u_i) h_j(u_i),$$

and similarly by using (3.4)

$$h_j(x) = \sum_{i=1}^{N} Q^x(Y_{T(Y)} = u_i) h_j(u_i).$$

Let $\gamma_i = P^x(X_{T(X)} = u_i) - Q^x(Y_{T(Y)} = u_i)$, $\gamma = (\gamma_1, \ldots, \gamma_N)^t$, $H = (h_j(u_i))$, $i, j = 1, \ldots, N$. Then $H\gamma = 0$. If the matrix H were not regular, there would exist $\beta_1, \ldots, \beta_N$ such that $h = \sum_{j=1}^{N} \beta_j h_j$ satisfied $h(u_i) = 0$, $i = 1, \ldots, N$. By Lemma 2, $h \equiv 0$ on K°, which is impossible, since h_j's are linearly independent. Thus, H is regular, and consequently, $\gamma = 0$. ∎

Let us now fix a point $p \in V_m^\circ$. For $j \geq m$, there are $w, \tilde{w} \in W_j$ such that $B_w \cup B_{\tilde{w}} = V_{j,p} \cup \{p\}$. Let, as in Section 2, $B_w = \{u_1^{(j)}, \ldots, u_{N-1}^{(j)}, p\}$ and $B_{\tilde{w}} = \{v_1^{(j)}, \ldots, v_{N-1}^{(j)}, p\}$. If $w = w_1 \ldots w_m$, let $w' = w_1 \ldots w_{m-1}$ (note that, with the same notation, $\tilde{w}' = w'$). Let $B_{w'} = F_{w'}(V_0) = \{p_1, p_2, \ldots, p_N\}$. We may assume that $p = 1/2(p_{N-1}p_N)$, and then $V_{j,p} \cap B_{w'} = \{p_{N-1}, p_N\}$. We note that the points are labeled in such a way that $p_{N-1} = u_{N-1}^{(j)}$ and $p_N = v_{N-1}^{(j)}$. For $N = 3$ see Figure 1.

Let $S^j(X)$ and $S^j(Y)$ denote the hitting times to $V_{j,p}$ for X and Y, respectively. Similarly, let $S(X)$ and $S(Y)$ denote the hitting times to $B_{w'}$. If X and Y were to have the same geometric trajectories, $X_{S^j(X)}$ and $Y_{S^j(Y)}$ would have to have the same distribution. For X, the distribution is given by (3.2). We are going to show now, that the same formulae hold for Y.

Figure 1:

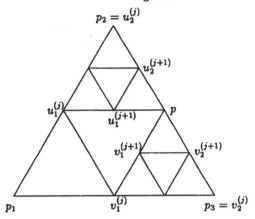

Let $\alpha_k^{(j)} = Q^p(Y_{S^j(Y)} = u_k^{(j)})$, $\beta_k^{(j)} = Q^p(Y_{S^j(Y)} = v_k^{(j)})$, $k = 1, 2, \ldots, N-1$. The system of equations relating above probabilities to the hitting distributions to $B_{w'}$ is the following one:

$$Q^p(Y_{S(Y)} = p_i) = \sum_{k=1}^{N-1} Q^p(Y_{S^j(Y)} = u_k^{(j)}) Q^{u_k^{(j)}}(Y_{S(Y)} = p_i)$$
$$+ \sum_{k=1}^{N-1} Q^p(Y_{S^j(Y)} = v_k^{(j)}) Q^{v_k^{(j)}}(Y_{S(Y)} = p_i), \qquad (4.2)$$

for $i = 1, 2, \ldots, N$. By Lemma 3, $Q^p(Y_{S(Y)} = p_i) = P^p(X_{S(X)} = p_i)$, $Q^{u_k^{(j)}}(Y_{S(Y)} = p_i) = P^{u_k^{(j)}}(X_{S(X)} = p_i)$, and $Q^{v_k^{(j)}}(Y_{S(Y)} = p_i) = P^{v_k^{(j)}}(X_{S(X)} = p_i)$. By scaling invariance of Brownian motion in K^o and by (2.2),

$$P^p(X_{S(X)} = p_i) = \begin{cases} \frac{1}{N+2}, & i = 1, \ldots, N-2 \\ \frac{2}{N+2}, & i = N-1, N \end{cases}$$

$$P^{u_k^{(j)}}(X_{S(X)} = p_i) = \begin{cases} \frac{1}{N+2}, & j = 1, 2, \ldots, N-2, \; j \neq i \\ \frac{2}{N+2}, & j = i \end{cases}$$

for $i = 1, 2, \ldots, N-2$,

$$P^{u_k^{(j)}}(X_{S(X)} = p_{N-1}) = \begin{cases} \frac{2}{N+2}, & j = 1, \ldots, N-2 \\ 1, & j = N-1 \end{cases}$$

$$P^{u_k^{(j)}}(X_{S(X)} = p_N) = \begin{cases} \frac{1}{N+2}, & j = 1, \ldots, N-2 \\ 0, & j = N-1, \end{cases}$$

and simmetrically when $u_k^{(j)}$ is replaced by $v_k^{(j)}$. Thus, (3.6) leads to the system

$$\sum_{k=1}^{N-2} \alpha_k^{(j)} + \alpha_i^{(j)} + \sum_{k=1}^{N-2} \beta_k^{(j)} + \beta_i^{(j)} = 1, \quad i = 1, \ldots, N-2$$

$$2\sum_{k=1}^{N-2} \alpha_k^{(j)} + (N+2)\alpha_{N-1}^{(j)} + \sum_{k=1}^{N-2} \beta_k^{(j)} = 2 \qquad (4.3)$$

$$\sum_{k=1}^{N-2} \alpha_k^{(j)} + 2\sum_{k=1}^{N-2} \beta_k^{(j)} + (N+2)\beta_{N-1}^{(j)} = 2$$

Since there are N equations with $2N - 2$ variables, the system is underdetermined (unless $N = 2$, when it trivially follows that $\alpha^{(j)} = \beta^{(j)} = 1/2$).

In order to show that $\alpha_k^{(j)} = \beta_k^{(j)} = 1/(2(N-1))$, $j = 1, \ldots, N-1$, we proceed as follows. First let $\gamma^{(j)} = (\alpha_1^{(j)}, \ldots, \alpha_{N-1}^{(j)}, \beta_1^{(j)}, \ldots, \beta_{N-1}^{(j)})^t$. We will relate $\gamma^{(j)}$ and $\gamma^{(j+1)}$ by conditioning Y on the first hitting to $V_{j+1,p}$. The following system of equations is obtained. For simplicity we write Y_{T^j} instead of $Y_{T^j(Y)}$ where $T^j(Y)$ is defined as $T^j(X)$ in (3.1).

$$Q^p(Y_{T^j} = u_k^{(j)}) = \sum_{i=1}^{N-1} Q^p(Y_{T^{j+1}} = u_i^{(j+1)}) Q^{u_i^{(j+1)}}(Y_{T^j} = u_k^{(j)})$$

$$+ \sum_{i=1}^{N-1} Q^p(Y_{T^{j+1}} = v_i^{(j+1)}) Q^{v_i^{(j+1)}}(Y_{T^j} = u_k^{(j)})$$

and similarly for $v_k^{(j)}$. This can be written as

$$\alpha_k^{(j)} = \sum_{i=1}^{N-1} \alpha_i^{(j+1)} Q^{u_i^{(j+1)}}(Y_{T^j} = u_k^{(j)}) + \sum_{i=1}^{N-1} \beta_i^{(j+1)} Q^{u_i^{(j+1)}}(Y_{T^j} = u_k^{(j)})$$

$$\hspace{9cm} (4.4)$$

$$\beta_k^{(j)} = \sum_{i=1}^{N-1} \alpha_i^{(j+1)} Q^{u_i^{(j+1)}}(Y_{T^j} = v_k^{(j)}) + \sum_{i=1}^{N-1} \beta_i^{(j+1)} Q^{u_i^{(j+1)}}(Y_{T^j} = v_k^{(j)}).$$

To compute $Q^{u_i^{(j+1)}}(Y_{T^j} = u_k^{(j)})$, we condition Y on the first hitting to $B_w = \{u_1^{(j)}, \ldots, u_{N-1}^{(j)}, u_N^{(j)}\}$, where $u_N^{(j)} = p$. Let τ denote the corresponding hitting time. It follows

$$Q^{u_i^{(j+1)}}(Y_{T^j} = u_k^{(j)}) = \sum_{l=1}^{N} Q^{u_i^{(j+1)}}(Y_\tau = u_l^{(j)}) Q^{u_l^{(j)}}(Y_{T^j} = u_k^{(j)})$$

$$= Q^{u_i^{(j+1)}}(Y_\tau = u_k^{(j)}) + Q^{u_i^{(j+1)}}(Y_\tau = p) Q^p(Y_{T^j} = u_k^{(j)}).$$

By Lemma 3, scaling invariance and symmetry of X, this is equal to $1/(N+2) + 2\alpha_k^{(j)}/(N+2)$ if $i \neq k$, and $2/(N+2) + 2\alpha_k^{(j)}/(N+2)$ if $i = k$. Similarly, by conditioning Y on the first hitting to $B_{\tilde{w}} = \{v_1^{(j)}, \ldots, v_{N-1}^{(j)}, p\}$, one gets

$Q^{v_i^{(j+1)}}(Y_{T^j} = u_k^{(j)}) = 2\alpha_k^{(j)}/(N+2)$, $i = 1, \ldots, N-1$. Hence, (3.8) can be written as

$$
\begin{aligned}
\alpha_k^{(j)} &= \sum_{\substack{i=1 \\ i \neq k}}^{N-1} \alpha_i^{(j+1)}\left(\frac{1}{N+2} + \frac{2}{N+2}\alpha_k^{(j)}\right) + \alpha_k^{(j+1)}\left(\frac{2}{N+2} + \frac{2}{N+2}\alpha_k^{(j)}\right) \\
&\quad + \sum_{i=1}^{N-1} \beta_i^{(j+1)}\frac{2}{N+2}\alpha_k^{(j)} \\
\beta_k^{(j)} &= \sum_{i=1}^{N-1} \alpha_i^{(j+1)}\frac{2}{N+2}\beta_k^{(j)} + \sum_{\substack{i=1 \\ i \neq k}}^{N-1} \beta_i^{(j+1)}\left(\frac{1}{N+2} + \frac{2}{N+2}\beta_k^{(j)}\right) \\
&\quad + \beta_k^{(j+1)}\left(\frac{2}{N+2} + \frac{2}{N+2}\beta_k^{(j)}\right)
\end{aligned}
\tag{4.5}
$$

Let $B^{(j)}$ denote the $(2N-2) \times (2N-2)$ matrix whose i-th row is $(\alpha_i^{(j)}, \ldots, \alpha_i^{(j)})$, for $i = 1, 2, \ldots, N-1$, and $(\beta_{i-(N-1)}^{(j)}, \ldots, \beta_{i-(N-1)}^{(j)})$, for $i = N, N+1, \ldots, 2N-2$. Let C_1 be the $(N-1) \times (N-1)$ matrix with all entries 1, and C_2 be the $(N-1) \times (N-1)$ matrix with all entries 0. Let C be the $(2N-2) \times (2N-2)$ matrix with blocks C_1 on the diagonal and C_2 off the diagonal. Next, let I denote the $(2N-2) \times (2N-2)$ identity matrix. Define the matrix $\Pi^{(j)}$ by

$$
\Pi^{(j)} = \frac{1}{N+2}(I + C + 2B^{(j)}).
\tag{4.6}
$$

Then the system (3.9) can be written as

$$
\gamma^{(j)} = \Pi^{(j)}\gamma^{(j+1)}.
\tag{4.7}
$$

Since the elements of $\gamma^{(j+1)}$ add up to 1, it holds $B^{(j)}\gamma^{(j+1)} = \gamma^{(j)}$. Thus (3.11) writes $\gamma^{(j)} = (1/(N+2))(I+C)\gamma^{(j+1)} + 2/(N+2)\gamma^{(j)}$. Finally, this can be written as

$$
\gamma^{(j)} = \frac{1}{N}(I + C)\gamma^{(j+1)}.
\tag{4.8}
$$

Let us denote $A = (1/N)(I + C)$. Note that A does not depend on the index j, hence (4.8) is valid for all $j \geq m$. Therefore,

$$
\gamma^{(j)} = A^n\gamma^{(j+n)},
\tag{4.9}
$$

for all $j \geq m$ and all $n \in \mathbf{N}$. An easy computation gives that $A^n = (1/N^n)(I + (N^{n-1} + \cdots + N + 1)C)$ and $\lim_{n \to \infty} A^n = C/(N-1)$.

On the other hand, all $\gamma^{(j)}$'s belong to the $2(N-1)$ dimensional simplex $S = \{x \in \mathbf{R}^{2N-2} : 0 \leq x_i \leq 1, \sum_{i=1}^{2N-2} x_i = 1\}$. Hence, there is a subsequence of $\{\gamma^{(j)}\}$ converging to $\gamma = (\alpha_1, \ldots, \alpha_{N-1}, \beta_1, \ldots, \beta_{N-1})^t \in S$. Taking the limit along this subsequence in (4.9) yields

$$
\gamma^{(j)} = \frac{1}{N-1}C\gamma,
\tag{4.10}
$$

for all $j \geq m$. But $C\gamma = (\sum_{i=1}^{N-1} \alpha_i, \ldots, \sum_{i=1}^{N-1} \alpha_i, \sum_{i=1}^{N-1} \beta_i, \ldots, \sum_{i=1}^{N-1} \beta_i)^t$. Together with (4.10) this implies that $\alpha_1^{(j)} = \cdots = \alpha_{N-1}^{(j)}$ and $\beta_1^{(j)} = \cdots = \beta_{N-1}^{(j)}$, for all $j \geq m$. Let us denote the common values by α and β. Using this in the system (4.3) gives $(N-1)\alpha + (N-1)\beta = 1$, $2(N-2)\alpha + (N+2)\alpha + (N-2)\beta = 2$, $(N-2)\alpha + 2(N-2)\beta + (N+2)\beta = 2$. A unique solution is $\alpha = \beta = 1/(2(N-1))$. Thus the following proposition is proved.

Proposition 1 Let $p \in V_m^\circ$, $j \geq m$ and $V_{j,p} = \{u_1^{(j)}, \ldots, u_{N-1}^{(j)}, v_1^{(j)}, \ldots, v_{N-1}^{(j)}\}$. If $S^j(Y)$ denotes the hitting time to $V_{j,p}$ for Y, then

$$Q^p(Y_{(S^j(Y))} = u_k^{(j)}) = \frac{1}{2(N-1)} = Q^p(Y_{(S^j(Y))} = v_k^{(j)}). \qquad (4.11)$$

Let $S_0^j(Y) = \inf\{t \geq 0 : Y_t \in V_j\}$, and $S_{i+1}^j(Y) = \inf\{t \geq 0 : Y_t \in V_j \setminus \{Y(S_i^j(Y))\}\}$, for $i = 1, 2, \ldots$. Let $Y_i^{(j)} = Y(S_i^j(Y))$, $i = 0, 1, \ldots$ denote the imbedded random walk on V_j. If the walk $X^{(j)}$ is defined in the same way from Brownian motion X, then Proposition 1 says that for all $p \in V_*$, $X^{(j)}$ under P^p and $Y^{(j)}$ under Q^p have equal one-step transition probabilities. Therefore, they have equal laws. Now it is a matter of routine to prove Theorem 1.

Proof of Theorem 1:

In proving the theorem, we use an idea from [4]. Let U be a relatively compact open set in K°. By U_n we denote the union of all sets in $\mathcal{W}_n$ contained in U (the family $\mathcal{W}_n$ was defined in Section 2). Then $U = \cup_{n \geq 0} U_n$. Let $T_n(X)$ and $T_n(Y)$ denote the hitting times to ∂U_n for X and Y, respectively. Note that $\partial U_n \subset V_n$. Let $\tau_U(X)$ denote the exit time from U for X. By continuity of paths $X(\tau_U(X)) \in \partial U$. Obviously $T_1(X) \leq T_2(X) \leq \ldots \leq \tau_U(X)$. It is easy to see that $\lim_{n \to \infty} T_n(X) = \tau_U(X)$. With analog notation, the same is valid for Y. Suppose that $x \in K^\circ$ and $m \geq 0$. If $x \in K^\circ \setminus V_m$, then by Lemma 3, distributions of $X(S_0^m(X))$ and $Y(S_0^m(Y))$ are equal. If $x \in V_n$ this is trivially so. Then Proposition 1 implies that $X(T_n(X))$ under P^x and $Y(T_n(Y))$ under Q^x have the same distribution. By continuity of paths, $X(\tau_U(X)) = \lim_{n \to \infty} X(T_n(X))$ and $Y(\tau_U(Y)) = \lim_{n \to \infty} Y(T_n(Y))$. Therefore, for a continuous function f on $\bar{U}$, and for $x \in K^\circ$

$$P^x[f(X(\tau_U(X)))] = \lim_{n \to \infty} P^x[f(X(T_n(X)))]$$
$$= \lim_{n \to \infty} Q^x[f(Y(T_n(Y)))] = Q^x[f(Y(\tau_U(Y)))]. \qquad (4.12)$$

Hence, $X(\tau_U(X))$ under P^x and $Y(\tau_U(Y))$ under Q^x have the same distribution, for all $x \in K^\circ$. The proof is finished by using the Blumenthal-Getoor-McKean theorem ([2],V-5.1).   ∎

2. The Function e

We prove that Theorem 1 follows from Theorem 2 by defining a function e on $\hat{K}_0$ and showing that it satisfies H7A) under the hypotheses of Theorem 1. In contrast to the proof of H7) under the hypothesis of Theorem 1 of [11], the present proof is much easier. e is defined in a similar way as in [11]. Choose a sequence of mappings $r \to F_k(r)$ from $[0,1]$ into D such that each $F_k(r) - \Delta$ is d-compact and for $r_1 < r_2, F_k(r_1) - \Delta \subset (F_k(r_2) - \Delta)^\circ$, (again, B° is the interior of B in (K, d)), and such that for all $x \neq \Delta$ and $\delta > 0$, there is k with $x \in F_k(0) - \Delta \subset F_k(1) - \Delta \subset B(x, \delta)$. Define for $x \in \hat{K}_0$

$$(2.1) \qquad e(x) = \sum_k 2^{-k} \int_0^1 [1_{F_k(r)-\Delta}(\hat{\pi}(x)) $$
$$+ 1_{(K-F_k(r))\cup\Delta}(\hat{\pi}(x))H_{F_k(r)}(x, F_k(r) - \Delta)]\, dr \, .$$

(Recall again the definition (1.3).) Note $H_{F_k(r)}(x, F_k(r) - \Delta)$ is increasing in r, as an easy consequence of (1.4). For $x \in K$

$$(2.2) \qquad e(x) = \sum_k 2^{-k} \int_0^1 H_{F_k(r)}(x, F_k(r) - \Delta)\, dr$$

which is the definition of e in [11]. Clearly, e is nonnegative $\hat{\mathcal{B}}_0$-measurable and $e(\Delta) = 0$.

Theorem 2.1. e *satisfies H7A).*

Proof. The proof of H7A.1) consists of the first 15 lines of the proof of [11], Theorem 2.1, using the stronger transience hypothesis H6A). The proof of H7A.2) is the same as H7.2); see [11], Theorem 2.5. H7A.3) is immediate since e is bounded. To prove H7A.4), we show that for $x \in \hat{K}_0 - K$, $D \in \mathcal{D}$

$$(2.3) \qquad e_D(x) = e(x) - \int H_D(x, dy)e(y) \geq 0 \, .$$

Since if $\hat{\pi}(x) \in F_k(r) - \Delta$ then

$$1_{F_k(r)-\Delta}(\hat{\pi}(x)) - \int H_D(x, dy)H_{F_k(r)}(y, F_k(r) - \Delta) = 1 - \text{2nd term} \geq 0 \, ,$$

References

[1]   Barlow, M.T., Perkins, E.A.: *Brownian Motion on the Sierpinski Gasket*, Probab.Th.Rel.Fields **79**, 543-623 (1988)

[2]   Blumenthal, R.M., Getoor, R.K.: Markov processes and potential theory, Academic Press, New York, 1968

[3]   Kigami, J.: *A Harmonic Calculus on the Sierpinski Spaces*, Japan Jour.Appl.Math., Vol.6, No.2, 259-290 (1989)

[4]   Øksendal, B., Stroock, D.W.: *A characterization of harmonic measures and Markov processes whose hitting distributions are preserved by rotations, translations and dilatations* Ann. Inst. Fourier, Grenoble, **32**, No.4, 221-232 (1982)

Department of Mathematics
University of Zagreb
P.O.Box 187
YU-41000 Zagreb
Croatia

Publications of Steven Orey

1. Formal development of ordinal number theory. *J. Symb. Logic* **20** (1955) 95–104 .

2. On ω-consistency and related properties. *J. Symb. Logic* **21** (1956) 246–252.

3. On the relative consistency of set theory. *J. Symb. Logic* **21** (1956) 280–290.

4. A central limit theorem for m-dependent random variables. *Duke Math. J.* **25** (1958) 543–546 .

5. Recurrent Markov chains. *Pacific J. Math.* **9** (1959) 805–827.

6. Model theory for the higher order predicate calculus. *Trans. Amer. Math. Soc.* **92** (1959) 72–84.

7. Strong ratio limit property. *Bull. Amer. Math. Soc.* **67** (1961) 571–574.

8. Sums arising in the theory of Markov chains. *Proc. Amer. Math. Soc.* **12** (1961) 847–856.

9. Change of time scale for Markov processes. *Trans. Amer. Math. Soc.* **99** (1961) 384–397.

10. Relative interpretations. *Z. Math. Log. Grundl. Math.* **7** (1961) 146–153.

11. A renewal theorem (with W. Feller). *J. Math. Mech.* **10** (1961) 619–624.

12. An ergodic theorem for Markov chains. *Z. Warsch. verw. Geb.* **1** (1962) 174–176.

13. 1-consistency and faithful representations (with S. Feferman and G. Kreisel). *Arch. Math. Log. Grundl.* **6** (1962) 52–63.

14. Non-differentiability of absolute probabilities of Markov chains. *Quart. J. Math. Oxford,* Ser. (2) **13** (1962) 252–254.

15. Absolute behavior of successive coefficients of some power series (with A. Garsia and E. Rodemich). *Ill. J. Math.* **6** (1962) 620–629.

16. Potential kernels for recurrent Markov chains. *Jour. Math. Anal. Appl.* **8** (1964) 104–132.

17. New foundations and the axiom of counting. *Duke Math. Jour.* **31** (1964) 655–660.

18. Construction of a Markov process from hitting probabilities (with F. Knight). *Jour. Math. Mech.* **13** (1964) 857–873.

19. Ratio limit theorems for Markov chains (with J.F.C. Kingman). *Proc. Amer. Math. Soc.* **15** (1964) 907–910.

20. Convergence of weighted averages of independent random variables (with B. Jamison and W. Pruitt). *Z. Wahrsch. verw. Geb.* **4** (1965) 40–44.

21. Tail events for sums of independent random variables. *Jour. Math. Mech.* **15** (1966) 937–951.

22. F-processes. *Proc. Fifth Berkeley Symp. Math Stat. and Prob., Vol. II,* 301–313. Univ. Cal. Press, 1967.

23. Polar sets for processes with stationary independent increments. *Markov Processes and Potential Theory.* Wiley, New York (1967) 117–126.

24. Markov chains recurrent in the sense of Harris (with B. Jamison). *Z. Wahrsch. verw. Geb.* **8** (1967) 41–48.

25. On continuity properties of infinitely divisible distribution functions. *Ann. Math. Stat.* **39** (1968) 936–937.

26. On the range of random walk (with N. Jain). *Israel J. Math.* **6** (1968) 373–380.

27. An optional stopping theorem (with B. Jamison). *Ann. Math. Stat.* **40** (1969) 677–678.

28. Growth rate of Gaussian processes with stationary increments. *Bull. Amer. Math. Soc.* **76** (1970) 609–611.

29. Subgroups of sequences and paths (with B. Jamison). *Proc. Amer. Math. Soc.* **24** (1970) 739–744.

30. Gaussian sample functions and the Hausdorff dimension of level crossings. *Z. Wahrsch. verw. Geb.* **15** (1970) 249–256.

31. Growth rate of certain Gaussian processes. *Proc. Sixth Berkeley Symp. Math. Stat. Prob., Vol. II,* 443–451. Univ. Cal. Press, 1971.

32. *Lecture notes on limit theorems for Markov chain transition probabilities.* Van Nostrand Reinhold Math. Studies, No. 34. Van Nostrand Reinhold, New York, 1971.

33. Sample functions of N-parameter Weiner process (with W. Pruitt). *Ann. Prob.* **1** (1973) 138–163.

34. Some properties of random walk paths (with N. Jain). *Jour. Math. Anal. Appl.* **43** (1973) 795–815.

35. Radon-Nikodym derivatives of probability measures: martingale methods. Dept. Found. Math. Sci., Tokyo University of Education, Tokyo, 1974.

36. How often on a Brownian path does the law of the iterated logarithm fail? (with S.J. Taylor). *Proc. London Math. Soc.*Ser. (3) **28** (1974) 174–192.

37. Conditions for the absolute continuity of two diffusions. *Trans. Amer. Math. Soc.* **193** (1974) 413–426.

38. Small random perturbations of dynamical systems with reflecting boundary (with R. Anderson). *Nagoya Math. Jour.* **60** (1976) 189–216.

39. Diffusions on the line and additive functionals of Brownian motion. *Proc. Conf. on Stoch. Diff. Equations and Appl.*, 211–230. Academic Press, New York, 1977.

40. The tail σ-field of one-dimensional diffusions (with B. Fristedt). *Stochastic Analysis* (Proc. Int. Conf. Northwestern Univ.), 127–138. Academic Press, New York, 1978.

41. Vague convergence of sums of independent random variables (with N. Jain). *Israel Jour. Math.* **33** (1979) 317–348.

42. Exterior Dirichlet problem and the asymptotic behavior of diffusions (with M. Cranston and U. Rosler). Stoch. Diff. Systems, 207–220. *Lecture Notes in Control and Information Science* **25**. Springer, New York, 1980.

43. Domains of partial attraction and tightness conditions (with N. Jain). *Ann. Prob.* **8** (1980) 584–599.

44. Some asymptotic results for a class of stochastic systems with parametric excitations (with P.R. Sethna). *Inter. Jour. Non-linear Mech.* **15** (1980) 431–441.

45. Stationary solutions for linear systems with additive noise. *Stochastics* **5** (1981) 241–251.

46. Probabilistic methods in partial differential equations. Studies in partial differential equations, 143–205. *MAA Studies Math.* **23**. Washington, D.C., 1982.

47. The Martin boundary of two-dimensional Ornstein-Uhlenbeck processes (with M. Cranston and U. Rosler). Probability, statistics, and analysis, 63–78. London Math. Soc. Lecture Note Series **79**. Cambridge Univ. Press, (1983).

48. Two strong laws for shrinking Brownian tubes. *Z. Wahrsch. verw. Geb.* **63** (1983) 393–416.

49. On the Shannon-Perez-Moy theorem. Particle systems, random media and large deviations, 319–327. *Contemp. Math.* **41** Amer. Math. Soc., Providence, (1985).

50. Large deviations in ergodic theory. *Seminar on Stochastic Processes 1984*, 195–249. Birkhäuser, Boston 1986.

51. Minimizing or maximizing the expected time to reach zero (with D. Heath, V. Pestien, W. Sudderth). *SIAM Jour. Control Optim.* **25** (1987) 195–205.

52. Reaching zero rapidly (with V. Pestien and W. Sudderth). *SIAM Jour. Control Optim.* **25** (1987) 1253–1265.

53. Large deviations for the empirical field of a Gibbs measure (with H. Föllmer). *Ann. Prob.* **16** (1988) 961–977.

54. Large deviation principles for stationary processes (with S. Pelikan). *Ann. Prob.* **16** (1988) 1481–1495.

55. Large deviations for the empirical field of the Curie-Weiss models. *Stochastics* **25** (1988) 3–14.

56. Weakly ergodic products of (random) nonnegative matrices. *Almost everywhere convergence*, 305–333. Academic Press, Boston, 1989.

57. Deviations of trajectory averages and the defect in Pesin's formula for Anasov diffeomorphisms (with S. Pelikan). *Trans. Amer. Math. Soc.* **315** (1989) 741–753.

58. Markov chains with stochastically stationary transition probabilities. *Ann. Prob.* **19** (1991) 907–928.

Progress in Probability

Progress in Probability is designed for the publication of workshops, seminars and conference proceedings on all aspects of probability theory and stochastic processes, as well as their connections with and applications to other areas such as mathematical statistics and statistical physics. It acts as a companion series to *Probability and Its Applications,* a context for research level monographs and advanced graduate texts.

We encourage preparation of manuscripts in some form of TeX for delivery in camera-ready copy, which leads to rapid publications, or in electronic form for interfacing with laser printers or typesetters.

Proposals should be sent directly to the editors or to:
Birkhäuser Boston, 675 Massachusetts Avenue, Cambridge, MA 02139, U.S.A.